Computer-Integrated Manufacturing

Second Edition

James A. Rehg

The Pennsylvania State University–Altoona College

Henry W. Kraebber

Purdue University

Upper Saddle River, New Jersey
Columbus, Ohio

Library of Congress Cataloging-in-Publication Data

Rehg, James A.

    Computer-integrated manufacturing / James A. Rehg, Henry W. Kraebber.—2nd ed.

       p. cm.

       Includes bibliographical references and index.

       ISBN 0-13-087553-8

       1. Computer integrated manufacturing systems. I. Kraebber, Henry W. II. Title.

    TS155.63. R45 2001

    670'.285—dc21

00-025596

Vice President and Publisher: Dave Garza
Editor in Chief: Stephen Helba
Executive Editor: Debbie Yarnell
Associate Editor: Michelle Churma
Production Editor: Tricia Huhn
Design Coordinator: Robin G. Chukes
Cover Designer: Tanya Burgess
Production Manager: Brian Fox
Illustrations: The Clarinda Company

This book was set in Zaph Calligraphic 801 by The Clarinda Company and was printed and bound by R.R. Donnelley & Sons Company. The cover was printed by Phoenix Color Corp.

10 9 8 7 6 5 4 3 2 1
ISBN: 0-13-087553-8

About the Authors

James A. Rehg, CMfgE, is an Assistant Professor of Engineering at The Pennsylvania State University–Altoona College, where he teaches automation controls courses in the B.S. program in Electromechanical Engineering Technology. He earned both a B.S. and M.S. in Electrical Engineering from St. Louis University and has completed additional graduate study at Wentworth Institute, University of Missouri, South Dakota School of Mines and Technology, and Clemson University. Before coming to Penn State–Altoona, he was the CIM coordinator and department head of CAD/CAM/Machine Tool Technology at Tri-County Technical College. Prior to that, he was the Dean of Engineering Technology and Director of Academic Computing at Trident Technical College in Charleston, South Carolina. He held the position of Director of the Robotics Resource Center at Piedmont Technical College in Greenwood, South Carolina, and was department head of Electronic Engineering Technology of Forest Park Community College in St. Louis, Missouri. In addition, he was a Senior Instrumentation Engineer for Boeing in St. Louis. Professor Rehg has authored five texts on robotics and automation, and has presented numerous papers on subjects related directly to training in automation and robotics. He has also been a consultant to nationally-recognized corporations and many educational institutions. He has led numerous seminars and workshops in the areas of robotics and microprocessors and has developed extensive seminar training material. In addition, he has received numerous state awards for excellence in teaching and was named the outstanding instructor in the nation by the Association of Community College Trustees.

Henry W. Kraebber, P.E., CPIM, is an Associate Professor of Mechanical Engineering Technology at Purdue University in West Lafayette, Indiana. He has fifteen years of experience and leadership in manufacturing operations, engineering, quality, and management. He has worked at the Collins Avionics and Missiles group of Rockwell International, the Plough Products Division of Schering-Plough Corporation, and Flavorite Laboratories. His work has supported the production of industrial, consumer, and military products in the food, consumer products, and electronics areas. Career highlights include developing and implementing manufacturing control systems (MRP2, including MAPICS and AMAPS) and implementing the concepts of total quality. He earned a B.S. degree in Industrial Engineering from Purdue University, and a M.E. degree in Industrial Engineering from Iowa State University. He is a senior member of the Institute of Industrial Engineers (IIE). As an active member of professional organizations, he is now Past President and the VP of Education for the Wabash Valley Chapter of APICS. In August 1989, Mr. Kraebber returned to Purdue University to become a faculty member in the School of Technology program in Computer Integrated Manufacturing Technology. He currently teaches courses in manufacturing operations, manufacturing quality control, and integrated systems. He is President of the CIM in Higher Education Alliance, a nonprofit corporation that supports CIM and manufacturing education.

This text is dedicated to two very special young men, Jim and Richard, and to our families, students and friends that have helped make this possible.

Contents

5 Design Automation: Computer-Aided Engineering   138

PART 3   MANAGING THE ENTERPRISE RESOURCES   197

6 Introduction to Production/Operations Planning   198

Preface

The new millennium brings many new issues and twists to the subject of computer-integrated manufacturing (CIM). It remains as broad as the complex manufacturing enterprises it attempts to model. Many would suggest that CIM is too broad for a single course or textbook. However, the essence of CIM is in the integration of the enterprise elements: physical integration through the linking of hardware and software systems, logical integration through shared common enterprise information and data, and philosophical integration based on a new sense of purpose and direction in every entity in the enterprise. Therefore, the integration so critical to a CIM implementation is best introduced in a single course so that links between the enterprise elements can be explored. This book was written to support such an introductory course.

To understand the operation of a comprehensive CIM solution requires some study of manufacturing as it was in the early 1990s, as it was at the end of the decade, and as we think it might look in the twenty-first century. The integration of basic product design techniques and manufacturing fundamentals and principles, along with a look at the changing operations and information systems that support CIM in the enterprise, make this book unique. In the book we:

- Describe the different types of manufacturing systems or production strategies used by industries worldwide. This description is important because no two CIM solutions are the same.

- Go beyond the description of automated machines and software solutions because a successful CIM implementation demands more than technology. In practice, ordering hardware and software is the last step in a CIM implementation; the preliminary work is what guarantees a successful CIM project.

- Include the impact of CIM on all the major elements in an enterprise: product design, shop floor technology, and manufacturing production and operational control systems.

- Provide a convincing argument for implementing CIM so that the enterprise will be competitive in the global market. In practice, the technologies available to manufacturers around the globe open every market to worldwide competition.

In addition, the second edition has the following significant changes: numerical problem examples have been added to Chapter 1; the introductory material presented in the old Chapter 3 has been incorporated into other chapters in the text; web addresses (universal resource locators, URLs) for companies offering CIM-related technology are listed at the end of many of the chapters; and a new chapter on enterprise resource planning (EPS) has been added.

To provide a complete overview of the computer-integrated enterprise, the book is divided into four parts. In the first part, Chapters 1 and 2, we provide an overview of global competition, describe an internal manufacturing strategy, discuss in detail the problem facing manufacturing and the development of an effective solution, and characterize the operation of different types of enterprises. In the characterization, we finish a classification and description of the manufacturing systems and production strategies used by manufacturing, an explanation of the product development and engineering change cycle, and an overview of the enterprise organization. At the end of Part 1, the need for change in manufacturing is made clear and a basic strategy for change in the organization is established. In addition, the description of the enterprise organization in Part 1 provides a framework for the CIM concepts introduced in the rest of the text. Part 1 provides the critical introduction of manufacturing and the enterprise that is necessary for a course designed to teach computer-integrated manufacturing.

In Part 2, which includes three chapters, we examine the three major design and engineering process segments that take a product from concept to production. Chapter 3 provides an introduction to design and production engineering concepts and issues. The use of CIM technology to design and produce world-class products with enhanced enterprise productivity is emphasized. The old design model is compared to a recommended new process that incorporates a concurrent engineering focus to product design. This part of the text concludes with an in-depth description of the functions in production engineering and the opportunities for productivity gains through CIM. Computer-aided design (CAD) is the focus of Chapter 4. A full chapter is devoted to CAD because it is one of the major building blocks in a CIM implementation. The topic includes a comprehensive definition and brief history of CAD, description of CAD systems and operation, classification of CAD hardware platforms and software systems from 2-D to solid modelers, and applications for CAD technology in the manufacturing systems described in Part 1. In Chapter 5, we explore the relationships between the concurrent engineering product design model and the computer-aided engineering (CAE) technology available to support every step of the design process and production engineering. We include a complete definition of CAE, design for manufacturing and assembly, finite-element and mass property analysis, rapid prototyping, group technology, computer-aided process planning, computer-aided manufacturing, production and process modeling and simulation, maintenance, automation, and product cost analysis. In the final section of Chapter 5, we describe the computer network used to tie the design and production engineering functions to the common enterprise database and other business functions.

Part 3 of the text shifts the focus to managing the enterprise resources. The concepts of material and manufacturing planning and the control systems used within the enterprise are addressed in Chapters 6 and 9. The function of manufacturing planning and the automation technology available for CIM implementations is recounted. The first chapter in the sequence, Chapter 6, provides an overview of the critical concepts that are explored in the following three chapters. In Chapter 7, we introduce the concept of manufacturing planning and control (MPC) with a model of a typical MPC system. In addition, three key elements in the MPC model—production planning, material requirements planning (MRP), and master production scheduling (MPS)—are covered. Finally, automation software used to implement CIM in this critical part of the enterprise is introduced and explained. Additional elements in the MPC model are defined in Chapter 8. The topics presented include inventory and data management, capacity management, production activity control, just-in-time manufacturing, and synchronous manufacturing. Software solutions for the manual MPC functions are included at the end of each section. At the conclusion of Chapters 6 through 8, the reader will understand the operation of an MPC system and will be able to perform the manual calculations for each function in MPC and describe application software capable of automating the MPC functions. Chapter 9, "Enterprise Resource Planning and Beyond," contains material completely new to the second edition. The chapter develops the links between the concepts from MRP and MRP II systems that are essential parts of the new ERP systems. The pace of change in technology and new systems at the end of the 1990s has been extraordinary. There is no way to predict the future, but it is clear that new systems will continue to be created. Technologies for design, processing and control, information systems, and communication are rapidly converging. The coming technologies will offer substantial new opportunities and risks for manufacturing enterprises.

Part 4 concentrates on the processes and systems that lay the foundation for modern manufacturing and enterprise-wide concepts critical to a successful CIM implementation. Chapter 10 covers the commonly used production process machines used in manufacturing. In addition, manufacturing systems composed for one or more machines, called flexible manufacturing cells and flexible manufacturing systems, are addressed in the chapter. Chapter 11 covers machines and systems that support production, including coverage of industrial robots, material-handling systems, automatic guided vehicles, and automatic storage and retrieval systems. The techniques used for the control of production systems is the focus of Chapter 12. The control systems discussed include cell control hardware and software, device control hardware and software, programmable logic controllers, and computer numerical controllers. The programming techniques used for CNC machines have been significantly expanded in this chapter. The operation and management of enterprise networks and common databases are also discussed. A successful implementation of any high technology requires a change in the management viewpoint on manufacturing management and human resource development. As a result, a discussion of a broad range of quality issues and the effective use of human resources is included in Chapter 13.

In summary, the chapter sequence starts in Part 1 with a global view of manufacturing. In the second and third parts, we focus on the activities required to convert raw material into finished goods and introduce technology to aid in the conversion and management of the enterprise. The last part of the text shifts back to systems that enable the enterprise to manufacture products competitively, with the discussion centered on the services and support functions required for successful CIM implementation. Common products (hardware, software and systems) are included throughout the book to demonstrate the technology and to stress the integration issues.

The logical order of topics and chapter content was tested in a series of workshops at Trident Technical College offered to college faculty and industrial employees. The insight gained through discussions with these workshop groups and the CIM team at the college was critical to the development of his book. We would especially like to thank Jerry Bell and Alan Kalameja for their help in the first edition with the design automation and control elements. Special thanks to Marci Rehg for her help in developing the CIM workshop material, where many of the presentation ideas were tested. And thanks to all the students who have helped us develop and test instructional materials related to CIM over the years.

Finally, thanks to the IBM Corporation, founders of the initial CIM in Higher Education Alliance program, for support in developing the CIM workshops and the CIM capability at two- and four-year colleges. The CIM in Higher Education Alliance is now an independent, nonprofit corporation that continues to encourage and support CIM and education for manufacturing. Thanks also to the reviewers: Don Arney (Ivy Tech State College (IN)), Dr. Michael Costello (Southern Illinois University at Carbondale), Dinesh Dhamija (Ohio University), Dave Hunter (Western Illinois University), and Herbert Tuttle (Kansas University, Ewards Campus).

James A. Rehg (jar14@psu.edu)

Henry W. Kraebber (hwkraebber@tech.purdue.edu)

Introduction to CIM and the Manufacturing Enterprise

After you have completed Part 1, it will be clear to you that:

- Manufacturing is vital to the economic health of the United States and our current high standard of living.
- Manufacturing must meet serious internal and external challenges to succeed in the face of increasing global competition.
- The operational philosophy described by computer-integrated manufacturing (CIM) is necessary at some level for the survival of most manufacturing companies.
- All manufacturing systems and production strategies will benefit from the application of the CIM philosophy.

The Manufacturing Enterprise

1–1 INTRODUCTION

This book is about the manufacturing enterprise and the operational principles that are necessary for manufacturers to be competitive in the present global market–driven economy. Defining *manufacturing* is a logical starting point.

> *Manufacturing is a collection of interrelated activities that includes product design and documentation, material selection, planning, production, quality assurance, management, and marketing of goods.*

The fundamental goal of manufacturing is to use these activities to convert raw materials into finished goods on a profitable basis. The ability to produce this conversion efficiently determines the success of the enterprise. History is rich with examples of manufacturing enterprises, large and small, unable to deliver products that were competitive in cost, quality, and timely delivery to the marketplace. Manufacturers never intentionally set a course that leads to product or company failure, so why do some succeed and prosper while others fade and die? No doubt, many reasons account for these failures, but studies of U.S. industries indicate that some common denominators influenced the failure. U.S.-based manufacturers were most successful in the captive markets associated with postwar years in the 1950s and 1960s. The successes in these glory years brewed complacency in U.S. companies that resulted in an inability to compete in the open markets of the 1970s and 1980s.

The lessons learned in the 1970s and 1980s resulted in changes across U.S. industries. The changes started in the multinational companies and trickled down to smaller industry groups. As a result of these improved manufacturing practices, U.S. industries reclaimed a leadership role by the mid-1990s and will continue that leadership role in the next millenium.

Foreign competition in Japan and Germany rose from the postwar ashes to build industrial giants that competed on an equal basis with their North American counterparts. When Japan began to penetrate world markets in the 1960s, U.S. manufacturers refused to make the necessary changes in manufacturing methodologies to retain their leadership position. A review of U.S. manufacturing over the last fifty years provides a useful base from which to view the changes necessary for manufacturing to remain strong and competitive in the global market.

Three Stages of Manufacturing Retreat

The first stage in the retreat of U.S. production dominance started during the Vietnam War with the emergence of small electronic consumer goods such as radios and tape decks. The retreat continued in the years that followed. Products with brand names such as Panasonic and Sony began to appear more frequently in U.S. markets and replaced products from General Electric and RCA. U.S. manufacturers mistakenly assumed that cheap labor and central government support were the reasons that Japan could produce these low-cost consumer goods. Although Japan did have a labor-cost advantage, its manufacturers were also developing efficient manufacturing methods and quality systems.

The next stage in the retreat was marked by the Japanese practice of copying successful U.S. products. During this period, the Japanese copied U.S. products and offered similar, higher quality products to U.S. customers at lower cost. In addition to the examples illustrated in Figure 1–1, the shift was especially evident in the automotive industry. Other countries followed this practice as a method of gaining entrance into the U.S. market. The oil shortage of 1973 opened up the automotive market to Japanese and German automotive manufacturers when U.S. customers turned to cheaper, more fuel-efficient imports. What consumers discovered, however, was a product that in many respects was far superior to the alternative built in the United States.

The third stage in industrial competition from offshore companies started in the late 1980s and featured rapid product development and manufacturing of upscale products. In producing these products, offshore companies continued to deliver world-class quality, design, and performance while extracting a premium price. The products, for example, include cars from Toyota and Mercedes-Benz, sound and video equipment from Sony and Phillips, medical and metalworking equipment—the list goes on. These competitive world-class companies proved that they could now play at any level in the production game.

Figure 1–1 Technology Lost: Invented Here, Made Elsewhere. In the Early 1970s, the United States Was the Leading Inventor and Manufacturer in the World.

(Courtesy of Council on Competitiveness, U.S. Department of Commerce.)

U.S.-invented technology	U.S. producers' market share (%)			
	1970	*1975*	*1980*	*1987*
Phonographs	90	40	30	1
Color televisions	90	80	60	10
CNC, DNC centers	99	97	79	35
Telephones	99	95	88	25
Semiconductors	89	71	65	64
Computers	N.A.	97	96	74

Return to Power

The United States regained the status of a manufacturing leader in the 1990s for several reasons, some business, some economic, and others political. A list of the most significant factors in the rebirth of U.S. manufacturing dominance follows.

Economic Factors

- Deregulation of energy and communications markets.
- Low inflation as a result of downward pressures on wages, the prices of raw materials, and the deregulated energy markets.
- Falling interest rates during the last half of the decade.
- The collapse of the Asian economy due to the excesses of financial institutions in managing real-estate portfolios and corporate loans.

Business Factors

- Consolidation of competitive companies and companies with complementary products in most markets.
- Restructuring of corporate America.
- New and expanding technological leadership.
- Partnerships between the United States and offshore companies.
- Adoption of computer integrated manufacturing (CIM) concepts in many industry groups.
- Increased productivity as a result of consolidations, restructuring, technology, CIM, and better labor-management relations.

Political Factors

- The consolidation of the European Union.
- Pressures to open closed markets.

Taken together, these factors permitted the U.S. industry base to enter the next millenium well prepared to compete in the global market.

Product Versus Process Goals

The success of U.S. manufacturers following World War II was due largely to the technology and industrial base spawned by the war and the captive markets associated with the postwar economy. In this period, two-thirds of U.S. research dollars were allocated to product development, while the balance was spent on developing technologies to improve the way in which products were manufactured. As a result, innovative products in transportation and consumer electronics flowed from U.S. industries. In contrast, most European and Japanese industries were neutralized by the war effort and could not compete with the innovative resources in the United States. As a result, the offshore countries, rebuilding their industrial base, focused the majority of their development efforts on improved manufacturing processes. Conceding product innovation to the United States, industries in Japan and Germany cultivated manufacturing technology and production systems that could produce products at higher quality and lower cost. At the same time, U.S. industry continued to emphasize product development research at the expense of better production processes and improved manufacturing technologies. As a result, few U.S. manufacturers were able to take product innovations to market dominance because they were not prepared to be the lower cost, higher quality producers when the product reached market maturity. The items listed in Figure 1–1 are good examples of products invented in the United States but "lost" to Japanese and other offshore industries because of inferior manufacturing technology and production systems in U.S. companies. The standards for world-class performance are based on excellence in production and manufacturing systems because innovation alone is insufficient to guarantee market dominance.

To be competitive a manufacturer must recognize and meet two challenges: one external and the other internal. If manufacturers fail to meet these challenges, the nation becomes a net user of consumable and durable manufactured goods instead of a net supplier. At that point, the standard of living falls in comparison to other superior industrialized nations.

1–2 EXTERNAL CHALLENGES

External challenges result from forces and conditions outside the enterprise. The major external challenges, illustrated graphically in Figure 1–2, include *niche market entrants, traditional competition, suppliers, partnerships and alliances, customers, global economy, cost of money,* and *the Internet.* Let's examine each item in more detail.

Companies have always had product lines called *cash cows.* These were high-revenue-generating products that carried the remainder of the product line. The success of the cash cow is a result of market dominance, technological superiority, patent protection, or the absence of competition. Through the late 1950s and 1960s, IBM's mainframe computer line is an example of this phenomenon. In the late 1960s and early 1970s, however, an employee of IBM recognized the profit potential of this market niche and set up a company, Amdhol, to compete with IBM in this market area. The Windows operating system from Microsoft is another example of a cash cow. In the late 1990s a small company, Red Hat Software, started shipping a competing operating system called Linux. While it is not clear that Linux will take significant market share from Microsoft, it is an example of niche marketing. The production and design technology available today and the global reach of the Internet make it possible for small companies, such as Amdhol and Red Hat Software, to develop a product in a market niche and compete with an established producer.

Traditional competition continues to provide a challenge to manufacturing. Manufacturers often counted on the traditional competitor's ignorance as a buffer; however, those days are gone. For example, a study conducted by a Midwest

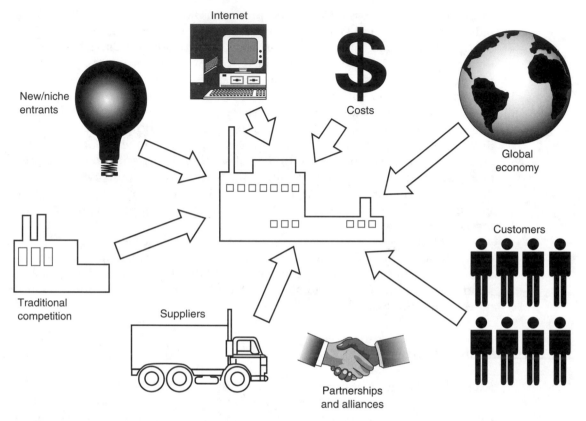

Figure 1–2 External Challenges.

product design and development company indicated that products they got to market six months after the competition reduced their product's profits after taxes by 33 percent. The technology to increase market share is affordable and available to everyone, so the challenge to stay ahead of the traditional competition is stronger today than ever before.

Another external challenge is presented by *suppliers*. Traditionally, many companies have purchased parts and subassemblies based on price variance with vendors selected on a low-bid basis. In this price variance model, the pool of suppliers can change frequently and is often large. Price continues to be a factor; for example, the same Midwest study indicated that a purchase price that was 9 percent higher than the competition resulted in a 22 percent decrease in profits over the product's life. However, in addition to competitive product cost, today's suppliers must *meet minimum defect levels,* have *predictable delivery,* and *shorten product design time.* The challenge for companies is to establish fewer but more collaborative purchaser–supplier relationships to satisfy the demands of the customers. The supplier challenge is often addressed through a process called supply chain management.

Some of the challenges facing manufacturing have been present for several years; the *partnerships and alliances* element is relatively new. Born from the intense global competition of the 1980s and 1990s, this external challenge results from companies with common interests joining forces to increase their competitive advantage and thus gain more market share. The mechanisms for combining assets and resources include mergers, friendly and hostile buyouts, partial ownership through stock purchases, and many types of business agreements. Examples of this challenge appear in every market sector. An example is the merger between Daimler Benz, the German automotive manufacturer, and Chrylser, the number 3 U.S. automotive producer, to create a powerful international company Daimler-Chrylser. With a partnership and alliance, two companies with lower market share can move ahead of the nearest competitor through their combination of customer accounts. The MCI and WorldCom merger is another example of this approach. The combined resources give the new organization a competitive advantage. In addition, a company can add technology faster through acquisition than by the traditional route of developing it in-house. Cisco Systems, the leading producer of network products, is an example. Cisco's numerous purchases of smaller communication companies add new products faster than Cisco could by using existing development resources. When companies within market sectors consolidate, the companies left without a strategic alliance are often not as competitive and run the risk of losing market share and profits.

The most difficult external challenge for manufacturing is provided by the *customers*. In the captive automotive market of the 1920s, Henry Ford could advertise that the public could have a Model T Ford in any color desired, as long as it was black. Today's sophisticated shopper buys on the basis of quality, service, cost, performance, and individual preference. If a manufacturer cannot meet all the needs of the customer, another manufacturer is ready to step in for the sale. The proliferation of the Internet provides access to products from across the world from the customer's home.

Another recent challenge is the business change created by the *Internet*. The development of the World Wide Web (WWW) is changing the way companies do business. The WWW is providing companies with a direct access to customer households. Through either the proliferation of the home computer or through the development of the *set-top box* for the cable TV market, the WWW has entered many homes around the globe. The challenge for companies is to determine the best strategy in this quickly changing market area. A good example is the electronic commerce (e-commerce) area. Before Amazon.com initiated the sale of books over the Web, the future for traditional bookstores appeared to be unchallenged. Now all of the major booksellers are scrambling for a Web presence because the percentage of books purchased over the Internet continues to increase. In addition to e-commerce, the Internet will affect manufacturing by providing windows into a manufacturer's operation from vendor sites and rapid communications and exchange of manufacturing data, and it will replace much of the paper transactions used in the past. Companies that embrace Internet technology intelligently will gain an advantage over their competitors who ignore the power of the WWW or misjudge their correct relationship with this new technology.

Most manufacturers recognize that the present *global economy* provides both an expanded base for marketing products and increased competition from every corner of the world. The challenge for companies is to provide products that meet world-class standards and to market the goods and services to the global economy just as effectively as the competitors both here and offshore.

Another external challenge presented to manufacturers is the *cost of money*. When money is borrowed at any interest rate, the company that uses capital resources the most effectively will be the long-term winner. The cost to introduce and sustain a product in the marketplace has reached staggering proportions. Manufacturing methods that reduce the cost of doing business, such as inventory reduction, become tools for survival. In addition, greater emphasis is placed on how money is used during the design phase to reduce production and delivery costs over the product's lifetime. For example, the results of the study by the Midwest design company indicate that a 100 percent overrun in development cost reduces profit for the product by only 4 percent. These data indicate that spending the money to ensure the best design is wise because money invested on product design up front can maximize profits over the product's lifetime.

Companies must respond to the external challenges (niche market entrants, traditional competition, suppliers, partnerships and alliances, customers, global economy, the cost of money, and the Internet) by:

- Recognizing that these challenges exist and admitting that the problems they create must be solved because they will never go away.
- Developing an internal manufacturing strategy to minimize the negative impact of the external challenges on the success of the business.

1–3 INTERNAL CHALLENGES

The internal challenge is to develop a *manufacturing strategy* that will conform to the following description:

> *A manufacturing strategy is a plan or process that forces congruence between the corporate objectives and the marketing goals and production capability of a company.*

Satisfying this definition is a major problem for many companies. In many organizations a *great manufacturing divide* separates the marketing and production sides of the enterprise (Figure 1–3). On one side of the divide, marketing views orders from a total dollar viewpoint, while on the production side, orders are judged from a volume and product-mix standpoint. The marketing areas seldom consider what the order product mix will do to production efficiency because they are not

Figure 1–3 The Great Manufacturing Divide.

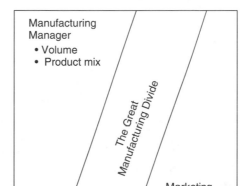

measured against that standard. In contrast, the production area is often judged on capacity utilization and shop productivity measures and not on total order value. As a result, manufacturing is frequently not aligned with the product needs of the marketplace, and the orders accepted by marketing provide a bad fit with the production processes available. A look at a typical corporate planning process (listed below) is the starting point for an understanding of this problem.

1. Define corporate goals.
2. Develop the necessary marketing strategies to satisfy the corporate goals.
3. Analyze the marketplace and determine how the products will fit with market conditions and competition.
4. Determine the process to be used for manufacturing the products.
5. Provide the manufacturing infrastructure necessary for the production.

Corporate planners use an iterative process in developing the first three planning steps because each step affects the other three. Too often, the last two are viewed as simply a linear extension of the first three. Experience indicates that steps 4 and 5 have a significant impact on the first three steps because the manufacturing process is often constrained and can change only over a limited range. In addition, the degree of flexibility in the production infrastructure has limits. However, including manufacturing (steps 4 and 5) in the planning iterations is frustrated by the complexity of the manufacturing processes and differences in the enterprise measurement systems. Read over the five planning steps above until you see the great manufacturing divide located between steps 3 and 4.

A procedure for recognizing both manufacturing and marketing in developing a corporate strategy while providing an ordered and analytical approach is needed. Such a procedure is provided by the Terry Hill model.

Terry Hill Model

The Terry Hill model furnishes a framework (Figure 1–4) that links manufacturing and marketing decisions so that a common executable strategic plan is possible. The five steps in the planning process are listed across the top of the matrix in Figure 1–4, but the third step is renamed *manufacturing order-winning criteria*. Note, for example, that marketing is interested in product markets and segments, range of products, standard versus custom products, and innovation. In contrast, manufacturing strategies focus on process choice and production infrastructure. The framework helps to stimulate corporate debate about the business so that manufacturing can assess the degree to which it can produce the products demanded by the marketplace. However, a common language understood by both marketing and manufacturing is necessary to debate current and future market needs. The common language is based on the order-winning criteria for the product.

Order-Winning Criteria

The order-winning criteria, between the marketing focus and manufacturing strategies, provide a vocabulary to describe product market requirements that can be translated into process choices and infrastructure requisites by manufacturing. Typical order-winning criteria include *price*, *quality*, *delivery speed*, and *innovation ability*. Therefore, *order-winning criteria* are defined as the minimum level of operational capabilities required to get an order. Note that the order-winning criteria provide direction to every unit or group in the enterprise. For example, if quality is an order-winning criterion, the manufacturing unit must provide that feature; however, when design innovation is the criterion, the design group is responsible. Order-winning criteria such as after-sales service, delivery reliability, and flexible financial policies touch areas outside manufacturing. The order-winning criteria provide every area with clear directions.

Although one criterion may be dominant, a mix of order-winning criteria are often present for a product. In addition, the order-winning criteria change over the life cycle of the product (Figure 1–5). The chart in Figure 1–5 indicates that products move through four stages: *introduction, growth, maturity,* and *decline.* Some products exhibit the characteristics of a commodity, and the decline phase levels to a relatively constant market share. Breakfast cereals, for example, exhibit that type of market performance. Note that product sales are at their maximum at maturity. In the early part of the product life cycle, innovation is frequently the dominant order-winning criterion. For example, in the early days of the transistor radio, RCA was one of the dominant companies because it had the research and design talent to satisfy the dominant order-winning criterion, innovation. When a product reaches maturity, price frequently becomes the key order-winning criterion. In the previous example, Japanese companies took over the transistor radio market during maturity because they could satisfy the dominant order-winning criterion, price.

Corporate goals	Marketing strategy	Manufacturing order-winning criteria	Manufacturing strategies	
			Structure	Infrastructure
What the company is going to do: • Growth • Profit margin • Other financial measures	How the company will reach desired goals: • Product markets and segments • Mix • Volumes • Standardization versus customization • Level of innovation • Leader versus follower	• Price • Quality • Lead time • Delivery/reliability • Flexibility • Innovation/ability • Size • Design Leadership	• Capacity • Facilities • Technology (processes used) • Vertical Integration (degree to which all parts are produced internally)	• Work force (pay, skill level) • Quality achievement process • Manufacturing, planning, and control system • Organization (control, measurement, motivation)
External Focus		Common Language	Internal Focus	

Figure 1–4 Order-winning Criteria Model.

Figure 1–5 Product Life-cycle Curve.

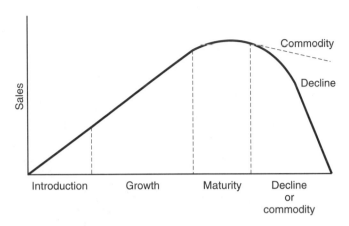

Changing the Product Life Cycle

Many U.S. manufacturers try to lengthen the maturity stage of the product life cycle in Figure 1–5 because that part of the curve provides maximum profits. Costs associated with product development are considered an *investment*, and a longer maturity stage represents a better return on that investment.

In contrast, some Japanese competition views the development cost of the initial product as *sunk cost*—just the cost of doing business. With this view, the business pressure for a long model life is not present. However, a shorter model life is not synonymous with design obsolescence; the product continues to have high quality standards for design and operational life. The model life is intentionally shortened by the design and introduction of a new and better model using one of three methods: *kiazen, leaping,* or *innovation*. The first method, *kiazen,* focuses on improvement of the current model. For example, the improvement team may be challenged to create a new model with enhanced features at a 10 percent reduction in cost within twelve months of the original product introduction. The second method, *leaping,* attempts to develop a different product with similar form, fit, and function characteristics to the initial product. The Sony Walkman is an example of this technique, with the initial product being a standard tape recorder. The last method, *innovation,* seeks to use genuine new product invention to identify follow-up merchandise.

This three-pronged approach to shorten model life is frequently accomplished with three cross-functional teams, each team working on one of the three methods. When this technique is executed successfully, three distinctly different models are introduced as a follow-up to the current product. As a result, profits grow through increased market share in the product area as consumers trade up from the original product or abandon an older model from a competitor.

Regardless of the length of the product life, the match between what marketing is selling over the life of a product, and the ability to design and manufacture the product desired, is strengthened by dialogue in a language both manufacturing and marketing understand. The order-winning criteria for the product provide this common language.

Order-Winning Versus Order-Qualifying Criteria

It is important to make the distinction between criteria that win product orders and those that qualify a product to compete in the marketplace. An example best describes this difference. In the 1970s the order-winning criterion for the color television market in the United Kingdom was price. The Japanese entered the market with products far superior in quality and reliability to the existing U.K. products. The British public set a new quality and reliability standard for television purchases based on the imported sets. The price could not drop low enough for the U.K. companies to sell sets that were below the new quality and reliability standard. Therefore, quality and reliability became the new order-qualifying criteria for selling televisions in the U.K. market. By the 1980s U.K. companies matched the quality and reliability standards set by the Japanese, and the order-winning criterion reverted to price. At this point the order-qualifying criterion became quality.

The enterprise must, therefore, provide the order-qualifying criteria to get into the market or maintain its market share. After the criteria are satisfied, the enterprise attends to the criteria on which orders are won. Market share is increased when the order-winning criteria are understood and executed better than the competition. Market leaders must also have the ability to anticipate when the order-winning criteria are changing during the product's life cycle.

The Terry Hill model provides a process and language that permit marketing and manufacturing to discuss the needs of the marketplace and the degree to which manufacturing process choices and infrastructure can satisfy these market needs. The intention in this text is not to describe the Terry Hill model in the detail required for implementation, but to emphasize that the internal challenge for every company is the development of a process that permits congruence between the corporate marketing goals and production capability.

Meeting the Internal Challenge

In simplistic terms the steps to achieve internal consistency are:

1. Analyze every product and agree on the order-qualifying and order-winning criteria for the product at the current stage in the product's life.
2. For every product, project the order-winning criteria for the future stages in the product's life.
3. Determine the *fit* between the required process capability or level and the existing capability in manufacturing.
4. Change or modify either the marketing goals or upgrade the manufacturing processes and infrastructure to force internal consistency.

A final word of caution is needed on setting the order-winning criteria necessary to meet marketing's goals. A tendency exists to select too many different order-winning criteria, thus failing to identify the criterion most critical to winning orders. Despite the imperfections in the process, it is critical for marketing to identify the order-winning criteria as prerequisites for the development of a corporate marketing strategy.

1–4 WORLD-CLASS ORDER-WINNING CRITERIA

A review of the order-winning criteria most often present indicates that the manufacturing unit in the enterprise has the responsibility to satisfy the criteria the majority of the time. Most of the criteria listed in the third column in Figure 1–4 must be addressed by the manufacturing unit. Manufacturing also shares the responsibility in overcoming the external challenges. For manufacturing to meet this responsibility, improvement in several shop floor standards is necessary. A table of manufacturing standards compiled by Coopers and Lybrand is given in Figure 1–6, with the U.S. average compared to the world-class standard. You can also view this comparison as the standard for the best manufacturers worldwide versus the average for all manufacturers. How well a company compares with the best in the world is an indication of how ready it is to provide the criteria necessary to win orders. The values used for the U.S. average are open for debate because they will change with industry group and geographic area studied. However, it is important for every company to determine how well it compares with the process capability of other world-class manufacturing units. In the following sections, we describe each attribute and indicate the order-winning criterion affected.

Setup Time

Setup time is the time required to get a machine ready for production. A long setup time causes the size of the minimum number of parts produced, or *lot size*, to be large. In addition, a long setup time results in higher manufacturing costs because of larger inventories. With more time allocated to setting up the machine, less time is

Figure 1–6   Internal Challenges.

Attribute	Manufacturing standards	
	World-class standard	U.S. average
Setup time		
System	< 30 minutes	24 hours
Cell	< 1 minute	
Quality		
Captured	1500 ppm	3–5%
Warranty	300 ppm	2–5%
Cost of quality	3–5%	15–25%
Manufacturing/total space	> 50%	25–35%
Inventory		
Product velocity	> 100 turns	2–4 turns
Material residence time	3 days	3 months
Flexibility	270 parts	25 parts
Distance	300 ft.	> 1 mile
Uptime	95%	65–75%

spent in production on the machine, so capital expenditures increase because more equipment and facilities are required. Calculating setup time cost requires a value for the burden rate. The *burden rate* is a cost rollup that includes: recovery of the original cost of the process machine and associated support hardware, such as robots and material handling equipment; maintenance costs (fault and preventive); labor costs (direct and indirect); programming for computer numerical control (CNC) machines and robots; and any other costs associated with having the machine in the production system. A typical range of burden cost for an automated production system might be $200 to $500 per hour, depending on the level of automation present. These costs accumulate every hour of the day. One part of the cost of manufactured items produced on the machine is set by the burden rate of the production machine. The following exercise describes the impact that setup time and burden rate have on production.

Exercise 1–1

A machine burden rate (the cost per hour to have the machine in production) is $150 per hour. Production requires 3 setups of 45, 90, and 50 minutes over 3 shifts. What is the annual cost of the setup time for a 6-day production operation?

Solution

$$\text{Total lost time per day} = 45 + 90 + 50 \text{ min}$$
$$= 185 \text{ min} = 3.08 \text{ hr}$$
$$\text{Lost time per year} = 3.08 \text{ hr} \times 6 \text{ days} \times 52 \text{ weeks}$$
$$= 961 \text{ hr}$$
$$\text{Annual cost} = 961 \text{ hr} \times \$150 \text{ per hr}$$
$$= \$144,150$$

The order-winning criteria affected include price, delivery speed, and flexibility.

Exercise 1–2

A production system has a burden rate of $300 per hour. The setup time for a machined casting is 3.6 hours, and the system can produce a finished part every 6 minutes.

a. For a lot size of 500 parts, how much of the part cost is associated with setup cost?

b. To reduce inventory levels, the ideal lot size is 100 parts. What new setup time is required to maintain the same setup cost on the parts?

c. The finished part cost is the sum of part production cost, setup cost, and a 30 percent margin. What percentage of the final cost results from the setup time for the 500 and 100 lot size production?

Solution

a.

$$\text{Setup time cost} = \text{setup time} \times \text{burden rate}$$
$$= 3.6 \text{ hrs} \times \$300$$
$$= \$1080$$

$$\text{Setup cost per lot size part} = \frac{\text{setup time cost}}{\text{lot size}}$$
$$= \frac{\$1080}{500}$$
$$= \$2.16$$

b. Determine the total setup time cost for a lot size of 100 at a per part setup cost of $2.16.

$$\text{Total setup time cost} = \text{lot size} \times \text{setup cost per part}$$
$$= 100 \times \$2.16$$
$$= \$216$$

$$\text{Setup time (new)} = \frac{\text{setup time cost}}{\text{burden rate}}$$
$$= \frac{\$216}{\$300 \text{ / hr}}$$
$$= 0.72 \text{ hrs}$$
$$= 43.2 \text{ mins}$$

c. Determine parts prod./hour, part production cost, and total part cost.

$$\text{Part production per hr} = \frac{60 \text{ min}}{\text{part production time (min)}}$$
$$= \frac{60 \text{ min / hr}}{60 \text{ min / part}}$$
$$= 10 \text{ parts/hr}$$

$$\text{Part production cost} = \frac{\text{burden rate}}{\text{parts production per hr}}$$
$$= \frac{300 \text{ / hr}}{10 \text{ parts / hr}}$$
$$= \$30/\text{part}$$

$$\text{Finished part cost} = (\text{production part cost} + \text{setup cost}) \times 1.3$$
$$= (\$30 + \$2.16)1.3$$
$$= \$41.81$$

$$\text{Setup percentage} = \frac{\text{setup cost}}{\text{finished part cost}} \times 100$$
$$= \frac{\$2.16}{\$41.81} \times 100$$
$$= 5.2\%$$

Quality

Quality is expressed as a percentage of defective parts produced or as a percentage of total sales in the three values in Figure 1–6. *Captured quality* is a measure of defects in units of parts per million and percentage that are found any time before the product is shipped to the customer. *Warranty quality* represents defective parts discovered after the product is shipped to the customer. *Total quality* represents the total cost in percentage of sales that quality costs the enterprise. A useful rule of thumb is that the total cost of poor quality in percentage of sales is usually three times the captured cost plus the warranty quality cost.

Exercise 1–3

Quality is often represented as a percentage and with units of part per million (ppm), as illustrated in Figure 1–6. Determine the number of bad parts for the following lot sizes using the U.S. average (use the small percent values in the range) and world-class standard values for captured quality listed in Figure 1–6.

a. 50,000 parts
b. 5000 parts
c. 500 parts
d. 50 parts

Solution

a. Determine bad parts for quality represented as a percentage of parts and in ppm notation.

$$\text{Bad parts} = \text{total parts} \times \frac{\% \text{bad parts}}{100}$$

$$= 50{,}000 \text{ parts} \times \frac{3\%}{100}$$

$$= 1500 \text{ parts (U.S.)}$$

$$\text{Bad parts} = \text{total parts} \times \frac{\text{ppm bad parts}}{10^6}$$

$$= 50{,}000 \text{ parts} \times \frac{1500}{1{,}000{,}000}$$

$$= 75 \text{ (world-class standard)}$$

b. Bad parts $= 5{,}000 \text{ parts} \times 0.03$
$= 150 \text{ parts (U.S.)}$

$$\text{Bad parts} = 5{,}000 \times 0.0015$$
$$= 7.5 \text{ parts (world-class standard)}$$

   c. Bad parts   = 15 parts (U.S.)
             Bad parts = 0.75 parts (U.S.) (world-class standard)
   d. Bad parts   = 1.5 parts (U.S.)
             Bad parts = 0.075 parts (U.S.) (world-class standard)

The order-winning criteria affected by this standard are quality and price.

Manufacturing Space Ratio

The *manufacturing space ratio* standard is a measure of how efficiently manufacturing space is utilized. The total footprint of the machines, plus the area of work-stations where value is added to the product, is divided by the total area occupied by manufacturing. The larger this ratio, the more efficient the production operation. Manufacturing shop floor space allocated to material handling, material transport, storage of raw materials and finished goods, in-process work queues, inspection, expediting, and supervision decrease this ratio. Therefore, the less floor space devoted to these non-value-added elements, the higher the ratio. The order-winning criterion affected is price.

Exercise 1–4
An automated work cell used by a manufacturer covers 300 square feet of factory floor space. The CNC process machine in the cell has a footprint that measures 75 inches by 80 inches. The remaining space includes pallets for raw materials and finished parts, an operator's desk, walking space around the machine, and an inspector's bench.

a. Calculate the manufacturing space ration for the work cell.
b. Calculate the total work cell area if the cell conforms to the world-class standard listed in Figure 1–6.

Solution:
a. Convert machine footprint to square feet. Then find ratio.

$$\text{Machine footprint (ft}^2) \quad = \frac{75 \text{ in.}}{12 \text{ in.}} \times \frac{80 \text{ in.}}{12 \text{ in.}}$$

$$= 6.25 \text{ ft} \times 6.67 \text{ ft}$$
$$= 41.69 \text{ ft}^2$$

$$\text{Manufacturing space ratio} = \frac{\text{process machine footprint}}{\text{total work cell area}} \times 100$$

$$= \frac{41.69 \text{ ft}^2}{300 \text{ft}^2} \times 100$$

$$= 13.9\%$$

b. Total work cell area $= \dfrac{\text{process machine footprint}}{\text{manufacturing space ratio (world - class)}}$

$$= \dfrac{41.69 \text{ ft}^2}{50\%}$$
$$= 83.4 \text{ ft}^2$$

Inventory: Velocity/Residence Time

Raw materials, partially completed parts, subassemblies, and finished goods that are waiting to be worked on or shipped to the customer represent inventory. Manufacturing *inventory* can be grouped into two broad categories: *movement* and *organization*. Movement or transport inventories, sometimes called *in-transit* or *pipeline*, are necessary because time is required to manufacture products or to transport goods from one location to another. Movement inventories in manufacturing are called *work-in-process* (WIP) inventories. The design and type of manufacturing system used dictates the level of movement inventory present in the system.

Organization inventories uncouple successive stages in the production and distribution systems to neutralize the disturbances in the system. Organization inventories are classified into the following three groups: cycle stock, safety stock, and anticipation stock. *Cycle stock* inventories exist when the quantity of produced or purchased parts is larger than the current requirement. The inventory costs associated with this excess number of manufactured or purchased parts are less than the costs of manufacturing the parts in smaller quantities or of purchasing small lot sizes. Using the calculation from Exercise 1–1, for example, you could argue that a larger than ideal lot size production is justified because the inventory costs are less than the costs associated with more frequent machine setup.

Safety stock provides protection against irregularities and uncertainties in the demand or supply stream. The most common example is the stocking of extra items of raw materials due to the uncertainty attached to production quality. In many manufacturing systems it is difficult to predict accurately how many production parts will be scrapped due to poor quality.

Anticipation stock is associated with products whose markets exhibit seasonal patterns of demand and whose supply stream is relatively constant. It is often not economically feasible to produce a sufficient number of products on a weekly or monthly basis to supply a seasonal demand. Examples are Christmas toys, calendars for the following year, and lawn mowers for the spring.

The inventory needed to reduce disturbances in manufacturing and in filling customer orders is a necessary cost. However, the goal is to reduce the inventory level continuously while maintaining continuous and smooth production plus on-time and complete order shipments to the customer. Inventory levels are measured in terms of inventory turns, or *velocity*, and in residence time. The number of *inventory turns* for a product is equal to the annual cost of goods sold divided by the average inventory value. The *residence time* is the average number of days that a part or raw material item spends in production. If the goal is reduced inventory, then the number of inventory turns should increase while residence time must decrease. If this standard is performed well, parts and raw materials spend less time in the plant. As a result, the order-winning criteria that benefit are usually price, quality, delivery speed, and delivery reliability.

Exercise 1–5

A manufacturer has an annual sales volume of $500 million with the cost of goods sold (COGS) at $0.90 per sales dollar. The average inventory value is $25,000,000. What are the inventory turns for this manufacturer?

Solution

$$\text{COGS} = \$500,000,000 \times \$0.90$$
$$= \$450 \text{ million}$$
$$\text{Inventory turns} = \$450,000,000/\$25,000,000$$
$$= 18$$

Flexibility

Flexibility is a measure of the number of different parts that can be produced on the same machine. Improved efficiency in this standard results from good part design, innovative fixture design, and well-planned production machine procurement. Excellence in this standard aids primarily the price order-winning criterion. Flexibility also measures the ability to produce new product designs in a short time. To force a shorter product life cycle, the enterprise must have flexibility as an order-winning criterion.

Distance

The *distance* standard measures the total linear feet of a part's travel through the plant from raw material in receiving to finished products in shipping. A high value for this standard indicates more non-value-added time for the part and decreased quality due to handling: therefore, higher cost result. Therefore, keeping the value of this standard low helps the price and quality order-winning criteria.

Uptime

Uptime is the percent of time a machine is producing to specifications compared to the total time that production can be scheduled. A reduction in the uptime standard means that more equipment is required for the same level of production. For example, industry data indicate that every 1 percent improvement in uptime is worth 10 percent in capital equipment cost. As a result, improvement in this metric helps the price order-winning criterion.

1–5 THE PROBLEM AND A SOLUTION

An enterprise must recognize the external challenges (niche market entrants, traditional competition, suppliers, partnerships and alliances, customers, global economy, the cost of money, and the Internet) that are present and develop a manufacturing strategy to win orders based on the criteria present in the marketplace. The problem is that the order-winning criteria drive the market. The enterprise cannot change these criteria, and the environment that creates the external challenges will not go away. Therefore, the enterprise must change.

The Cost of Doing Nothing

Figure 1–7 illustrates the consequences to a product for a manufacturer who chooses to ignore the forces present in the marketplace. The next product is less competitive than its predecessor, so the margins continue to drop. The southeastern United States had numerous examples of textile manufacturers that followed this downward spiral in the 1960s and 1970s. One case is that of a mill that produced first-quality cloth in the 1950s and could sell in markets of its own choosing. Over time the production machines and processes aged and did not stay even with world competition; as a result, the manufacturer was relegated to producing only low-market value muslin before closing its doors in the 1970s. A similar story can be told in countless other market sectors, steel production being one example. The turbulence in manufacturing from the mid-1960s to the early part of the 1980s

Figure 1–7 Cost of No Action.

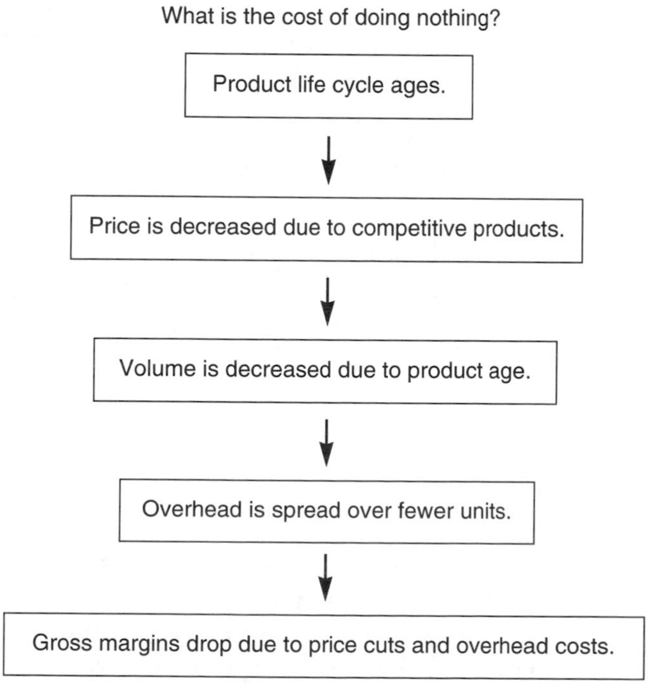

What is the cost of doing nothing?

Product life cycle ages.

↓

Price is decreased due to competitive products.

↓

Volume is decreased due to product age.

↓

Overhead is spread over fewer units.

↓

Gross margins drop due to price cuts and overhead costs.

caused a dramatic change in U.S. manufacturing. In general, the pressures created a stronger manufacturing base in most market sectors. However, the challenge from on- and offshore competition requires that a company be alert. With the life of the enterprise at risk, what should be done?

A Solution

The following seven world-class measures must be addressed by manufacturers.

1. Design and manufacturing lead time by product.
2. Inventory turns by product.
3. Setup times on production equipment.
4. Production efficiency by product.
5. Employee output/productivity by product.
6. Total quality and rework.
7. The number of product improvement suggestions per day per employee.

In general, these seven basic standards are often not adequately addressed by companies. Manufacturers looking to improve market share or reduce product cost frequently improve one area with an *island of automation*. Improvements in design lead time using computer-aided design (CAD) software are a good example of an island of automation that is frequently built. However, fixing one problem while ignoring others often reduces overall productivity. Manufacturing presents an integrated set of problems that require an integrated solution. That solution is *computer-integrated manufacturing*. The late Joseph Harrington Jr. recognized the value of integration and in 1973 introduced the term *computer-integrated manufacturing* in his book of that title. Although he intentionally avoided developing the acronym CIM (pronounced "sim"), the abbreviation was in common use by the early 1980s.

Computer-Integrated Manufacturing Defined

The Computer and Automation Systems Association (CASA) of the Society of Manufacturing Engineers (SME) defined *CIM* as follows:

> *CIM is the integration of the total manufacturing enterprise through the use of integrated systems and data communications coupled with new managerial philosophies that improve organizational and personnel efficiency.*

An appendix to this chapter, the *Industry Week* survey titled "The Benefits of a CIM Implementation," supports the use of CIM to deliver a host of order-winning criteria.

The Manufacturing Enterprise wheel (Figure 1–8) illustrates the integration called for in the definition and shows the interrelationship among all parts of an enterprise. Take a few minutes to study all the segments present in the wheel, and then read the definition again.

CIM describes a new approach to manufacturing, management, and corporate operation. Although CIM systems can include many advanced manufacturing technologies such as robotics, computer numerical control (CNC), computer-

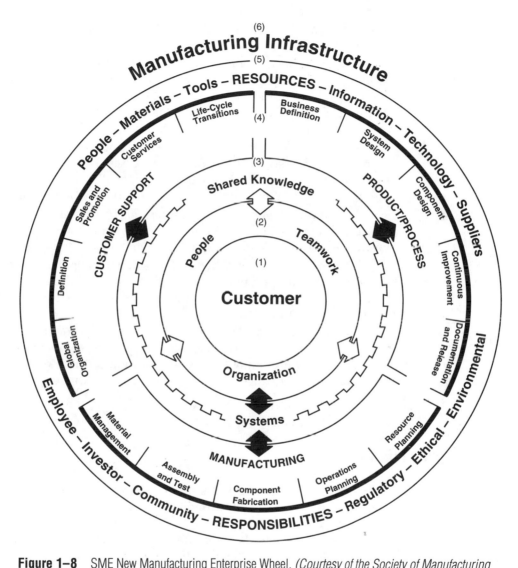

Figure 1–8   SME New Manufacturing Enterprise Wheel. *(Courtesy of the Society of Manufacturing Engineers, Deerborn, Michigan, copyright 1993, Third Edition.)*

aided design (CAD), computer-aided manufacturing (CAM), and just-in-time (JIT) production, it goes beyond these technologies. CIM is a new way to do business that includes a commitment to total enterprise quality, continuous improvement, customer satisfaction, use of a single computer database for all product information that is the basis for manufacturing and production decisions in every department, removal of communication barriers among all departments, and the integration of enterprise resources. The CIM concept model in Figure 1–8 is an update from the original SME/CIM Wheel and has the following defined areas:

1. The hub of the wheel, titled *Customer,* is the primary target for all marketing, design, manufacturing, and support efforts in the enterprise. Only with a

clear understanding of the marketplace and the customer can the enterprise be successful.

2. The next layer on the wheel focuses on the means of organizing, hiring, training, motivating, measuring, and communicating to ensure teamwork and cooperation in the enterprise. The techniques used to achieve this goal include self-directed teams, teams of teams, organizational learning, leadership, standards, rewards, quality circles, and a corporate culture.

3. This section focuses on the shared corporate knowledge, systems, and common data used to support people and processes. The resources used include manual and computer tools to aid research, analysis, innovation, documentation, decision making, and control of every process in the enterprise.

4. The three main categories of processes, *product/process definition, manufacturing,* and *customer support,* make up this section of the wheel. Included in this group are fifteen key processes that form the product life cycle.

5. The enterprise has resources that include capital, people, materials, management, information, technology, and suppliers. It also has responsibilities to employees, investors, and the community, as well as regulatory, ethical, and environmental obligations.

6. The final part of the wheel is the manufacturing infrastructure. This infrastructure includes customers and their needs, suppliers, competitors, prospective workers, distributors, natural resources, financial markets, communities, governments, and educational and research institutions.

The CIM principles covered in this text address most of the areas in CASA/SME New Manufacturing Enterprise Wheel.

1–6 LEARNING CIM CONCEPTS

Just as the Enterprise wheel divides the operational aspects of CIM into major segments, the study of CIM in the following chapters is divided into three corresponding parts. The first part introduces and defines CIM. In the second part, the three process segments are described and the automation options are explored. Finally, in the third part, implementation is addressed together with management of the information systems and human resources. Learning often starts with the process segments.

Process Segments

Figure 1–9 illustrates another view of the CIM structure, including three process segments with overlapping areas that indicate shared data and resources. The product design data, for example, is needed in manufacturing planning and control (MPC) to plan process routings. The degree of the overlap varies because some segments share more information than do others. In addition, the three process segment circles are supported by the resources present in the circle that encloses them. The development of a new product usually starts in the design circle and moves in the direction of the arrows in Figure 1–9. The CIM concepts

Figure 1–9   The Enterprise Areas.

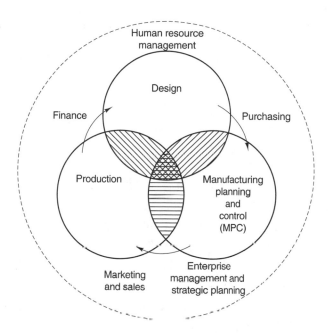

required for each segment are covered in the same order in this book. As a result, Part 2 starts with the design and documentation work required when a new product is produced. In the CIM wheel this is called "product/process" and includes *business definition, system design, component design, continuous improvement,* and *documentation and release*. Discussions about this area in the literature usually reference computer-aided design (CAD), the automation software frequently used to improve design and drawing productivity. However, implementing CIM in the product/process segment requires more than just using CAD software. In Chapters 3 through 5 we describe the CIM requirements and benefits in this segment of the CIM wheel and enterprise.

The second process segment required for product development, labeled *manufacturing planning and control* (MPC) in Figure 1–9, includes all the process planning, scheduling, inventory management, and capacity planning required for efficient manufacturing. In the Enterprise Wheel (Figure 1–8), the segment called manufacturing includes elements from MPC and from the production circle in Figure 1–9. The elements include *material management, assembly and test, component fabrication, operations planning,* and *resource planning*. Manufacturing resource planning (MRP II) concepts and software are frequently applied to manage this area of the production process. Implementation of MPC software such as MAPICS or Fourth Shift does not guarantee that this segment complies with the CIM definition, however. The requirements for a CIM implementation in this area of manufacturing are detailed in Chapters 6 through 8.

The last process segment, labeled *production* in Figure 1–9, includes all the activity associated with the production or shop floor. This includes some of the elements found in the manufacturing segment of the wheel. The application of CIM principles for the shop floor or production areas is covered in chapters 9 and 10. Frequently, manufacturers confuse the installation of flexible manufacturing cells (FMCs), computer numerical control (CNC) machines, smart conveyor systems, automatic tracking, or industrial robots with the implementation of CIM. However, a careful examination of the CIM definition and wheel indicates that CIM is *not* just:

- Automated hardware and software.
- A manufacturing system bought from a vendor, installed in the enterprise.
- Manufacturing strategies such as just-in-time (JIT) and simultaneous engineering.

An enterprise that embraces CIM concepts has a managerial philosophy that:

- Uses customer satisfaction as the basis for decisions.
- Espouses total quality (TQ) principles.
- Values the ideas of every employee.
- Does not accept the status quo but works toward continuous improvement.

In addition, a successful CIM implementation:

- Has a common database with data shared across all the departments.
- Uses automation hardware and software to integrate enterprise operations effectively so that product data are created only once and used many times.

However, the process segment produces a world-class product only when the enterprise resources are well-managed.

1–7 GOING FOR THE GLOBE

The ultimate goal of the enterprise is to develop an internal strategy that leads to world-class levels of manufacturing performance. Order-winning criteria are used to identify and rank the world-class performance measures most critical for manufacturing success in a given market. The bar graph of CIM benefits in the appendix at the end of the chapter implies that CIM is synonymous with world-class measures such as lower manufacturing cost, higher product quality, better production control, better customer responsiveness, reduced inventory, greater flexibility, and smaller lot-size production. Therefore, a CIM implementation starts an enterprise on a journey toward acquiring these world-class standards. A three-step process for CIM implementation follows.

The CIM Process: Step 1

Implementation of a successful CIM system follows a three-step process.

STEP 1: Assessment of the enterprise in three areas:

- *Technology.*
- *Human resources.*
- *Systems.*

Building an enterprisewide CIM system requires extensive planning, many months of hard work, and a substantial investment in people, hardware, and software. Assessment must be the first step because it prepares the enterprise for step 2, simplification, and step 3, implementation. The assessment of the enterprise technology, human resources, and systems includes a study to determine:

- The current level of technology and process sophistication present in manufacturing.
- The current state of employee readiness for the adoption of CIM automation across the enterprise.
- The reason *why* the production systems function as they do.

In each of the three areas the capability, strengths, and weaknesses are checked and documented. The assessment process consists of an internal self-study with a large educational component. The critical nature of education is illustrated in Figure 1–10 by a survey of 139 companies planning a CIM implementation. Note that 55 percent of the respondents listed *lack of in-house technical expertise* as a major obstacle to CIM implementation. In-house expertise in the design and construction of automation systems is often not present in smaller companies. To overcome this deterrent, the education component for every employee must focus on:

- The necessity for enterprise change to remain competitive in a national and world marketplace.

Figure 1–10   Obstacles to a CIM Implementation.

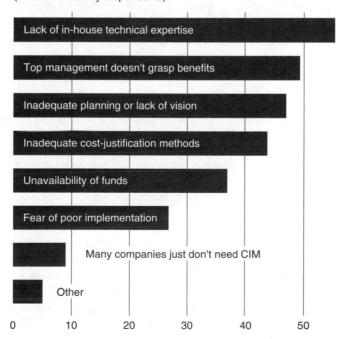

(Percent of survey respondents)

- The need to support a new model in enterprise operations that includes teamwork, total quality, improved productivity, reduced waste, continuous improvement, common databases, and respect and consideration for all ideas regardless of the level from which they are initiated.
- The hardware and software necessary to implement a CIM system, and the management strategy required to run the system successfully.

CIM is not hardware and software; CIM is a way to manage the new technologies for improved market share and profitability. From the start of the implementation, everyone in the organization must understand how CIM relates to his or her job. As a result, assessment and education must be first.

The CIM Process: Step 2

In step 1 the enterprise was studied department by department, process by process, and activity by activity. Personnel were educated about the company and the technology planned for the CIM implementation. With this base established, step 2 is begun.

STEP 2: Simplification—elimination of waste

In most departments, CIM moves the manual or functionally automated operations into an integrated enterprisewide solution. For example, the manual record-keeping process for tracking the location and quantity of some inventory moves to a computerized system, with all inventory information stored in a common electronic database. Installing automation, for example, without first eliminating the unneeded operations in the inventory process would just automate many of the poor practices present in the manual system. *Simplification* is defined as follows:

Simplification is a process that removes waste from every operation or activity to improve the productivity and effectiveness of the department and organization.

What is waste, and where is it found? *Waste* is every possible operation, move, or process that does not *add value* to the final product. If an activity log was made for raw material as it passed through manufacturing, the material time would be divided into five general categories: *moving, waiting in queue, waiting for process setup, being processed,* and *being inspected.* Only one of the five, *being processed* by the production machine, adds value to the part, and then only if the process performs within specifications. The remaining are classified as *cost-added* operations because the part is *not* increasing in value as a result of the activity, and the enterprise continues to pay overhead costs during that time. In short, any operation or activity that does not add value to a product is cost added and is waste. Removing all waste from operations is often not possible; however, all avoidable cost-added processes must be eliminated.

The search for waste often focuses on direct labor cost. Yet in most manufacturing operations, direct labor usually accounts for only 2 to 10 percent of cost of the goods sold. Therefore, eliminating waste by direct labor cost alone would not provide significant savings. However, case studies show that cost-added operations account for as much as 70 percent of all activities in an enterprise. Eliminating as many of the cost-added operations as possible requires significant change in the daily business and manufacturing process. The rules in Figure 1–11 offer a process designed to stimulate enterprise change and reduce the level of cost-added operations significantly. Armed with a comprehensive list of cost-added operations for an area, the system or manufacturing operation is changed to eliminate half of the waste, a 50 percent reduction. Reducing waste in these significant amounts requires that the enterprise *attack waste fundamentally.* For example, a plant uses a forklift to move material between production machines in a batch manufacturing operation. To reduce the transit time between machines, an identified cost-added activity, management links the production machines with a material-handling conveyor. *Can you recognize the problem with this waste reduction solution?* The process is quicker, but material is still moved, work-in-process inventory is not less, and the batch sizes have not changed significantly. The conveyor is a *superficial improvement.* A *fundamental improvement,* elimination of transportation, is achieved by moving the production machines close together. Identifying fundamental waste reduction requires that every employee be trained in the fundamentals of the business during the first step in the CIM implementation process.

The reduction of waste by 50 percent in Figure 1–11 the first step in the process of eliminating 90 percent of the identified cost-added processes. As the figure indicates, the 50 percent of the waste still remaining is reduced by 50 percent, so the total waste present is only 25 percent of the original amount. The process continues until no more than 10 percent of the original waste is present. At this point, the manufacturing process is sufficiently stripped of unnecessary cost-added operations so that a successful CIM implementation is possible. However, the war on waste is never over because employees must use the *continuous improvement* technique to continue searching for cost-added operations to eliminate.

An aggressive simplification exercise on the shop floor to reduce cost-added activity ensures that what gets automated by the shop floor hardware and software

Figure 1–11   Rules for Elimination of Cost-added Operations.

Three-Step Rule for Eliminating Waste

Reduction	Total (%)
Reduce by 50%	50
Reduce by 50% again	75
Make it 10% of what it originally was	90

A world-class company works the three-step rule to arrive at a 90 percent reduction in waste.

is not bad production processes. When the simplification process is applied in every office and department in the enterprise, performance improves and installed automation is effective. World-class manufacturing requires the elimination of as much of the cost-added activity from the entire enterprise as possible during the simplification step in the CIM implementation.

The CIM Process: Step 3

The first two steps prepared the enterprise for CIM implementation; step 3 builds the system.

STEP 3: Implementation with performance measures

Implementation with *performance measures* has two requirements: (1) measuring the success of CIM implementation at regular intervals, and (2) recording the changes in key manufacturing and business parameters. The objective for the enterprise is to become world class, so the enterprise must track the level of achievement in moving to the new world-class performance. One tracking technique uses four key standards from Figure 1–6, along with checks on productivity and continuous improvement efforts. The seven key measurement parameters are:

- Product cycle time.
- Inventory turns by product.
- Production setup times.
- Manufacturing efficiency.
- Quality and rework.
- Employee output/productivity.
- Employee continuous improvement suggestions.

Initially, operational data from all seven measurements are recorded to establish a baseline before starting implementation of any automation or process improvements. In enterprises with successful CIM programs, the performance measurements are reviewed monthly to record changes in the performance. However, for the initial CIM implementation, companies often start by measuring just two or three of the first five measurement parameters listed previously. They focus first on the parameter(s) that would provide the greatest initial payback. Inventory turns is often selected first because most manufacturing operations have some excess inventory, and dollar savings are easy to identify. As control of the processes is established, additional performance measures are added to the improvement project. The results listed in Figure 1–12 show the changes achieved over a 6-month period by a valve manufacturer with $25 million in sales. The remarkable improvements in *floor space* (manufacturing efficiency) are a result of a shift from job shop and repetitive production environments to a *structured flow* system that groups machines by product. The case study at the end of the chapter illustrates a similar improvement process for another company. Compare the final results for the valve manufacturer in Figure 1–12 and the company described in the case study with the world-class standards shown in Figure 1–6.

Valve Manufacturer's Performance Report Card Case History

	Baseline	*Six months*	*Eighteen months*
Cycle time	18 weeks	6 weeks	1 week
Inventory turns	4	8	48
Quality (finished part)	85%	95%	99.8%
Floor space	800 ft^2	400 ft^2	80 ft^2

Figure 1–12   Results of Improved Business Operations.
(Source: Stickler, Going for the Globe Part III, P and IM Review, November, 1989, p. 42. Reprinted by permission of APICS.)

Another performance measurement technique, called the *ABCD checklist*, was developed by the Oliver Wight Companies. The checklist measures enterprise progress toward world-class performance in the following five basic business functions: strategic planning processes, people/team processes, total quality and continuous improvement processes, new product development processes, and planning and control processes. A company uses the rating scale in Figure 1–13 to answer a series of sixty-eight questions. Based on the average of the numeric score, the company is classified as class A, class B, class C, or class D. The overview questions are included in Appendix A at the end of the book. *Take several minutes and review the questions.* Determining the response for some overview questions is difficult without additional clarification. This clarification is provided by a group of detailed questions provided for some of the overview questions.

The audit process to determine the ABCD classification level is usually performed by a team of company managers who know current production performance and a representative of Oliver Wight Companies, who facilitates completion of the ABCD checklist instrument. The general characteristics of companies operating at the Class A, B, C, and D levels is also provided in the appendix. *Review the characteristics of a Class A company.* As you can see, achieving a class A designation implies world-class performance. Establishing the class of operation is not the critical factor; however, identifying the areas in the enterprise that caused a score of less than class A is valuable. Based on the overview question results, processes can be improved, required training for personnel can be identified, and continuous improvement can be tracked. Reread the case study at the end of the chapter and note the journey to class A status and world-class performance.

The final step in the CIM implementation process consists of the acquisition and installation of hardware and software to the specifications developed in steps 1 and 2. When assessment and simplification are complete and a performance measurement system is in place, successful implementation of hardware and software is assured. Unfortunately, many companies implement CIM automation starting with step 3. As a result, they automate all the disorder of the poor production processes and produce waste at a record rate.

Level	Planning and control processes	Continuous improvement processes
Class A	Effectively used companywide; generating significant improvements in customer service, productivity, inventory, and costs.	Continuous improvement has become a way of life for employees, suppliers, and customers; improved quality, reduced costs, and increased velocity are contributing to a competitive advantage.
Class B	Supported by top management; used by middle management to achieve measurable company improvements.	Most departments participating and active involvement with some suppliers and customers; making substantial contributions in many areas.
Class C	Operated primarily as better methods for ordering materials; contributing to better inventory management.	Processes utilized in limited areas; some departmental improvements.
Class D	Information inaccurate and poorly understood by users; providing little help in running the business.	Processes not established.

Scoring process from overview questions:

Class A: average greater than 3.5
Class B: average between 2.5 and 3.49
Class C: average between 1.5 and 2.49
Class D: average less than 1.5

Figure 1–13   ABCD Scoring Process.
(Courtesy of Oliver Wight Publications, Inc.)

Managing the Resources

Meeting internal and external challenges with a strategy built on CIM requires change throughout the enterprise. The installation of new automation hardware and software, the development of communications networks, and the establishment of a common enterprise database are some of the significant changes under CIM. For the CIM implementation to be successful, however, the most significant change must be in the human resources area. A new awakening to the power of the employee must occur. Traditional corporate managers must make the difficult but necessary transition to team concepts, horizontal management structures, and the idea of a workforce empowered to make decisions and solve problems. Chapters 11 and 12 provide a view of the new role for managers and the complex systems they manage.

Manufacturing has changed significantly since 1973 when Dr. Harrington coined the phrase *computer-integrated manufacturing*. Computer technology, especially, made exponential advances in speed, function, and computation power.

Despite the changes in world markets and technology, the principles outlined by Harrington continue to ring true. His arguments for a system-oriented approach to the enterprise and against highly fragmented manufacturing operations that produce only localized optimization are still valid. Manufacturing continues to evolve and technology continues to progress, but the need for an integrated solution to the many problems across the enterprise remains.

1–8 SUMMARY

Much has been written and said about the value of a manufacturing base in the United States. Our current standard of living depends on our manufactured goods remaining competitive in world markets. To be competitive today requires that manufacturers meet the external and internal challenges and produce products to world-class standards. Meeting these standards requires a manufacturing strategy that forces internal consistency between marketing and manufacturing. One model, offered by Terry Hill, uses product order-winning criteria as a common denominator for all marketing and manufacturing decisions. Application of the model frequently exposes an integrated set of problems present in the organization. Computer-integrated manufacturing (CIM) offers a solution for this integrated problem set. The dependence on CIM concepts for survival of manufacturers in world markets becomes clearer every day.

In basic terms, CIM is the integration of all enterprise operations and activity around a common corporate database. Although the CIM concept is simple, application and implementation are difficult and complex. Application and implementation always start with education, and in the chapters that follow, the basic concepts associated with computer-integrated manufacturing will be explained in detail.

REFERENCES

BARRIS, R. R., *Justifying Automation—The New Realities.* Burlington, MA: Coopers and Lybrand, 1990.

CLARK, P. A., *Technology Application Guide MRP II Manufacturing Resource Planning.* Ann Arbor, MI: Industrial Technology Institute, 1989.

GOODARD, W., *The Oliver Wight ABCD Checklist for Operational Excellence,* 4th Ed. Essex Junction, VT: Oliver Wight Limited Publications, Inc., 1993.

GROOVER, M. P., *Automation, Production Systems, and Computer-Integrated Manufacturing,* 2nd ed., Englewood Cliffs, NJ: Prentice Hall, 1987.

HARRINGTON, J., Jr., *Computer-Integrated Manufacturing.* New York: Industrial Press, 1973.

HILL, T., *Manufacturing Strategy.* Homewood, IL: Richard D. Irwin, 1989.

OWEN, J. V., "Flexible Justification for Flexible Cells." *Manufacturing Engineering,* September 1990.

Rowen, R. B., *A Manufacturing Engineer's Introduction to Supply Chain Management*, CASA/SME Blue Book Series, 1999.

Sheridan, J. H., "Toward the CIM Solution." *Industry Week*, October 16, 1989.

Shrensker, W. L., *CIM Computer-Integrated Manufacturing: A Working Definition*. Dearborn, MI: CASA of SME, 1990.

Sobczak, T. V., *A Glossary of Terms for Computer Integrated Manufacturing*. Dearborn, MI: CASA of SME, 1984.

Stickler, M. J., "Going for the Globe Part II." P and IM Review, December 1989, pp. 32–34.

Stickler, M. J., "Going for the Globe Part III." P and IM Review, November 1989, pp. 41–43.

Vollmann, T. E., W. L. Berry, and D. C. Whybark, *Manufacturing Planning and Control Systems*, 2nd ed. Homewood, IL: Richard D. Irwin, 1988.

QUESTIONS

1. Define *manufacturing*.
2. Describe the types of economic conditions that made U.S. manufacturers most successful in the 1950s and 1960s.
3. Describe the retreat of U.S. production from the 1950s to the present day.
4. What are the external challenges faced by manufacturers worldwide?
5. Identify the external challenge that is the most difficult to overcome, and describe why.
6. How should companies respond to external challenges?
7. What is the internal challenge faced by manufacturers?
8. What causes the great manufacturing divide?
9. What is the difference between order-winning criteria and order-qualifying criteria?
10. How does the Terry Hill model help to overcome the internal challenge faced by manufacturers?
11. Describe how the life cycle of a product affects the order-winning criteria.
12. What steps are necessary to achieve the internal consistency required for a congruence between marketing/management and manufacturing operations?
13. What are the world-class order-winning criteria compiled by Coopers and Lybrand?
14. What are inventory turns, and how are they defined?
15. Why does a 1 percent increase in machine uptime translate into a 10 percent saving in capital equipment?
16. What are seven requirements for a solution to the primary problem facing manufacturing?

17. What contribution did Joseph Harrington Jr. make toward solving the manufacturing problem?
18. How is computer-integrated manufacturing defined?
19. Identify and describe the segments that form the CIM wheel.
20. Compare the CIM wheel in Figure 1–8 with the CIM representation in Figure 1–9. What are the similarities and differences?
21. What is the CIM managerial philosophy?
22. What constitutes a successful CIM implementation?
23. What resource in an enterprise must change most when a CIM system is implemented? Why is that necessary?
24. Describe the three-step process for implementing CIM.
25. What three areas are studied in the assessment stage, and what information is gathered?
26. Describe the focus of CIM education during the implementation.
27. Define *manufacturing waste* and *simplification*.
28. Compare fundamental process improvements with superficial improvements.
29. What are performance measures, and how are they used?
30. Give one example of a performance measure that could be used to determine the effectiveness of a CIM implementation.
31. What is the ABCD checklist, and how is it used?

PROBLEMS

1. True Bore Machine Company has annual sales of $120 million and an average annual inventory valued at $5 million. With a COGS of $0.85 per sales dollar, what is the inventory turn value for the company? What average inventory turns value would produce a turns ratio of 300?
2. The plastic injection molding machines at Great Plastics Molding Inc. are down due to setups for an average of 16.7 percent of the time in 3 shifts. If machine downtime cost is $135 per hour, how much does setup time cost for each machine on an annual basis?
3. Metal Enclosures Inc. has machine footprints that total 5780 square feet and 8240 square feet of value-adding assembly area. If the plant has 18,420 total square feet of product and inventory area, what is the manufacturing space ratio?
4. A production system has a burden rate of $200 per hour and a setup time of 4.4 hours. What is the setup cost per part for a lot size of 200 parts?
5. The system described in Problem 1 must move to a 75-part lot size. How much must the setup time be reduced to keep the setup cost per part constant?
6. A casting costs $15, the burden cost on the machine that finishes the casting is $300 per hour, time for the part on the machining center is 15 minutes, and the

setup time is 3 hours. If the castings are machined in lot sizes of 50, what percentage of the part cost is due to setup time?

7. A study to replace the operator in Problem 3 with automation indicates that the burden rate for the cell would increase to $345 per hour, the part machining time would drop to 10 minutes, and the setup time would drop by 0.5 hours. Would this additional automation decrease part cost? What is the part cost for the ideal lot size of 25 on this new cell?

8. Repeat the calculations in Exercise 1–3 using the warranty quality standard values from Figure 1–6.

9. Compare the results of the defective part calculations from Example 1–3 and Problem 5. What is the significance in manufacturing when the projected defect rate in a lot size production is less than 1 part?

10. A work cell includes two 48-in. square pallets for raw casting and finished castings, an operator's work area used to insert bushings into the machined casting that is 30 in. by 48 in., a tool rack that covers 12 square feet, wide walkways and machine clearances that cover 45 square feet, and the production machine that measures 58 in. by 97 in. Determine the manufacturing space ratio for the work cell.

11. A company has 10 work cells like the one described in Problem 7. Due to increased demand, another cell must be added but there is no factory floor space. What manufacturing space ratio would be needed in each cell to put eleven work cells in the space that now supports ten? What two areas in the work cell could most likely be reduced to allow for an eleventh work cell?

12. What is the significance to product cost or manufacturing profit of the two inventory turns values in Problem 1 when the cost of money is considered?

PROJECTS

1. Use primary and secondary research techniques to make a list of the names, addresses, phone numbers, and products of manufacturers in your city, region, or state using the following guidelines: five manufacturers with more than 500 employees in the plant location, five manufacturers with more than 100 but fewer than 500 employees at the location, and five manufacturers with 100 or fewer employees at the location.

2. Determine the current order-winning criteria and stage in the life cycle for one product from each group of manufacturers identified in Project 1. Identify and describe an order-qualifying criterion for the products examined.

3. Prepare a report on the concept of order-winning criteria described in the book *Manufacturing Strategy* by Terry Hill.

4. Prepare a report on the definition of CIM provided by Harrington in his book *Computer-Integrated Manufacturing.* Compare this with the SME definition given in this chapter.

5. Identify up to five local or national companies or products that experienced loss of market share due to the process illustrated in Figure 1–7. What was the significant mistake made in each case?
6. View the SME video "CIM: Focus on Small and Medium-Size Companies" and prepare a report that describes how either the Ex-Cell-O Corporation or the Modern Prototype Company used CIM to meet the external and internal challenges faced.

APPENDIX 1–1: THE BENEFITS OF A CIM IMPLEMENTATION

Industry Week conducted a survey of managers and executives in manufacturing industries to determine the status of CIM implementations. When asked what the benefits of a CIM implementation meant for their companies, the 139 respondents, including forty-nine CEOs and presidents and fifty-three vice presidents, indicated that the biggest payoff is in manufacturing cost and quality. A bar graph that shows the survey responses for this question appears in Figure 1–14. An important

Figure 1–14 CIM Benefits.

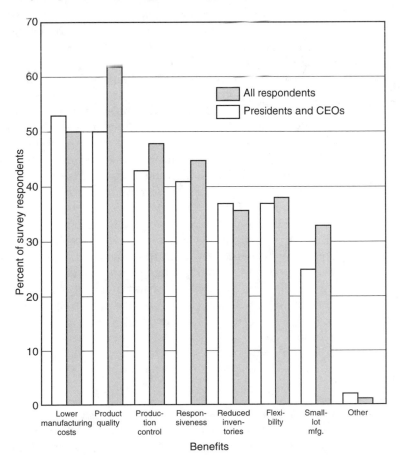

point is that the manufacturing industry's list of benefits from CIM includes a high percentage of the key order-winning criteria just identified. Stated more directly, if a firm needs to improve on the standard order-winning criteria, CIM will deliver results.

APPENDIX 1–2: A TALE OF TWO FARMERS—A CIM PARABLE

(Courtesy of Melnyk and Narasimhan, Computer Integrated Manufacturing—Guidelines and Applications from Industrial Leaders. *BusinessOne Irwin: Homewood, II. 1992.)*

The Parable Told

Once upon a time, in a land not too far from here, there lived two farmers. These two farmers were neighbors. They lived on two adjacent pieces of property that were almost identical. They were about the same size; they shared the same general terrain; in both, the soil was good, the water plentiful, the sun always shone, and the weather was temperate. In short, the land was good for growing.

Over time the two farmers had become very prosperous. Their produce was always good and it was always in demand. With the money earned from the sales of their produce, both farmers had hired good, conscientious hands to help with the task of working their land.

It was almost inevitable that news of the success of these two farmers would eventually reach farmers in the other valleys who were not as successful. In some cases the reason lay in their land. The soil was not as rich. In other cases the problem lay in the climate. In many instances the problem lay in how these other farmers managed their own plots of land. When they heard about the success of the two farmers, many of these farmers decided that they should visit the two farmers so that they could try to learn the secret of their success.

The two farmers prospered. As they prospered, they found themselves visited by farmers from different valleys. Each of these foreign farmers (that is the way that the two farmers referred to those who came from other valleys) had the same request. All they wanted was to see how the two farmers worked their land. They simply wanted to learn from those who really best knew how to farm the land. The more the two farmers prospered, the more farmers came to visit them. After a time the two farmers began to take great pride in their work. They began to think of themselves as the best farmers in the country. They also began to believe that their ways of managing the land were indeed the best and that these methods could not be improved on. Whenever one of their farmhands approached them with a suggestion for improving yields on the farm, the answer was always the same—no. "That's not the way we've done things here," would reply one farmer. "If it ain't broke, then why fix it," said the other.

The Beginning of the End   Then it happened. At first it took place gradually. The two farmers started to notice that the demand for their crops was falling. When in the past they would sell all that they could harvest, the two farmers now noticed that they were returning with some of their crops. It wasn't that much at first. "The economy," ventured one farmer. "Who knows and who really cares," commented the other farmer. Yet every year it seemed that the amount of crops being returned was getting a little bit larger.

Then one year the situation got worse. The two farmers found themselves with a great deal of produce that no one seemed to want. They first reacted by cutting price or by offering buyer incentives (two carrots for the price of one). That seemed to help somewhat but it couldn't stop the slide in sales.

The farmers then reexamined the crops and the methods that they were using to grow the crops. The crops seemed to be as good as ever. The farming methods had not changed. The next step that the farmers took was to tell their hands that they had to work harder. After all, reasoned the two farmers, with all of the success that their farms had enjoyed in the past, it shouldn't come as a surprise that the workers had gotten lax. However, this announcement was not well received. The hands felt that they were already working hard enough. Besides, noted some of the more vocal hands, the farmers should not blame the workers for problems caused by poor management

The two farmers looked around for other explanations for their problems. They talked to the village people who had bought their produce in the past. The two farmers were surprised to learn that the demand for produce of all types had increased, not decreased. The farmers then asked the villagers whether they knew whose produce they were buying. "Of course," answered many of the villagers. "We are now buying our carrots, potatoes, beans, and such from these farmers living in the nearby valleys. "Their harvest is just as good as yours. In fact, it is often better. Their produce is also cheaper and fresher."

The two farmers were shocked when they heard that the same farmers that had once visited them were now raising better crops. The first reaction of the two farmers was to believe that these foreign farmers were somehow cheating. How could they be cheating? According to one farmer, the answer was simple: "They're probably selling their produce for less than it's costing them to raise it. That's the ticket." "You're wrong," said the other farmer. "You know what they're doing? I'll tell you. Either they got low quality seeds or else they got cheap labor. That's the only way."

When the two farmers raised these concerns with the village elders, they were surprised by their response. To the charge that the foreign farmers were selling their crops for less than their cost, the two farmers were told by one of the elders who had visited these farmers, "It ain't so. I've been there and I can tell you that they're selling their goods to us at a fair price. Not only that, they're making a good buck on top of it all."

To the charge that their labor was cheaper or that they were using inferior inputs, the two farmers were again told that they were wrong: "'The labor ain't that much cheaper. Furthermore, the seeds that they are using are even better than yours."

One of the elders then talked to the two farmers and said, "You know what your problem is? You've had it your own way for such a long time that you think that you are the best. Well, it looks like that isn't the case now. Those farmers are selling everything they bring to market because they got better and cheaper vegetables and fruits. Why are their vegetables and fruits better and cheaper? Well, as I see it, they work harder and smarter. If you want my advice, I'd see what they were doing and then do the same. Until then, the only thing you're going to get out of me is sympathy."

Both farmers had to admit that the last suggestion was a good one. After they left the village elders, they made arrangements to see several of the more successful foreign farmers.

When they arrived at these farms, they saw several obvious differences. The first was that the farming equipment was much newer than what they had on their farms. Second, the seeds were different. Many of the seeds were newer and were more resistant to such problems as weeds, insects, long periods of drought, and excessive moisture. The hands on these farms seemed to work more diligently. When the two farmers asked the foreign farmers to explain the reason for their sudden success, they heard references to a new way of farming, a way that these other farmers called *Scientific Seeds Farming*, or SSF. One foreign farmer went as far as saying that the only reason he could compete effectively was because his farm used SSF.

Returning home, the two farmers were confused. Here was something that they had never heard of before. They didn't really understand SSF, yet it did seem to account for a good deal of the success now enjoyed by the foreign farmers. SSF seemed to involve seeds and farm equipment. They both decided to visit the farming implement and seeds seller in the village to see if he knew anything about SSF.

The Move to SSF   Of course, the seller had heard about SSF: "SSF, do I know anything about it? Damn right I do. SSF is the newest method for farming. It works. It merges the new technology of agro-engineering with the practices of farming. It has a great payback period. As I like to tell others, if you don't use SSF, then you might as well sell your farm. You know that if you want to have an SSF farm, you got to put your money in better seeds and in better equipment. Unless you have these two things, you really cannot do much with this system. Now if you are interested, I can show you what I've got that just might help you with putting in an SSF system on your farms. Heck, I'll even show you how to use everything."

The two farmers listened to everything that the seller had to tell them. They looked at the various ploughs, spreaders, drills, and harvesting machines. They studied the various types of seeds and listened as the seller described their various properties. It wasn't long before both farmers felt that they had so much information that they simply couldn't absorb anything else. They were even having trouble remembering everything that the agent had just told them.

The two farmers returned home. Each looked at the facts as he knew them. Each thought about how best to correct the worsening situation facing them. Each realized that something had to be done soon since the very survival of their farms was at stake. The two farmers reacted to the crisis facing them in very different ways.

Two Approaches to SSF   The first farmer decided that the key to SSF lay in getting the best farm equipment and the best seeds. He returned to the village implement and seeds seller. The farmer didn't really know what he needed. He did know two things. First, he had to do something to improve operations on his farm - now. Second, he had money. If he had to he would get himself the best and newest farm implements and seeds.

When the first farmer entered the store, he approached the seller and told him that he wanted to become a class A SSF user (Class A was the term that the seller used when describing those farms where SSF was used successfully and effectively). "I want to implement SSF and I want to do it now," announced the farmer.

"Well then, let's start by doing it right and looking at the latest and best," said the seller. That day, the seller showed the first farmer the newest seeds and the newest farm equipment. He demonstrated the features of vehicles considered to be the state of the art.

"Consider how envious your neighbors will be when they see this equipment in your fields. Of course, it does cost a few more pieces of gold but look at what you're getting. First of all, you are getting the newest equipment. Nobody else has it. Second, the equipment that you are buying offers a lot of extra features that you wouldn't find in the older equipment. Think of all the uses to which you can put it. Isn't it worth it to spend the extra money?" noted the seller.

The first farmer really didn't know what he would do with those few extra features. He would probably find a use or two for them. In the end he couldn't really disagree with what the seller had told him. It seemed to make sense.

After the first farmer looked at everything that the seller had shown him, he really didn't know what to buy. So he asked the seller to recommend some seeds and the equipment that he thought would be best for the farmer. The seller thought for a moment and pointed out that since he really wasn't familiar with the farmer's crops and needs, the only thing he could do would be to recommend what he thought was the best overall. He did so. The result was a list consisting of the most expensive and most advanced equipment and seeds. The first farmer looked at the list. When he saw the total cost, he was shocked. He knew it would be expensive to improve operations at his farm, but this was unbelievable!

After the initial shock, the farmer decided that there was nothing else to do but buy them. After all, it was only a one-time expense. If it worked, reasoned the first farmer, he could save money in several ways. First, operations would become more efficient. Costs would fall and profits would improve. If operations became more efficient, reasoned the first farmer, the number of workers needed might be reduced. Now here was a way to save money by getting rid of the lazy workers. It all seemed to make sense.

He left with many things to carry to the farm. The first farmer also left much, much poorer.

The approach taken by the second farmer was very different. He knew that he would have to get new equipment and seeds. Yet before he invested any money in making these purchases, he decided that he first had to study how he was farming now.

He looked at the land and the crops he was growing. He talked to his farm-hands. He talked to his customers. All of this activity took a great deal of time, but it was very revealing. Among the various things he had learned were the following:

- Many of his fields had been overworked and needed to lie fallow for at least a season.
- Some of the crops that he was growing were not really suitable for his land.
- Some of the crops were not well placed. He had been growing crops requiring dry soil in the clay areas (where it was always wet). He was also growing wet crops in the sandy soil.
- The way he was using his farm help was very inefficient. Often, the men working in the field didn't want to do what they had to do. Instead of weeding the fields (something that they often didn't do), the farmhands would be working on threshing.
- His existing farm equipment was in poor condition. It was out-of-date and poorly maintained.
- He really didn't know what customers he was serving. He was growing too many different crops and not doing a good job of meeting the needs of any one group of customers.
- Each worker had his own way of doing things. There seemed to be no set standards. As a result, the farmer had difficulty in determining whether the problem lay in the methods, the crops, the equipment, the customer, or the land.

The second farmer decided that, before buying any new seeds or equipment, he first had to correct the existing problems. Correcting the various problems was not an easy task. It took him two seasons to put most of the changes in place. He had dropped certain crops. He rearranged his crops. He improved the use of his farm hands. Weeding was now done regularly. Every day something different was being done.

At the end of the second season, the second farmer visited the village implement and seed seller. Unlike his neighbor, the second farmer knew exactly what he wanted. He was not interested in the newest or the best. He wanted something that worked and did exactly what he needed. He also wanted something that could be easily used by his field hands. To him, the new equipment really didn't replace his workers as much as it made them more effective. It didn't seem to bother him that he wasn't buying the best. "Why buy something that I don't really need or that might not work," said the second farmer. Like the first farmer, the second farmer replied that he too was now implementing SSF.

The Results with SSF Compared   While the second farmer was quietly making changes to his farm, the first farmer was telling everyone about what he was doing. He invited people to see the new equipment that he had bought. He told people that he was now well on the way to becoming a Class A SSF system. He showed the reductions in manpower that had resulted from the introduction of the new equipment He showed people how the new equipment worked. The people who did visit the first farmer were impressed by the new equipment and the amount of money that he had invested. They marveled at the operation of the new equipment (when it worked, which was not often). "Surely," they said to themselves, "here is an example of how to do SSF right."

In using the new equipment and seeds, the first farmer had not really changed any of his past practices. He had simply replaced the old seeds with the new seeds and the old equipment with the new equipment. He replaced manpower with technology. Everything else stayed the way it had always been.

At the end of three seasons, the two farmers met at the local village market. They discussed their experiences with scientific seeds farming.

To the first farmer, SSF was a failure: "I've spent a fortune on new equipment and seeds. It hasn't been worth it. We really don't want anything to do with this new equipment. Much of the time, it doesn't really work. The seeds aren't really that much better than the old. I'm not selling any more than I used to at these markets. You know, I thought that SSF was supposed to really reduce my costs. Well, it hasn't. If anything, with my investments, my costs are now higher than ever. If you ask me, SSF is a crock. It's just another way for sellers to unload seeds and equipment that we don't need."

The second farmer looked at the first farmer. He didn't know what to say. For him SSF did work. The benefits of SSF were not that evident in costs. Instead, SSF seemed to enhance his ability to offer better quality produce. His crops were better than ever. While not as cheap as some of the produce offered by the foreign farmers, it was now in demand. He was making more gold than ever before. His farm hands were happy. His customers were happy. Even the foreign farmers admitted that they couldn't compete successfully with him. The second farmer felt good because he now knew that the crops he was growing and selling really met his customers' needs.

As the two farmers walked away from each other, the second farmer couldn't help but wonder why SSF was a success for him and such an expensive failure for his neighbor. One thought kept reoccurring to him:

**If you cannot do the basic things right,
all of the technology in the world won't help you.**

Manufacturing Systems

Manufacturing was defined in Chapter 1 as "activities that eventually lead to the marketing of goods." In the process, raw materials are converted into finished products. The complete input/output model is described in Figure 2–1. The five inputs required are *raw materials, equipment, tooling and fixtures, energy,* and *labor.* The traditional outputs are finished goods and scrap.

A visit to manufacturing sites confirms that no two companies or manufacturing operations are the same. Even plants built by the same corporation to produce the same products have variations. The reason for the major differences between manufacturing sites is both *product* and *technology* based. In the first case, the *product* being manufactured dictates the manufacturing process required. For example, assembling cars is very different from refining gasoline. As a result, manufacturing automobiles requires a production facility different from the refinery used for gasoline.

Changes in *technology* cause significant differences in manufacturing plants as well. The differences are evident in production facilities producing the same product. For example, an automotive assembly plant built five years ago would be different from one built today. The new plant would take advantage of any technological advancement in the last five years that improves production efficiency. If every production facility is different, it naturally follows that the CIM systems installed in

Figure 2–1 Production Model.

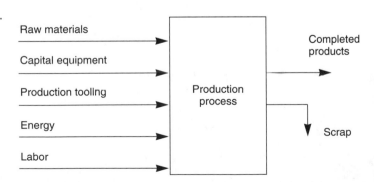

these production facilities will each be unique. Therefore, the variation in CIM implementations is also *product* and *technology* based. For example, production scheduling software for the automotive assembly plant would not be effective for the refinery. Thus the ability to recognize the different types of production facilities is critical for the correct recommendation of a CIM solution. Learning the description of the various types of manufacturing facilities currently in use is the first step and the subject covered in this chapter.

Technology change also influences the CIM solution. Using the same example, production control software installed five years ago would not have the features available in similar software installed today. The continuous improvements in technology are tolerable only if they can be easily incorporated into the current system. Passing product data between manufacturing software packages from two different vendors is a good example. If both software solutions transfer the data to common database using an industry standard format for the data, future enhancements to the software are acceptable from a system integration standpoint. However, increased software capability may require upgrading of the hardware platform where the software resides. Learning the detailed operation of current hardware and software applications is just a temporary solution. It is more important to understand *why* different CIM technologies are used and *how* they affect the total integrated system. After the manufacturing systems are studied in this chapter, in the remainder of the book we answer the *why* and *how* questions regarding the CIM technology required to integrate these systems.

2–1 MANUFACTURING CLASSIFICATIONS

Manufacturing is classified using two criteria:

- How merchandise is produced in the *manufacturing system.*
- How customer demand is satisfied by a *production strategy.*

Classification of manufacturing operations is necessary for CIM implementation. Many of the solutions are tailored to a specific type of manufacturing system.

Manufacturing System Classification

Classification by manufacturing system divides all production operations into the following five groups: *project, job shop, repetitive, line,* and *continuous.* Overlap cannot be avoided between some of the categories, and most manufacturers use two or more of the manufacturing systems in the production of an entire product line. Classification of companies into these groups requires a detailed analysis and evaluation of the production operations. As a result of the classification process, the activity in the three major process segments (design, manufacturing planning and control, production) in Figure 1–9 is identified and understood. Therefore, success in matching enterprise needs with automation hardware and software is easier to achieve. Distinguishing characteristics of each classification category are described below.

Project.   The most distinguishing characteristics of this category is that products are complex, with many parts, and are most often *one of a kind.* For example, project-type companies build oil refineries, large office buildings, cruise ships, and large aircraft. In each case the products may be similar but usually are not identical. The plant layout, another discriminating factor for this category, is called *fixed position* (Figure 2–2a). Because of their size and weight, products such as ships and large aircraft remain in one location, with the equipment and parts being moved to them. In addition, the design drawings are complex, the lead time is long, the customer is identified before production starts, and production scheduling usually uses project management techniques.

Job Shop.   Job shops are also distinguished by low volume and production quantities, called *lot sizes,* that are small. However, compared to the products in the project category, the size and weight of the parts in the job shop group are very small. As a result, the parts are moved or routed between fixed production work cells for manufacturing processing. The classic machine shop with lathes, mills, grinders, and drill presses is the example most often cited for this category. The plant layout for the job shop (Figure 2–2b) is frequently called a *job shop* or *process layout.* Other distinguishing features include less than 20 percent repeat production on the same part, noncomplex products, and intensive scheduling and routing on the shop floor. In addition, the raw material is usually purchased as needed for each project.

Repetitive.   The repetitive manufacturing system has the following unique characteristics: orders for repeat business approach 100 percent, blanket contracts with customers for multiple years occur frequently, moderately high volume with lot sizes varying over a wide range, and fixed routings for the production machines. The plant layout could be either the process layout shown in Figure 2–2b or more like the product-flow layout shown in Figure 2–2c. The production machines are frequently special-purpose machines, called *transfer machines,* built to produce a specific product or family of products. Automotive subcontractors are representative of this type of manufacturing. A supplier to Ford, for example, could have an order to supply 3000 water pumps per week made up from three different model types. The weekly quantity for each model may vary over some range, and the contact might span three years.

Line.   The line manufacturing system has several distinguishing characteristics: (1) the delivery time (often called *lead time*) required by the customer is often shorter than the total time it takes to build the product, (2) the product has many different options or models, and (3) an inventory of subassemblies is normally present. Car and truck manufacturing is an example of this category. If a customer or dealer order for a car triggered the start of production for all of the car's parts, it would take months to build the car. For example, if the bolts were not made, the seat material not woven, the raw material for the tires not produced, the engine not built, and cassette player not assembled when the order arrives, the car could

Figure 2–2 Types of Plant Layout: (a) Fixed-Position Layout; (b) Process Layout; (c) Product-Flow Layout.

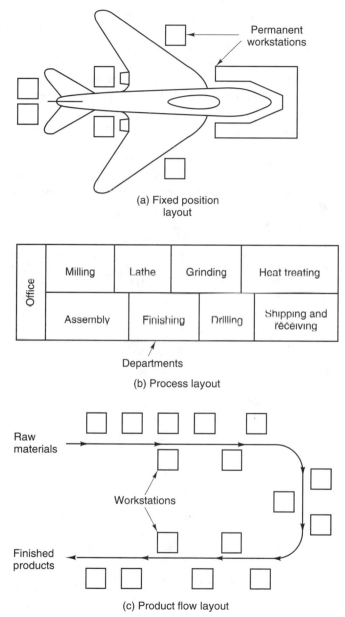

(a) Fixed position layout

(b) Process layout

(c) Product flow layout

not be built as fast as the customer or dealer requires. Manufacturing facilities in this group usually use the product-flow (Figure 2–2c) plant layout.

Continuous.   Continuous manufacturing systems have the following characteristics: Manufacturing lead time is greater than the lead time expected by the customer, product demand is predictable, an inventory of finished products is maintained, volume is high, and products have few options. This type of production system always uses the product-flow plant layout (Figure 2–2c) with the production line limited to one or just a few different product models. Examples of this industry type include the production of nylon carpet yarn, the production of breakfast cereal, and the production of petroleum and chemical products. In products like these, the production process is often continuous with raw material, such as chemical compounds, entering on one end of the system and a finished product, nylon filament thread, flowing out the other end.

Figure 2–3 offers comparative data on the characteristics of the manufacturing systems just described. Two of the items, *labor content* and *design component*, may be misunderstood without additional clarification. Labor content refers to the size of the human labor component or the difficulty that automation faces in

	Project	Job shop	Repetitive	Line	Continuous
Process speed	Varies	Slow	Moderate	Fast	Very fast
Labor content	High	High	Medium	Low	Very low
Labor skill level	High	High	Moderate	Low	Varies
Order quantity	Very small	Low	Varies	High	Very high
Unit quantity cost	Very large	Large	Moderate	Low	Very low
Routing variations	Very high	High	None	Low	Very low
Product options	Low	Low	None	Very high	Very low
Design component	Very large	Large	Very small	Moderate	Small

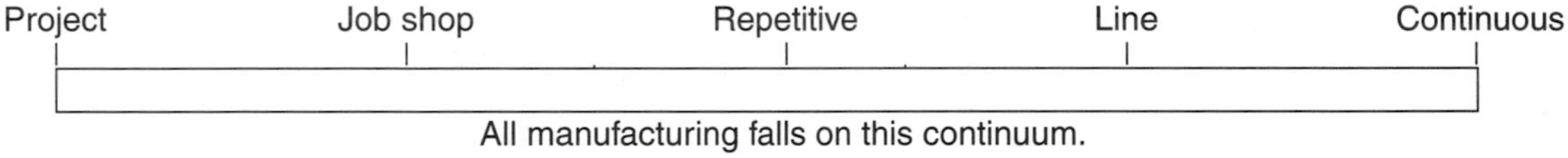

Project	Job shop	Repetitive	Line	Continuous

All manufacturing falls on this continuum.

Figure 2–3   Manufacturing Characteristics.

replacing the human element. The design component element indicates the relative number of hours devoted to product design. Take the time to study the chart. You should begin to see how the type of manufacturing operation dictates where the productivity gains will occur as automation is introduced with CIM. For example, consider computer-aided design (CAD) technology, which helps when the design component is large. The last element on the chart, design component, indicates that in a project type, the manufacturer has an excellent opportunity for an impact from CAD, but CAD's effect on a continuous manufacturing operation would probably be minimal because the design component (the need for CAD) is small.

Exercise 2–1
Use your general knowledge about the following products to classify (P, project; J, job shop; R, repetitive; L, line; C, continuous) each by the manufacturing system(s) used.

a. Nylon carpet yarn
b. Gillette Plus razors
c. Space shuttle
d. Personal computers
e. Pontiac Sunbirds
f. Cheerios
g. Electric fan motors

h. Replacement pump part
i. Automotive alternators
j. Oil refinery
k. Hard disk drives
l. Stainless steel dinnerware
m. Televisions
n. Special metal bracket

Solution:
The first letter represents the most common type of system used. (a) C; (b) C; (c) P; (d) L, R; (e) L; (f) C; (g) R; (h) J; (i) R; (j) P; (k) R, L; (l) R, L; (m) L; (n) J.

The number of different groups in the manufacturing classification system is not critical. Some literature lists only three categories: job shop, batch production, and mass production. In that case the five categories just described would be mapped into these three.

The number of categories is not critical; understanding what makes automotive production different from nylon production is, however. The continuum at the bottom of Figure 2–3 represents the different type of manufacturing systems just described. "Project" and "Continuous" are opposite in function and represent the two possible extremes on the left and right, respectively. The other three fall somewhere between. Every manufacturing system fits on the continuum, with some companies using several production technologies.

It should be clear by now that a single manual process or software package can be used to handle the production scheduling requirements of the five manufacturing system categories. Selecting a process or software application requires knowledge about the production operation. Implementing a CIM solution requires a complete knowledge of the manufacturing system to be improved or automated, and classification is the first step.

Production Strategy Classification

The *production strategy* used by manufacturers is based on several factors; the two most critical are *customer lead time* and *manufacturing lead time*. Knowing the definition of each is important.

> *Customer lead time identifies the maximum length of time that a typical customer is willing to wait for the delivery of a product after an order is placed.*

For example, the consumer expects preferred brands of commodities, such as toothpaste, to be available on the store shelf whenever a purchase is desired. Rarely will the consumer wait for a delivery if the brand selected is not available. When the brand is not available, a competitor's product is chosen or the product desired is purchased at a different store. In this example, immediate delivery or satisfying a customer lead time of zero is the order-winning criterion.

> *Manufacturing lead time identifies the maximum length of time between the receipt of an order and the delivery of a finished product.*

Manufacturing lead time and customer lead time must be matched. When a new car with specific options is ordered from a dealer, for example, the customer is willing to wait only a few weeks for delivery of the vehicle. As a result, automotive manufacturers must adopt a production strategy that permits the manufacturing lead-time to match the customer's needs.

The production strategies used to match the customer and manufacturer lead times are grouped into four categories: *engineer to order* (ETO), *make to order* (MTO), *assemble to order* (ATO), and *make to stock* (MTS). A description of each follows.

Engineer to Order.   A manufacturer producing in this category has a product that is either in the first stage of the life-cycle curve or a complex product with a unique design produced in single-digit quantities. Examples of ETO include construction industry products (bridges, chemical plants, automotive production lines) and large products with special options that are stationary during production (commercial passenger aircraft, ships, high-voltage switchgear, steam turbines). Due to the nature of the product, the customer is willing to accept a long manufacturing lead time because the engineering design is part of the process.

Make to Order.   The MTO technique assumes that all the engineering and design are complete and the production process is proven. Manufacturers use this strategy when the demand is unpredictable and when the customer lead-time permits the production process to start on receipt of an order. New residential homes are examples of this production strategy. Some outline computer companies make personal computer to customer specifications, so they followed MTO specifications.

Assemble to Order.   The primary reason that manufacturers adopt the ATO strategy is that customer lead time is less than manufacturing lead time. An example from the automotive industry was used in the preceding section to describe this situation for line manufacturing systems. This strategy is used when the option mix for the products can be forecast statistically: for example, the percentage of four-door versus two-door automobiles assembled per week. In addition, the subassemblies and parts for the final product are carried in a finished components inventory, so the final assembly schedule is determined by the customer order. John Deere and General Motors are examples of companies using this production strategy.

Make to Stock.   The last strategy, MTS, is used for two reasons: (1) the customer lead time is less than the manufacturing lead time, (2) the product has a set configuration and few options so that the demand can be forecast accurately. If positive inventory levels (the store shelf is never empty) for a product is an order-winning criterion, this strategy is used. When this order-winning criterion is severe, the products are often stocked in distribution warehouses located in major population centers. This option is often the last phase of a product's life cycle and usually occurs at maximum production volume.

Figure 2–4 compares the production activities, such as design, to the four production strategies. The manufacturing lead times are set by the three major

Figure 2–4 Production Strategies and Manufacturing Lead Times.

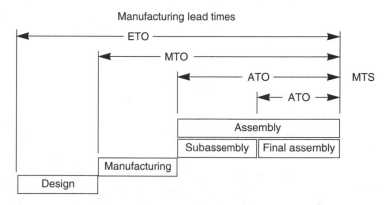

activities in product development: design or engineering, manufacturing, and assembly. Assemble to order (ATO) has two lead times; the use of subassemblies produces the shortest ATO time. Make to stock (MTS) has a zero manufacturing lead time because the customer is not willing to wait and delivery must be immediate. Compare the figure with the definitions of the four production strategies.

The relationship between the product life-cycle curve described in the last chapter and the production strategies just described is illustrated in Figure 2–5. When a product enters the market, the demand is usually low, so a make-to-order strategy does not stress the resources of the company. As demand builds, an assemble-to-order strategy keeps the product delivery time even with competitive products without carrying inventories of finished goods. At peak demand a company often chooses a make-to-stock strategy to meet customer orders and maintain market share. The cost of the finished-goods inventory is reduced by the efficiencies of scale due to higher volume and is more advantageous than increasing production capacity.

Figure 2–5 Production Strategies.

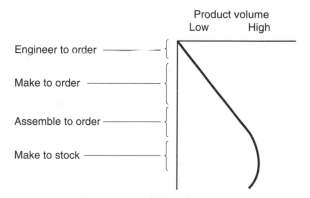

The relationship between the manufacturing systems described earlier and the production strategies just covered is illustrated in Figure 2–6. Study the chart for a few minutes before continuing. Customer demand and lead time still deter-

	Engineer to order	Make to order	Assemble to order	Make to stock
Project	x			
Job shop		x		
Repetitive		x		x
Line			x	x
Continuous		x		x

Figure 2–6   Comparison of Manufacturing Systems and Strategies.

mine the production strategy shown in Figure 2–6, but in two cases the manufacturer has a choice of the type of manufacturing system to be used.

The key point from the discussion of classification of manufacturing systems and production strategies is that industries are all different. The classification process simply helps you understand some of the similarities and differences that are present. Armed with that knowledge, it becomes much easier to meet the external challenges, identify the order-winning criteria, implement the CIM principles, adjust management philosophy, integrate the hardware and software systems, and compete in the world marketplace.

Exercise 2–2
Use your general knowledge about the following products to classify (E, engineer to order; MO, make to order; A, assembly to order; MS, make to stock) each by the production strategy(ies) used.

a. Nylon carpet yarn

b. Gillette Plus razors

c. Space shuttle

d. Personal computers

e. Pontiac Sunbirds

f. Cheerios

g. Electric fan motors

h. Replacement pump part

i. Automotive alternators

j. Oil refinery

k. Hard disk drives

l. Stainless steel dinnerware

m. Televisions

n. Special metal bracket

Solution:
The first letter represents the most common production strategy used.
(a) MS, MO; (b) MS; (c) E; (d) MS, A; (e) A, MS; (f) MS; (g) A; (h) MO; (i) MS, A; (j) E; (k) MS; (l) MS; (m) MS; (n) E.

2–2 PRODUCT DEVELOPMENT CYCLE

Despite the differences that exist across manufacturing, the product development cycle is generally the same. The cycle, illustrated in Figure 2–7, shows the linear process used to bring a product to market. Remember, however, that the customer continues to set the order-winning and order-qualifying criteria that influence every step in the process. Also, the type and level of activity in each block is

Figure 2–7   Product Development Cycle.

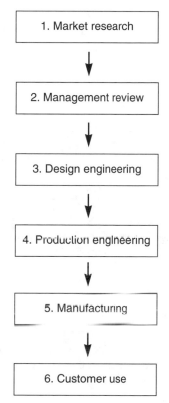

1. Market research

2. Management review

3. Design engineering

4. Production engineering

5. Manufacturing

6. Customer use

directly proportional to the type of manufacturing system and production strategy present. For example, the level of activity in the design engineering block would be high in a project system using an engineer-to-order strategy.

Do the names in blocks 3, 4, and 5 look familiar? They are the three major process segments found in Figure 1–9. As a result, the product development cycle is consistent with the CIM model described earlier. Two types of product development are supported by the cycle: new product development and existing product changes.

New Product Development

For a new product, the *customer support* area in Figure 1–8 determines the form, fit, and function that the product must satisfy, along with the order-winning criteria necessary to succeed in this new market sector. The form, fit, and function concept is described in detail in the next chapter, but for now it is defined as the technique used to specify a product. The *management review* process (Figure 2–7) looks at all the market data, product specifications, and current order-winning criteria and makes a *go* or *no go* decision on the new product initiative. With the project approved, *design engineering* starts adding detail to the new product with an eye focused on the order–winning criteria present in the current market. In the next step, *production engineering* initiates the manufacturing planning and control functions. For example, processes, machines, and routings are selected that will ensure that a product is consistent with the order-winning criteria. The last enterprise activity in the cycle is *manufacturing* the product.

The process appears to be linear with manufacturing at the end getting the demands from decisions made far upstream. The order-winning criteria should force congruence between the development of new products and a production system with the order-winning attributes identified by marketing. However, use of order-winning criteria is no guarantee that the design, for example, is optimized to the machines that must make the product in manufacturing. In Chapter 3 a process called *concurrent engineering* is superimposed on the product development cycle to overcome the problems created by this totally linear process.

Existing Product Changes

The process of handling changes to existing products effectively is almost identical to the procedure just outlined for a new product. The only exception is the point where the activity starts in the development cycle (Figure 2–7). With a product change the process can start at any block and then must ripple through the system in an orderly fashion.

The product development cycle identifies the need for an orderly flow of product data that the CIM system must support for the departments and the design and manufacturing process. New product development emphasizes the need for a top-to-bottom data interface to support information flow. However, changes to existing products require an interface that works from any starting point in the cycle. Since design and product changes are inevitable, the CIM system must accommodate product changes, initiated at any point in the development cycle, in an orderly and controlled fashion.

2–3 ENTERPRISE ORGANIZATION

The enterprise model presented in Figure 2–8 shows the functional blocks found in most manufacturing organizations. The lines connecting the areas indicate formal communications that occur regularly between enterprise functions. There are as many different representations of manufacturing organizations as there are books discussing them. This model has sufficient detail to demonstrate the information and data flow normally present and provides the framework to discuss the functions of the various areas involved in getting a product to the customer. Study the model until you are familiar with the names in each box and compare the block names with the element list in the Enterprise wheel in Figure 1–8.

A CIM implementation affects every part of an enterprise; as a result, every block in the organizational model is affected. Therefore, a successful CIM implementation across the enterprise requires an understanding of the functions performed by each block in the model. The basic operation of the enterprise is described in the following sections.

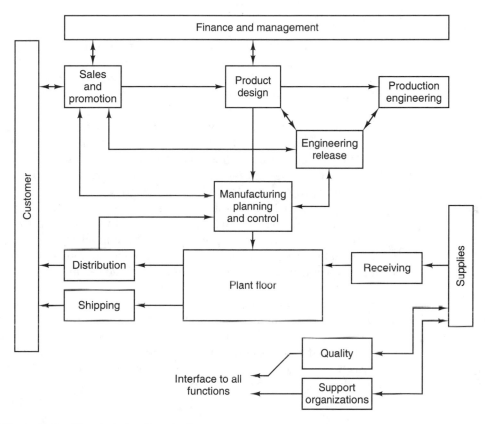

Figure 2–8 Manufacturing Organization.

Sales and Promotion

The fundamental mission of *sales and promotion* (SP) is to create customers. To achieve this goal, nine internal functions are found in many companies: *sales, customer service, advertising, product research and development, pricing, packaging, public relations, product distribution,* and *forecasting.* An analysis of the enterprise model in Figure 2–8 indicates that sales and promotion interfaces with several other areas in the business. The *customer services* interface supports three major customer functions: *order entry, order changes,* and *order shipping and billing.* The order change interface usually involves changes in product specifications, change in product quantity (ordered or available for shipment), and shipment dates and requirements.

Sales and marketing provide strategic and production planning information to the finance and management group, product specification and customer feedback information to product design, and information for master production scheduling to the manufacturing planning and control group. The interface with engineering release ensures that any proposed changes to the product are reviewed and approved by sales and promotion before implementation.

Finance and Management

Finance and management has the responsibility for setting corporate goals, performing financial functions, and performing medium- and long-range planning. In the financial area four major internal functions are performed: *cash management, financial planning, financial analysis,* and *strategic planning.* The financial planning is long-term planning for the operations of the business. The financial analysis unit generates a direction for the future that is based on past and current financial conditions. They usually draw data from three sources: *balance sheet, income statement,* and *statement of change* in the financial position. The strategic planning unit works with two time frames: *medium term* and *long range.* The medium term includes yearly budget projections for production objectives, corporate structure, and manufacturing infrastructure. The long-range planning includes long-term objectives for the corporation and the strategies necessary to achieve these objectives.

The *accounting* unit is responsible for three areas: *general accounting, cost accounting,* and *related functions.* The areas supported by the accounting unit include accounts payable, accounts receivable, general ledger, and the three cost accounting areas: manufacturing, product, and overhead.

If this is a book about CIM, why study finance and management or sales and marketing? The answer is that every part of the enterprise uses the common database and must be considered in the integration process. As a result, the CIM integrater must understand how these units function in the organization. This reference to the function or operation of these two units is the last because the operation of these functions is well documented in business books. However, when production control software is discussed in a subsequent chapter, the links to these units will be explored.

Product/Process Definition

The units that share a formal interface with the *product* and *process definition* unit are illustrated Figure 2–9. Keep in mind that this unit is one of the three process segments in the Enterprise wheel (Figure 1–8) and is the design component in Figure 1–9. The unit includes *product design, production engineering,* and *engineering release.* The product design provides three primary functions: (1) product design and conceptualization, (2) material selection, and (3) design documentation. The production engineering area establishes three sets of standards: *work, process,* and *quality.* The engineering release area manages engineering change on every production part in the enterprise. Engineering release has the responsibility of securing approvals from departments across the enterprise for changes made in the product or production process. The operation and automation of this area of the enterprise model are covered in considerable detail in later chapters.

Manufacturing Planning and Control

The *manufacturing planning and control* (MPC) unit (Figure 2–10) has a formal data and information interface with several other units and departments in the

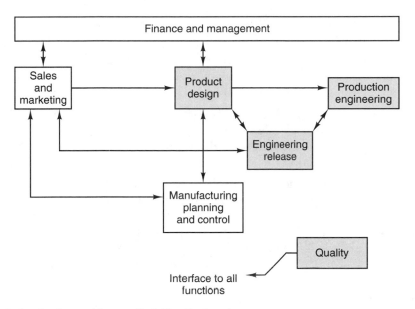

Figure 2–9   Product and Process Definition Engineering.

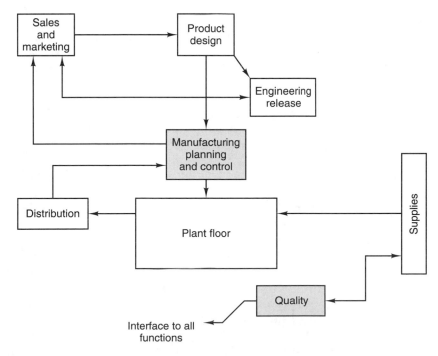

Figure 2–10   Manufacturing Planning and Control.

enterprise. This unit is the second process segment identified on the Enterprise wheel (Figure 1–8) and in Figure 1–9. The MPC unit has responsibility for:

- Setting the direction for the enterprise by translating the management plan into manufacturing terms. The translation is smooth if order-winning criteria were used to develop the management plan.
- Providing detailed planning for material flow and capacity to support the overall plan.
- Executing these plans through detailed shop scheduling and purchasing action.

This unit is the most complex in the enterprise, based on the quantity of data and the large number of interfaces into and out of the area. As a result, the data and information flow is expressed through an MPC model such as that shown in Figure 2–11. Study the illustration and learn the names of the various functions listed. Each function is explored in detail in later chapters.

Figure 2–11 MPC Model for Information Flow.

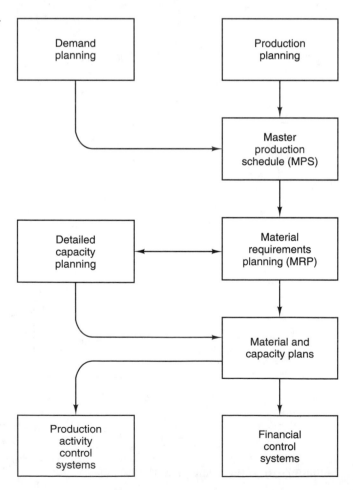

Shop Floor

Interfaces with the *shop floor* unit are illustrated in Figure 2–12. This unit is the third process segment and is identified in Figure 1–9 as production. Although the use of different terms to identify this area may seem confusing, several terms are used in the literature to refer to the production unit. Shop floor activity often includes *job planning and reporting, material movement, manufacturing process, plant floor control,* and *quality control.* The operation and automation of this area of the enterprise model are covered in considerable detail in later chapters.

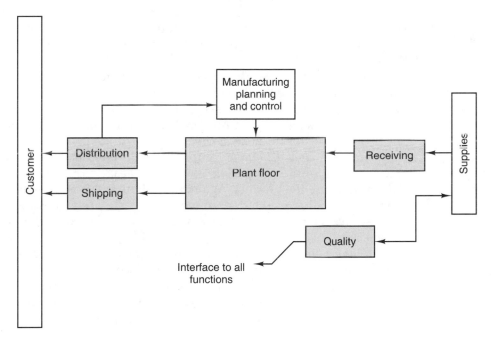

Figure 2–12   Shop Floor.

Support Organization

The *support* organizations, indicated by the box in Figure 2–8, vary significantly from firm to firm. The functions most often included are *security, personnel, maintenance, human resource development,* and *computer services.* Basically, the support organization is responsible for all of the functions not provided by the other model elements.

In the last six sections we described the general organization of an enterprise engaged in manufacturing. In addition, we provided an overview of the functions performed in each of the six major units. In the remainder of the book, we investigate the CIM management and automation issues present in these units. Before starting this more detailed study, however, we need to look at how production flow occurs in this enterprise model (Figure 2–8).

2–4 MANUAL PRODUCTION OPERATIONS

The previous discussion, centered on the enterprise organization, identified the links between each unit and the functions provided. However, we have not yet provided a sense of how production flows. Figure 2–13 shows one possibility for the flow required to bring a product to a customer. Study the figure and note the symbol for a paper operation and the symbol for computer-driven processes. Refer frequently to the illustration when the product flow through the system is described below.

Activity enters the system as either a *design* or a *request for engineering action* (REA). An REA could come from marketing and result from a product change requested by a customer, or the REA could be a change requested by an internal department to correct a problem. Note that currently both are usually paper-based operations.

The *product design* area, using computer-aided design (CAD), makes the drawings to describe the design or REA. Although this unit is usually automated with CAD, the drawing file is saved on the computer in the product design area. The paper drawing of the design or REA is thrown over the office wall, figuratively speaking, for the next activity.

In the *product definition* group, all the different parts in the design or REA are listed in a bill of materials so that it is easy to understand what it takes to manufacture the product. Again, even though this stage may be a computer-based application in the department, the work rarely goes to the engineering release management group for approval as an electronic file or as computer-based data.

The *manufacturing definition* group starts to work after the design or REA is approved through the engineering release process and is delivered to production. Here a manufacturing bill of materials is produced that separates the parts in the design into different categories, such as those that are purchased and those that are manufactured. Internally, this process is frequently manual, so the solution is passed to the next area in a paper format.

Manufacturing process planning determines the type of machines required to process the parts and the production sequence, called the *routing*, to be used. Establishing the routing is most often a manual process; as a result, information is passed to the production area in a paper format.

The next step, *business production planning*, frequently is automated with a computer and scheduling software. The production process and machine sequence information from the handwritten manufacturing process plan is keyed into the software, and production schedules are produced. Shop orders are released across the manufacturing area and production activity control is established to control the shop floor.

The production flow just described had the following characteristics:

- There was unidirectional flow from design or REA to the shop floor. The system did not provide feedback to the department on the status of the

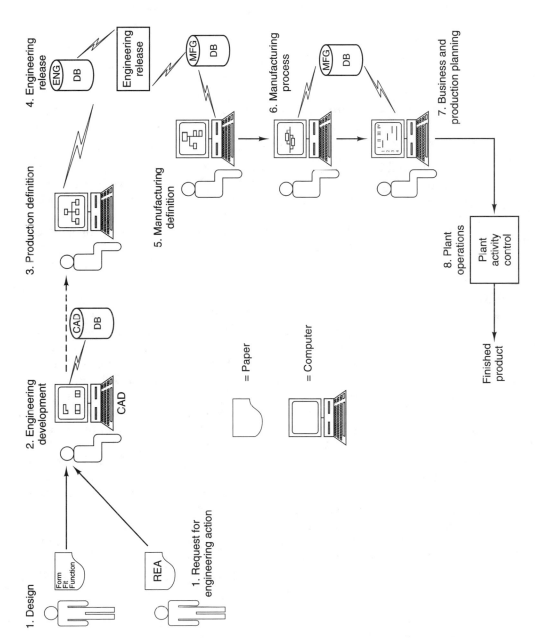

Figure 2–13 Production Sequence.

1. Design

2. Engineering development

CAD

3. Production definition

4. Engineering release

ENG DB

Engineering release

MFG DB

5. Manufacturing definition

6. Manufacturing process

MFG DB

7. Business and production planning

8. Plant operations

Plant activity control

Finished product

1. Request for engineering action

REA

Form Fit Function

= Paper

= Computer

product after it left a unit. Most critically, the status of the part in production was available only by going to the shop and checking.

- A mixture of manual operations and islands of automation prevented communication with other units.
- Productivity was lost from the same product information being reentered in two or more units.
- The system is inflexible.
- There is a sense that the operation was a collection of independent units, each operating on its own with little regard for how that operation affected the remainder of the enterprise.

Unfortunately, this description fits some manufacturing organizations. Intervention is required to bring these organizations in line with the production strategies that win orders and overcome the external challenges described earlier. The intervention to bring about the desired change comes from the implementation of computer-integrated manufacturing principles. A description of that process forms the remainder of this book.

2–5 SUMMARY

Each company is unique, due to the product manufactured or the technology level used in the production process. The differences among the companies require that the CIM implementation for each be different. To specify the type of CIM solution that is appropriate, you must know where a company fits in the manufacturing systems and production strategy classification scheme. Classification under manufacturing systems divides all production operations into the following five groups: *project, job shop, repetitive, line,* and *continuous.* The production strategies in use are grouped into four categories: *engineer to order, make to order, assemble to order,* and *make to stock.* The production strategy changes at different points in the product life cycle. Knowing where a company fits into the relationship between the manufacturing system and production strategies helps in identifying successful CIM solutions.

The product development cycle (Figure 2–7) describes the activities required for the development of a new product and for making changes in existing products. Analysis of the development cycle reveals the interfaces that the CIM system must support in the enterprise. It becomes apparent that the CIM system must control product changes in an orderly and controlled fashion regardless of the point in the development cycle where they are initiated.

The enterprise organization needs a model to support an understanding of the functions performed by every department and group. The model in Figure 2–8 is used to describe these functions. The production flow of a product through the enterprise is provided by Figure 2–13. A design or request for engineering action (REA) triggers the activity, which then moves through the groups in sequential fashion.

REFERENCES

CLARK, P. A., *Technology Application Guide: MRP II Manufacturing Resource Planning.* Ann Arbor, MI: Industrial Technology Institute, 1989.

GROOVER, M. P., *Automation, Production Systems, and Computer-Integrated Manufacturing,* 2nd ed., Upper Saddle River, NJ: Prentice Hall, 1987.

SHRENSKER, W. L., *CIM Computer-Integrated Manufacturing: A Working Definition.* Dearborn, MI: CASA of SME, 1990.

SOBCZAK, T. V., *A Glossary of Terms for Computer-Integrated Manufacturing.* Dearborn, MI: CASA of SME, 1984.

VOLLMANN, T. E., W. L. BERRY, and D. C. WHYBARK, *Manufacturing Planning and Control Systems,* 4th ed., Homewood, IL: Richard D. Irwin, 1997.

QUESTIONS

1. Why is every CIM implementation different?
2. Describe the two classification techniques used to differentiate between manufacturers.
3. Define the following manufacturing systems: project, job shop, repetitive, line, and continuous.
4. Describe why the comparative data in Figure 2–3 support the fact that every CIM implementation is different.
5. Use your general knowledge about the following products to classify (P, project; J, job shop; R, repetitive; L, line; C, continuous) each by the manufacturing system(s) used.

   a. Antifreeze
   b. Bic Lighters
   c. Cruise ships
   d. Electric typewriters
   e. Jeep Cherokee
   f. Nachos
   g. Automotive water pumps

   h. Special coupling
   i. Automatic transmission fluid
   j. Concrete
   k. Computer power supplies
   l. Plastic picnic forks
   m. CD players
   n. Special motor mount

6. Define the following production strategy terms: engineer to order, make to order, assemble to order, and make to stock.
7. Describe how the production strategies change over the life cycle of a product.
8. Use your general knowledge about the products listed in Question 5 to classify (E, engineer to order; MO, make to order; A, assemble to order; MS, make to stock) each by the production strategy(ies) used.

9. What is the product development cycle, and how is it supported by the CIM wheel?
10. What is the primary distinction between new product development and product changes in the development cycle?
11. Study the product development cycle in Figure 2–7 and list some of the data and information interfaces that CIM must provide.
12. What is the function of the enterprise organization model in Figure 2–8?
13. What is the function of each block in the organizational model?
14. Briefly describe the production flow through the manual production system.
15. What is the difference between a design and REA?
16. What problems in the manual production system described in Question 14 must CIM solve?

PROJECTS

1. Using the list of companies developed in Project 1 in Chapter 1, determine the manufacturing system and production strategies used for each product produced.
2. Compare the enterprise model (Figure 2–8) with the structure used by one company from the small, medium-sized, and large groups and describe how their operations differ from that of the model.
3. Compare the production operation sequence described in the text with that used by the companies selected in Project 2. Describe the differences that are present.
4. View the SME video "CIM: Focus on Small and Medium-Size Companies" and prepare a report that describes the manufacturing system and production strategy used by the Ex-Cell-O Corporation and the Modern Prototype Company.

APPENDIX 2–1: CIM AS A COMPETITIVE WEAPON

How important is CIM as a competitive weapon for U.S. industries? That question was answered by 139 respondents, including company presidents, CEOs, and vice presidents Their responses to this *Industry Week* survey are illustrated by the pie charts in Figure 2–14. Two key indications can be drawn from these results: (1) a large majority, 85 percent, consider CIM *essential* or *very important;* and (2) the small number of *no answers* indicates that management understands the question in sufficient depth to have an opinion. The second indication is significant because it says that the decision makers are committed and willing to listen to recommendations from the organization.

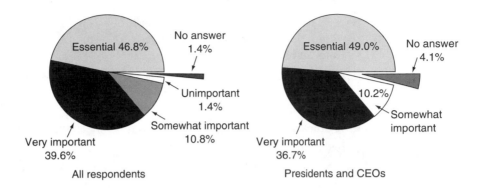

Figure 2–14   Importance of CIM.

CASE STUDY: EVOLUTION AND PROGRESS—ONE WORLD-CLASS COMPANY'S MEASUREMENT SYSTEM

A manufacturing company with over $450 million in annual sales was losing market share due to the following problems:

- The delay in shipping customer orders was getting larger.
- The production time on new orders was increasing.
- Development time on new products was increasing.
- Inventory levels for most products was high.
- Product rework was high due to poor production quality.

The company had an MRP II software system, but the accuracy of the inventory and product parts list along with poor vendor delivery performance made the system dysfunctional. While the company use a cycle-count system to keep track of inventory, it was not enforced across all areas in the company. In addition, the company carried higher inventory levels as a buffer against a vendor on-time delivery of only 30 percent. To regain the competitive edge, management developed a CIM plan designed to achieve the following aggressive business goals:

- Class A MRP II greater than 95 percent with a continuous increase in standards and levels of performance.
- Average order turnaround of less than 6 days from the receipt of order to shipment, with a process in place to ensure that this level of performance could be sustained.
- Delivery performance greater than 95 percent to promised delivery and greater than 80 percent to requested delivery with a shift in philosophy from "delivery when we say we can" to "delivery when the customer requests it."

- A quality performance of fewer than 7500 defective parts per million, with outgoing product quality measured in terms of the customers' expectations and standards.
- A 7 percent reduction in overall real cost for the current fiscal year with continuous improvement in material, labor, and overhead costs.
- A run-rate inventory turnover of five or greater per year.
- A time to market of less than fifteen days measured from the receipt of the job by operations to the release date.

Then the company developed a performance measurement system to track and report progress toward these ambitious goals. Baseline readings were taken in the current year, and improvement was rapid, as indicated by the consolidated report card, which follows in Figure 2–15.

	Current Year	Three months later	Six months later	Nine months later	One year later
1. Class A MRP II	D– (51%)	B (84%)	A (93%)	A (95%)	A+ (97%)
2. Order turnaround time	18 days	10 days	10 days	10 days	7.6 days
3. Delivery performance (weekly)	26%	90%	92%	95%	96%
4. Order backlog (orders)	11,000	446	486	191	269
5. Production operations management performance (slips)	325	0	0	1	1
6. Direct ship performance	72%	94%	89%	95%	95%
7. New products released on schedule (weekly)	40%	95%	100%	100%	100%
8. Shortage area (orders)	4,800	271	570	220	174
9. Vendor delivery performance	31%	81%	98%	94%	96%
10. Inventory record accuracy	87%	91%	98%	96%	98%
11. Customer service posture	Fire fighter	Maytag repairperson	Minimum class A	Class A	Class A
12. Morale	Rotten	Positive	Good	Better	Even better

Figure 2–15 Documentation of Enterprise Improvements.

(Source: Stickler, "Going for the Globe Part III," P and IM Review, November 1989, p. 43. Reprinted by permission of APICS.)

The Design Elements and Production Engineering

After you have completed Part 2, it will be clear to you that:

- A successful CIM solution must emphasize the design of products and the processes that will be used to produce them.
- Graphics-based systems are the new language of design.
- Computer-based systems supporting the engineering design process are essential elements of modern CIM systems.
- CIM design processes can help enable true concurrent design for manufacturing and assembly.

Product Design and Production Engineering

To produce a world-class product, the Product Process segment must (1) have a design model that takes the product through all the critical design steps; (2) use computer software and other analytical tools to create, analyze, and test design options; (3) support a *concurrent engineering* process that brings everyone involved in the design and production process together during the design; and (4) use design principles to create a product that can be manufactured defect free. In the next three chapters, we provide an overview of the design process and the automation used to support all stages in the product development.

3–1 PRODUCT DESIGN AND PRODUCTION ENGINEERING

The product design and production engineering area is an appropriate starting point for a detailed study of CIM for the enterprise for two reasons. First, these departments have embraced and encouraged the use of technology to reduce the many tedious manual tasks present in the design and documentation of products and product systems. Although the automation gain in the design area is good news, success is tempered by the fact that the technology is not always implemented effectively. In many cases the design automation solution fell far short of the standards set for CIM because the implementation had only a departmental focus.

Functional automation, the narrow view of automation isolated to a single function or departmental area, appears to provide significant productivity gains; however, under careful analysis the benefits for the total enterprise are often negative. For example, a case study of a manufacturer producing riding lawn mowers uncovered three locations where functional automation was used to produce product and part drawings: initial design (part and product design drawings), marketing (service manual drawings and customer information), and production (drawings to aid in the programming of metal-cutting machines). The three areas, working in isolation with no enterprise-wide plan, chose different automation hardware and software drawing solutions with *no compatibility* between the computer files created in each department. As a result, the benefit for the enterprise was negative for several reasons: Time was wasted entering redundant data, three separate computer images of the same product had to be maintained, the number of drawing errors for the product increased by a factor of 3, product quality suffered due to the drawing errors, and production costs were increased due to more part list errors and obsolete parts.

The second reason for starting with product design and production engineering is because the initial creation of product data starts there. Generation of design drawings for a new product is one of the first activities in the design area. However, the format for creation and storage of the design must be consistent with a basic CIM axiom that demands the establishment of a *single enterprise database with a single image of all product information.* The implementation of a central database for all products requires an enterprise CIM network where employees use computer technology to access and share product data electronically.

A CIM local area network (LAN) to support product design is illustrated in Figure 3–1. Study the network and find the storage location for the common data. Note that the common database is not restricted to the enterprise mainframe computer. Current network technology permits archiving of common part files on storage devices across the network. In this example, product design drawings would be saved on a computer in the LAN where most of the design activity occurs. In addition, the network technology permits all nodes or users on the network to access the common data distributed around the enterprise system through the bridged LANs. In Figure 3–1, the group responsible for documenting the design through development of working drawings uses a LAN separate from the group performing initial product design and design analysis (CAD/CAE). That situation would not be necessary every implementation because all of the computers could be on the same LAN.

When this principle of common enterprise data is applied to the riding-lawn-mower case study, the three different drawing systems are affected. Each system used to develop new mowers must be able to share a single drawing file in the common enterprise database. Typically, the product drawing originates in the design area and is used or modified by other departments. Building CIM systems around a common data precept changes *functional automation* into *enterprise data integration,* the topic of the remainder of the book.

3–2 ORGANIZATIONAL MODEL

It is important to start with an understanding of the fit between product design and production engineering in the organization model presented in Figure 2–8. The product design and production engineering portion of the organizational model is illustrated in Figure 3–2. The formal interfaces, illustrated by the connecting lines and arrows, represent information and data flow that occur on a regular basis and are dictated by the information flow for normal operations. No formal interface is indicated between the design and production engineering blocks and the shop floor. Certainly, informal communications links between

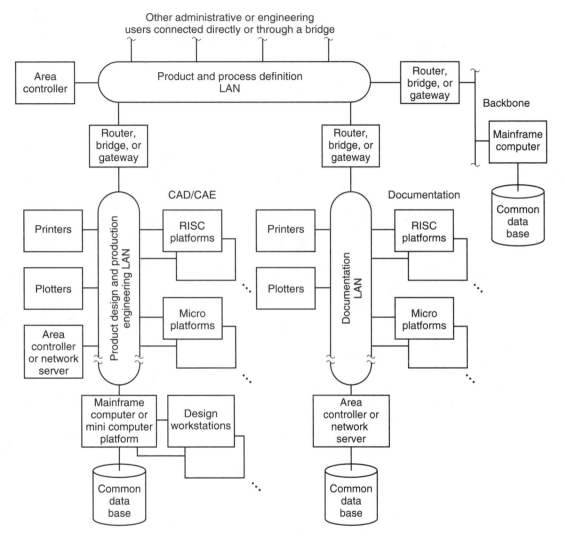

Figure 3–1   Product and Process Definition LAN.

these units exist, and in some organizations a formal link is established between the production engineering area and the shop floor.

Design Information Flow

The design activity starts with either a *new design creation* or a *request for engineering action* (REA) in the product design block in Figure 3–2. Most frequently, the new design information comes from sales and marketing. An internal REA can originate from any department in the enterprise that identifies a product problem; external REAs originating from customer problems are usually initiated by the

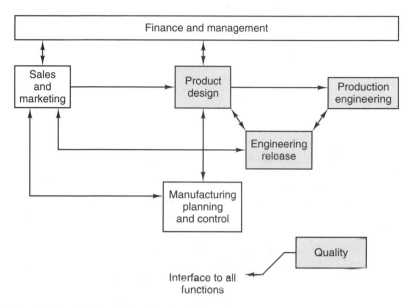

Figure 3–2   Product Design and Production Engineering.

sales and marketing area. The product design area is responsible for *product design and analysis, material selection,* and *design and production documentation.*

The production engineering area adds *production standards* for labor, process, and quality to the product data from the design area. Labor standards identify the time required to produce each part in the product. If in-house production is the result of the buy versus make analysis, the process standards required to make the product are developed. Finally, quality standards and verification techniques for the product and production process are established.

Engineering release is responsible for *product change control.* All changes to existing products and all new designs are normally reviewed by the four enterprise areas highlighted in Figure 3–2. The review process ensures congruence between customer needs and production capability for the proposed product design or change.

The separate block for quality does not imply that quality is external to the design and production engineering departments in the enterprise. In fact, world-class companies practice *quality at the source* to move toward a goal of defect-free operation for every action performed in the organizations. Quality at the source implies that all areas implement defect-free management in a unique way; for example, product design must have no design or documentation errors, and the product must be designed in compliance with the order-winning criteria and manufacturing capability.

In the remainder of this chapter, we develop the design process in greater detail and describe why the application of automation to the design segment is critical for world-class manufacturing.

3–3 THE DESIGN PROCESS: A MODEL

The general process for design is characterized by five basic steps:

1. Conceptualization
2. Synthesis
3. Analysis
4. Evaluation
5. Documentation

These five steps are modified from a six-step process described by Groover (1987) and Shigley (1983). In this new five-step model conceptualization replaces *recognition of need* and *definition of the problem* in the six-step process. The *conceptualization* process, developed in the next section, defines the problem based on the stated need and then divides it into two categories: typical and atypical. In the *synthesis* step, detail is added to the aggregate solution produced in conceptualization. When the product or part leaves the synthesis step, the design has sufficient detail to determine how it will perform. In the next step, *analysis*, the design is tested and the performance data are collected on as many phases of operation as possible. *Evaluation* of the performance data occurs in the fourth step. If evaluation of a product indicates that any part of the product did not meet performance and design specifications, then alternatives to the design are considered. The last step is *documentation;* here the final part details are added that permit manufacturing to produce a product that matches design specifications. The design process, just described, is used for all five manufacturing systems (*project, job shop, repetitive, line,* and *continuous*) and the four production strategies (*engineer to order, make to order, assemble to order,* and *make to stock*) described in Chapter 1.

The design process is illustrated by a sequence of blocks in Figure 3–3. The illustration implies a linear process with reiteration between the blocks. The top three blocks are highly interactive, especially in the early stages of a product design. The new product concept is divided into subsystems or parts. Potential solutions for the product and each part whirl through the conceptualization, synthesis, analysis, and evaluation steps with iterations generated whenever a better alternative is uncovered. Eventually, the best alternatives for the product emerge and documentation is prepared to support the manufacturing requirements. Integrated design/synthesis, analysis, evaluation, and documentation software packages are used by CIM-driven companies in the design model in Figure 3–3. The software has integrated applications to support the entire design model and works from a common electronic database of the product.

The origin of product and part information is the design process. For a CIM implementation, product design is the start of the enterprise-wide central database. Studies in automated assembly indicate that up to 70 percent of a product's cost are set by the time a product leaves the evaluation stage. Therefore, revolutionary decreases in product cost can come only from the initial design group. As a result, the design process is the starting point for improving the cost order-

Figure 3–3   Design Model.

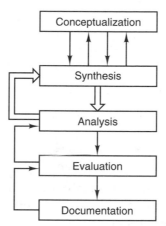

winning criteria. The development of an effective design process is addressed in the remainder of this section.

Form, Fit, and Function

The design process starts with the characteristics of the product defined in terms of *form, fit,* and *function.* The desired shape, style, and character of the product are defined by the term *form.* The *fit* characteristic refers to the marketing fit or the order-winning criteria necessary to be successful in the market. The fit also describes the relationship of the desired product to other products in the company's line, and the degree to which the product matches the target population. The *function* character-istic defines the product in terms of performance, reliability, maintainability, and any other specific order-winning criteria present. The form, fit, and function stan-dards for the desired product establish the limits for the finished design. The data for these three characteristics for new products often come from the marketing area. However, changes to the form, fit, and function for existing product come from a host of internal and external sources including customer marketing, sales, and design. With the three characteristics identified, the design problem is defined and the design process is started.

Conceptualization

The first step in the design process, *conceptualization,* defines the design problem as either typical or atypical. A typical design problem is one that is similar to pre-vious product designs. The atypical design statement defines a product need that is totally new or different from any previous product. The atypical problems require a concept design approach; the typical problems are handled with a process called repetitive design. *Repetitive design* is defined as follows:

> *Repetitive design is the application of the design process to a new product by using pieces of previously designed items or small variations from previous designs.*

Repetitive design applies when the designs are primarily collections of some standard parts or have large sections that are similar from design to design.

Conceptualization Using Repetitive Design

The repetitive design process is described by the following three-step sequence.

1. Establish all the form, fit, and function information and data for the desired product.
2. Categorize current products with similar form, fit, and function characteristics and then apply the following guidelines:
   - Apply parametric analysis to families of parts or assemblies that are similar in form and function but vary in size and detail, and/or
   - Develop a set of standard parts for use across similar products.
3. Model the product design graphically and analytically to communicate the design configuration effectively.

Parametric analysis, referenced in the second step in the repetitive design sequence, is a powerful repetitive design tool offered with some computer-aided design (CAD) programs. A simple example of a roller manufacturer illustrates the parametric design concept. The company produces rollers with the general shape shown in Figure 3–4. Dimensions A, B, C, and D vary from order to order, with about 1500 sizes produced. To manufacture the rollers without a CAD system and parametric design, a drawing for all 1500 standard sizes must be produced. To add a new size roller, a draftsperson must draw the complete new part. When a parametric-based CAD solution is used, the part, shown in Figure 3–5a, is drawn once in the CAD system, with the dimensions that change represented by variables. The variable values for the standard rollers are listed in a table (Figure 3–5b) in the parametric design software. When a standard size is selected, the CAD software

Figure 3–4   Roller Shape.

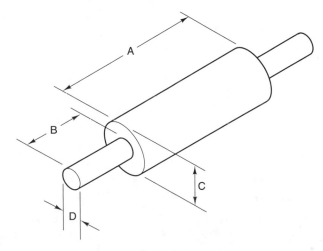

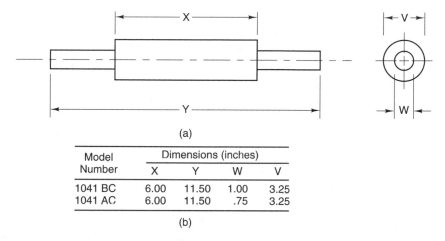

(a)

Model	Dimensions (inches)			
Number	X	Y	W	V
1041 BC	6.00	11.50	1.00	3.25
1041 AC	6.00	11.50	.75	3.25

(b)

Figure 3–5   (a) Part Dimensions; (b) Parameter Table.

creates the roller to the correct dimensions, plots a drawing of the part, and creates a program file to produce the part on a computer-controlled production machine. To create a new non-standard-size roller, the desired values for the variables are entered into the table, and the CAD software produces the drawings and other necessary files for the new roller. The power of parametric design software extends beyond this simple example. For example, the input data for the new roller could be a required minimum force on the roller surface; the parametric software would calculate the minimum dimension for W in Figure 3–5a based on the force and the tensile strength of the roller metal.

The case study at the end of the chapter illustrates how repetitive design using the second step in the repetitive design sequence increases the profitability of a power tool manufacturer. Study the case carefully.

Conceptualization Using Concept Design

The *concept design* process is defined as follows.

> *Concept design is the application of the design process for the creation of a new product that is unique, with no similarity to any product currently produced.*

The concept design process draws heavily on the creative nature of the designer, who must create, in the mind's eye, a totally unique solution to the present problem. Therefore, the concept design process is just a two-step sequence.

1. Establish all the form, fit, and function information and data for the desired product.
2. Model the product design graphically and analytically to communicate the design configuration effectively.

History has many examples of products that resulted from true innovation on the part of the designer. One such product was the Sony Walkman cassette player.

While concept design principles were used by the engineering team that conceived the first wearable entertainment cassette player, they borrowed many ideas from other existing products. Therefore, at the start of every concept design process, the potential for using standard parts, parametric parts, and results from previous designs must be considered along with the requirement for *original* design work dictated by a unique form, fit, or functional requirement. Data indicate that concept design is required less than 10 percent of the time, with the remaining 90 percent divided between repetitive design and some combination of the two design processes.

Synthesis

The second step in the design process is *synthesis.* The synthesis activity in Figure 3–3 includes the specification of material, addition of geometric features, and inclusion of greater dimensional detail to the aggregate design emerging from conceptualization. However, at this point the original product design model needs some modification. The new model (Figure 3–6) identifies the second step as a synthesis filter. The synthesis filter is analogous to a paper filter that removes impurities from a liquid. The synthesis design filter removes geometric features and material specifications that add cost to the product but not market value. As the model illustrates, the filter uses two processes, *design for assembly* and *design for manufacturing,* to ensure a good design. Conceptualization and synthesis are closely tied and highly iterative, with activities in these two processes often inseparable. As preliminary product ideas are enriched with features and details early in the design process, the design engineer uses both conceptualization and synthesis skills. However, as the product design becomes firm, more time is spent in synthesis adding and verifying product features and details. With 70 percent of the

Figure 3–6 New Design Model.

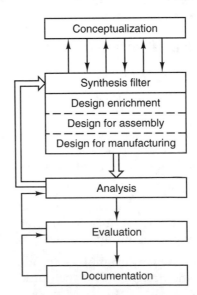

manufacturing cost fixed in these early activities, the time spent getting the product *right* in the synthesis step is well justified. The procedures used to get the product *right*, design for assembly and design for manufacturing, are covered in Chapter 5.

Analysis

Analysis is a method of determining or describing the nature of something by separating it into its parts. In the model of the design process (Figure 3–6), the analysis stage studies a single design solution or several alternative design choices using mathematical and other scientific procedures. In the process the elements, or nature of the design, are analyzed to determine the fit between the proposed design and the original design goals. The highly interactive nature of the synthesis and analysis stages is best illustrated with a simple example. The aluminum bracket in Figure 3–7a was part of a larger design. In the synthesis stage, the thickness of the metal was specified. However, when each part of the bracket (horizontal plate, junction, and vertical plate) was analyzed, the strength of the junction was not adequate for the typical bracket load. After several iterations between synthesis and analysis, the correct thickness at the junction was determined (Figure 3–7b), and the optimum design for the bracket was produced.

The types of analysis frequently used fall into two categories: *mass properties* and *finite element*. While the analysis calculations can be performed manually, the use of a computer increases the analysis capability and significantly reduces the time required. In mass properties analysis, solid object features such as volume, surface area, weight, center of mass, and center of gravity are calculated. In addition, mechanical tolerance requirements are established and assembly interference and fit analysis are performed to ensure that the individual parts can be assembled properly even if all the tolerances go to worst-case condition. In *finite element analysis* (FEA), typical calculations are (1) the limits of stress and strain for a part, (2) the heat transfer capability of the part for a specific material, and (3) the theoretical limits of operation of an electrical or electronic circuit. FEA and mass property analysis often require that the product and/or parts in the product be represented electronically by a solid model. This solid model has all the characteristics of the actual product or part, so the FEA or mass properties software uses the model to perform the analysis.

Figure 3–7   Aluminum Bracket.

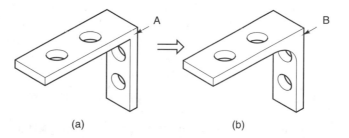

(a)   (b)

Study the design model shown in Figure 3–6 and note how the relationship between the top three blocks is illustrated. The many arrows in both directions indicate that conceptualization and synthesis are highly interactive and often inseparable. However, the line between synthesis and analysis is quite clear. A design is developed and then analysis is performed. Analysis is a critical link to ensure that the product meets the order-winning criteria required by the customer.

Evaluation

The *evaluation* step (Figure 3–6) checks the design against the original specifications. The optimum design delivered from the synthesis and analysis process is compared to the form, fit, and function requirements for the new product. Additional evaluation is performed to determine the match between the manufacturing and assembly requirements of the design and the capability of the manufacturing facility. If the design satisfies the evaluation criteria in every case, the product is passed to documentation. If any areas fall short, however, the design is returned to either the conceptualization, synthesis, or analysis process for more work.

Evaluation often requires the construction of a prototype to test for conformance with critical order-winning criteria such as operational performance, reliability, compatibility, user-friendly operation, and other criteria. *Rapid prototyping*, a technique to produce a sample product quickly, is frequently used in this design step. Evaluation and rapid prototyping techniques are described in detail in Chapter 5.

Documentation

The last step in the design process is the presentation of the design through a *documentation* process. The following items are frequently part of the documentation process.

- Creating all necessary product and part views in the form of *working drawings* and *detail and assembly drawings*
- Addition of all design details, such as standard components, special manufacturing notes, and all dimensions and tolerances
- Creation of required engineering documents, such as part number assignments, detailed part specifications, design bill of materials (BOM), and a part and part number *where-used list*
- Creation of product electronic data files used by the following departments: manufacturing planning and control, production engineering, marketing, and quality control

Documentation examples are provided in Figures 3–8 through 3–12.

This concludes the description of the process (Figure 3–6) required to bring a product design from initial concept to completed design. A problem remains with

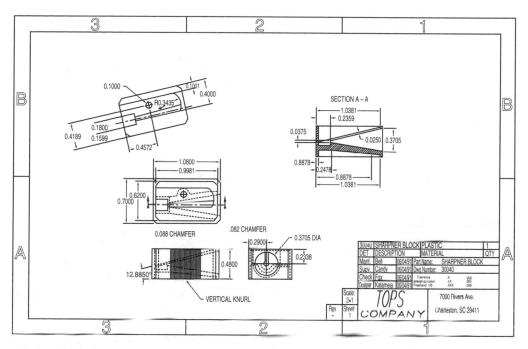

Figure 3–8   Working Drawing.

this design process, however. The process was conducted in isolation because representation from manufacturing, external parts and equipment vendors, and other areas in the enterprise affected by the design were not integrated into the design process. To ensure that the design is optimum for both the initial form, fit, and function requirements and the manufacturing capability of the enterprise, the design process must be enlarged. The required new model has several names:

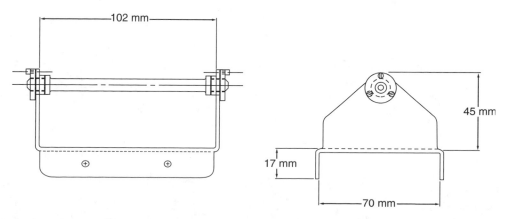

Figure 3–9   Assembly Drawing.

Figure 3–10 Bill of Materials for Spindle Housing Assembly.

Bill of Materials			
Quantity	Part number	Description	Material
1	23301	Sheet metal base	Stainless steel
1	23302	Spindle	Stainless steel
2	23303	Bushings	Nylon
6	23304	Screws	Purchased

simultaneous engineering, early manufacturing involvement, and *concurrent engineering.* In this book, the term *concurrent engineering* is used.

3–4 CONCURRENT ENGINEERING

Concurrent engineering is defined as follows.

> *Concurrent engineering implies that the design of a product and the systems to manufacture, service, and ultimately dispose of the product are considered from the initial design concept.*

Concurrent engineering (CE) is not a new concept; it was used in Japan in the early 1960s. As the definition indicates, CE is a design philosophy dealing with the process that a company uses to bring a new product to market. A comparison

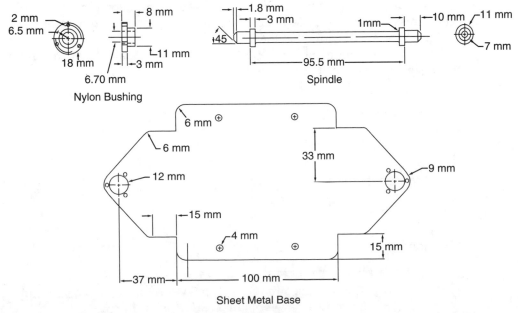

Figure 3–11 Working Drawing for Spindle/Housing Assembly.

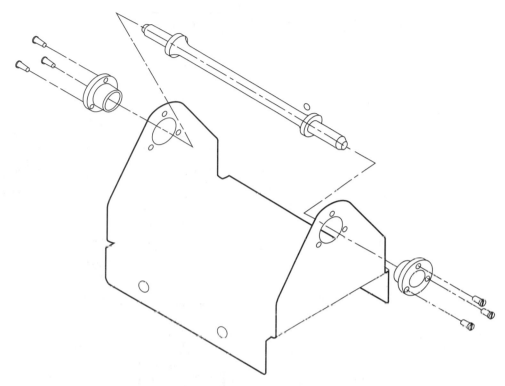

Figure 3–12   Exploded Assembly View of Spindle/Housing Assembly.

between CE and the traditional approach produces some interesting conclusions. First, a look at the traditional process.

The Traditional System

A traditional design system still used by some companies is illustrated in Figure 3–13. The information flow in the sequence is linear, with little interaction between the many groups responsible for new product development. In the traditional system, the product design process (Figure 3–6) is often completed by design engineers without input from any other group. As a result, production engineers get product information after the product design is completely finished and documented on engineering drawings. The production system is frequently designed by production engineers in the same type of isolation. Personnel on the shop floor frequently receive completed product drawings and production system specifications from the production engineers shortly before production must begin. The group responsible for customer documentation and product service is often not involved with the product until shortly before the product is shipped from the factory. In most cases the disposal of the product after its useful life is never considered.

This type of product development creates several problems, including late entry into a market segment, poor quality, poor customer service, and high

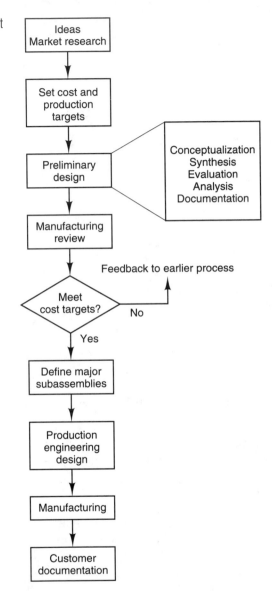

Figure 3–13 Traditional Product Development Process.

product cost. The problems occur because the design produced in isolation by engineering cannot be manufactured in the shop as specified on the drawings. The manufacturing divide described in Chapter 1 has been created. As a result, the merchandise produced usually does not have the order-winning criteria necessary for a successful product.

What is needed is a process that brings design engineering, marketing, production engineering, external vendors, production control, and the shop floor together throughout the design process. The design that results from this collective endeavor is usually optimum for the enterprise and matched to the order-winning criteria present in the marketplace. The design process described in Figure 3–6 must be modified to include concurrent engineering.

The New Model for Product Design

The design process model shown in Figure 3–14 includes concurrent engineering. With this change, participation in the design process is not limited to product design engineers but includes all the specialties listed in the right side of the figure. Study the new model until you become familiar with all the participants. The width of the concurrent block indicates the level of participation from the initial concept design through documentation. Notice that the level of participation from across the enterprise drops as the product design becomes firm. As the crucial design features are locked in at the conceptualization, synthesis, and analysis stages, there are fewer critical decisions that require broad input from across the enterprise.

The traditional product development chart shown in Figure 3–13 is changed to include concurrent engineering, and the result is illustrated in Figure 3–15. Note that the concurrent engineering design process shown in Figure 3–14 is embedded in the process, and the CE process includes product design, manufacturing system design, and customer documentation.

Concurrent Engineering: An Operational Model

There are many opinions about how CE should be managed and executed, but one of the clearest processes was proposed by James Nevins, Daniel Whitney, and

Figure 3–14   New Model for Product Design.

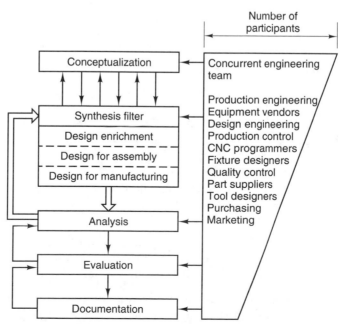

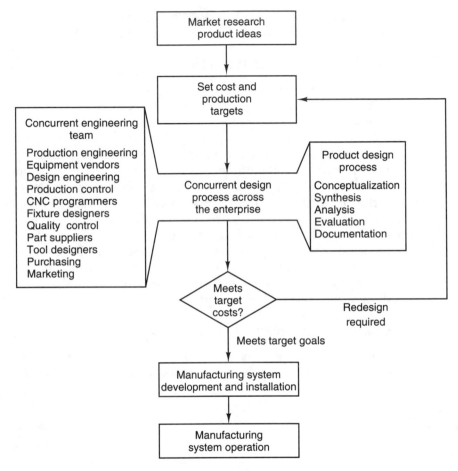

Figure 3–15   Improved Product Design Process.

four collaborators from the Draper Laboratory. The process is general enough to apply to any type of manufactured product. The process steps are consistent with the design process model shown in Figure 3–14 and the product development process shown in Figure 3–15.

The product design begins with a market plan in which rough concepts are developed and price and production targets are set. Concurrent engineering follows the approval of the market plan and proceeds in five phases.

1. *Concept phase.* In this phase most of the major decisions about the form, fit, and function of the product are tentatively established. Example areas include style, materials, performance, and energy consumption.

   ■ Decisions in this phase account for 70 percent of the overall product cost and impact marketability.

- The best designers and the most sophisticated design tools are required.
- Analysis of alternative designs requires the use of performance simulation software and cost-estimating models.

2. *Major subassembly design phase.* Following approval of the conceptual design, the product is subdivided into major subassemblies for more detailed design specification.

   - Model variations in the product are identified and isolated to the fewest possible subassemblies. Modular design practices that reduce the number of different subassemblies reduce cost and permit a quick response to changing market conditions.
   - The number of different departments involved in the design increases significantly in this phase.

3. *Single-part design phase.* The design of all the parts of the main product and all subassemblies is completed in this phase.

   - The number of people participating in the design is at a maximum.
   - Major design changes at this time cause major delays in design completion and are expensive.

4. *Design of part-pairs phase.* Following the design of detailed parts, the emphasis shifts to manufacturing processes instead of product performance. For mechanical designs, individual parts are analyzed for fit by employing part-mating theory (geometric tolerances), tolerance analysis, and tolerance simulation. For electronic products, the development of test programs and acceptance criteria for individual integrated circuits is an example of the activity.

5. *Grouping of parts and subassemblies.* The final phase focuses on additional refinements for efficient assembly. The assembly sequence is determined and the assembly equipment, jigs, and fixtures are designed.

   - Frequently, previous designs for individual parts and subassemblies must be modified. For example, tolerances may be refined; lifting eyes and gripping surfaces may be added.
   - Test strategies and programs are developed.

At several steps throughout the concurrent design process, the development of ISO 9000 process descriptions would be developed to ensure that the production process satisfies this important standard. The process just described may require the commitment of more time and money to the design process than in the traditional method. However, concurrent engineering will reduce total time to market because the production tooling and manufacturing systems are ready sooner.

Automating the Concurrent Engineering Process

The concurrent engineering process is well supported by product design software running on the mainframe computers, engineering workstations, and PCs. The products permit two or more designers located in different offices or even different cities to work jointly on the same part or product drawing. In some cases video feedback between the sites is provided with audio links. Using this technique, every person involved in the selection of certain design features can witness the changes and can provide input to achieve a better, lower-cost product. Another feature of the design software, called *full associativity*, links all the related drawings files that have been created for a product. For example, the working drawings, solid model, 3-D drawings, and section views are linked by this feature. As a result, any change made on any of the drawings is changed automatically on every drawing of the part.

Concurrent Engineering Success Criteria

Companies experienced in the application of the concurrent engineering model have identified some important general guidelines.

- Construction of tooling and manufacturing systems does not begin until all five phases of the Nevims and Whitney model of concurrent engineering are completed.
- The concurrent engineering philosophy suggests that prototype tooling and production tooling be as similar as is practical so that production tooling produces no unexpected surprises.
- Each step in the concurrent engineering process involves experienced designers of manufacturing systems, service systems, and disposal systems, not just product designers.

The involvement of participants requires more than simply a design review. Production engineers may need to begin preliminary design of manufacturing systems to ensure that the product can be made for the amount budgeted. Such a design may include:

- Outside suppliers of tooling and production machinery
- Designs of repair and maintenance systems to evaluate serviceability
- Preliminary design and cost estimates of tools for repair, assembly, and disassembly
- Design of ultimate disposal techniques, called the *cradle-to-grave process,* consistent with environmental regulations for the product area

This definition and operational model for concurrent engineering is broad because of the *cradle-to-grave* emphasis placed on the design process. CE is also demanding because it requires exceptionally good, thorough engineering throughout the design process. However, the demands of the marketplace for world-class products require a concurrent engineering process.

3–5 PRODUCTION ENGINEERING

Production engineering (PE) has the responsibility for developing a plan for the manufacture of the new or modified product that was generated through the concurrent engineering process. The PE plan has many elements:

1. Process planning
2. Production machine programming
3. Tool and fixture engineering
4. Work and production standards
5. Plant engineering
6. Analysis for manufacturability and assembly
7. Manufacturing cost estimating

In some organizations these production engineering activities are grouped under the industrial engineering department; in others, they would be assigned to the manufacturing engineering area. The name of the department where these functions are located is not critical. However, the use of concurrent engineering to bring these seven areas together in the product design team is crucial for a successful product. When the classical method is used, the design is passed through one or more groups responsible for these seven activities. If this occurs, the design becomes less and less optimum as it passes through the groups. For example, if tooling to produce the product is not considered until last, it may not be possible to hold the material during some manufacturing processes. Regardless of the location of these activities, it is important to understand the function they perform on the product. Therefore, the remainder of this chapter is devoted to a description of these activities.

Process Planning

Process planning is often called manufacturing planning, material processing, and machine routing by different industry groups. The people who carry out this process are called process planners, material processors, or just planners. *Process planning* is defined as follows.

> *Process planning is the procedure used to develop a detailed list of manufacturing operations required for the production of a part or product.*

Figure 3–16 indicates the order of events in process planning in the traditional company. The process starts with a part or a family of parts for a completed product arriving in planning along with the detailed product documentation listed in the upper left box in Figure 3–16. On each part a make-versus-buy decision is made by production engineering. Every part to be made and assembled in-house has a *routing sheet* prepared.

Routing sheets, sometimes called process plans or operations sheets, describe the sequence of operations or manufacturing processes required to produce the finished product. A typical set of operations and machines for the machining of surfaces is illustrated in Figure 3–17. Take a few minutes to study the chart. An example routing sheet for 100 of the stainless steel spindles shown in Figure 3–11 is illustrated in Figure 3–18. Note that in Figure 3–18 the work center (a specific production facility that includes one or more people and/or machines) is specified, along with the ID number and brief description of the operation. The time data provided for each operation often include setup time, unit run time,

Figure 3–16 Process Planning Sequence.

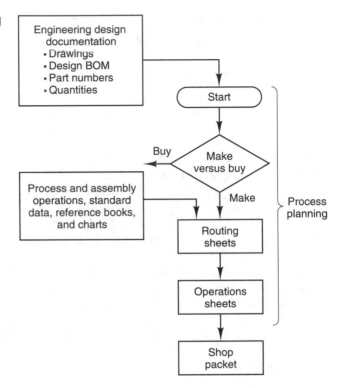

queue time, and other standard times. Identify the operations listed in the routing sheet that appear on the process chart shown in Figure 3–17.

The *operations sheet* (Figure 3–16) describes in detail each of the operations on the routing sheet for the part. These operations often include tooling, jigs/fixtures needed to hold the part, sketches of setups, semifinish dimensions, machine settings, assembly instructions, handling requirements, inspection and testing requirements, and operator skill levels. Everything necessary to produce the part to design specifications not covered in the working drawing is included. Note in Figure 3–16 that the planner has process, assembly, and standards data reference books and charts to use as a guide in setting up the routing and operation sheets. In some organizations, the routing sheet and operation sheet are combined into one document. All the information and sheets prepared in process planning go into a *shop packet*. The shop packet, a package of documents used to plan and control the movement of an order on the shop floor, normally includes a *manufacturing order* called a traveler, *operation and routing sheets, engineering documentation, pick list, move tickets, inspection tickets, time tickets,* and others. The process used to develop routing and operation sheets for products is similar for all five manufacturing systems (*project, job shop, repetitive, line,* and *continuous*) and the four production strategies (*engineer to order, make to order, assemble to order,* and *make to stock*) described in Chapter 1.

In the concurrent engineering model, the process planners are working with the CE design team to produce routing information during the synthesis stage. The

Operation	Block diagram	Most commonly used machines	Machines less frequently used	Machines seldom used
Shaping	Tool / Work	Horizontal shaper	Vertical shaper	
Planing	Tool / Work	Planer		
Milling	Slab milling, Tool, Work / Face milling, Work, Tool	Milling machine		Lathe (with special attach-ment)
Facing	Work / Tool	Lathe	Boring mill	
Turning	Work / Tool	Lathe	Boring mill	Vertical shaper Milling machine
Grinding	Tool / Work	Cylindrical grinder		Lathe (with special attach-ment)
Sawing	Tool / Work	Contour saw		
Drilling	Work	Drill press	Lathe	Milling machine Boring mill

Figure 3–17 Manufacturing Processes.

(Reprinted with the permission of Macmillan College Publishing Company from Materials and Processes in Manufacturing, *Seventh Edition, by E. Paul DeGarmo, J. Temple Black, and Ronald A. Kohser. Copyright ©1988 by Macmillan College Publishing Company, Inc.)* ***(cont.)***

Operation	Block diagram	Most commonly used machines	Machines less frequently used	Machines seldom used
Boring		Lathe Boring mill Horizontal boring machine		Milling machine Drill press
Reaming		Lathe Drill press Boring mill Horizontal boring machine	Milling machine	
Grinding		Cylindrical grinder		Lathe (with special attachment)
Sawing		Contour saw		
Broaching		Broaching machine		
ECM		ECM machine		
Laser		CO_2 laser YAG laser		

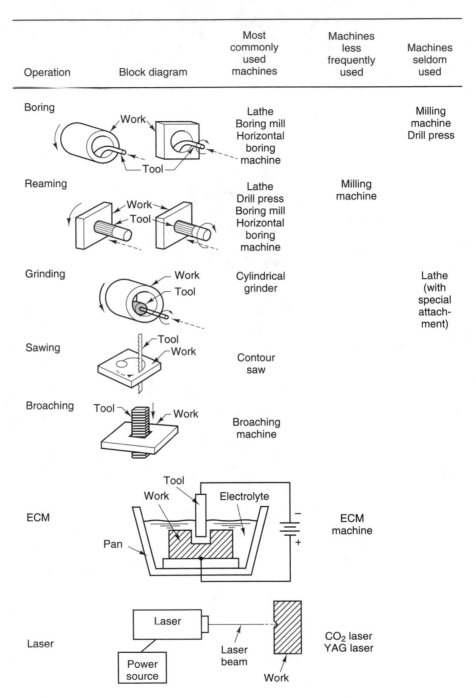

Figure 3–17 Continued.

Figure 3–18 Route Sheet for
Stainless Steel Spindle in Figure 3–9.

Part number: 252601   Lot size: 100
Part description: Spindle—stainless steel

Work center	No.	Operation Description	Setup time (min)	Unit run time (min)
10	01	Saw 3/4-in. stainless steel bar to length	10	0.47
30	01	Deburr	5	0.25
20	01	Turn shaft ends	30	1.17
20	02	Turn center	10	0.74
30	02	Deburr	5	0.15
40	01	Grinding	25	0.37
50	01	Degrease	5	0.20
60	01	Inspection	10	0.10

procedure just described does not change, but the process planning is moved up into the design process. As a result, designs that are optimized for the production area are generated. In CIM enterprises, computer software can be used to generate most of the data in the packet. In addition, the packet need not move physically from work center to work center in paper form; rather, operators view the data files containing packet information using computer terminals in the work center.

In the fully manual planning system, the process plan is produced by the planner, who studies the part drawings and then establishes the appropriate procedures to manufacture the part. It requires a person knowledgeable in the processes listed in Figure 3–17, who has many years of experience in planning production. It is as much an art as it is a science.

Production Machine Programming

In the past, when there were only manually operated machines in the production area, this activity was not required. Automation of production machines started about the same time in the discrete and continuous production industries. In the discrete manufacturing area, automation started with metal-turning and metal-cutting machines such as the lathe and mill illustrated in Figure 3–17. At the same time, the process industries, producing products such as gasoline and nylon, started introducing pneumatic process controllers to regulate the continuous production processes.

As automation on the plant floor evolved to computer-driven machines and processes, the need for machine programming developed. Today, the use of computer-controlled production machines, often called numerical control (NC) or computer numerical control (CNC), in discrete- and continuous-process manufacturing is the rule rather than the exception. In many organizations, the responsibility for the programming of the production machines falls in the production engineering area in the enterprise. As a result, the automation of the factory floor created a new term, *computer-aided manufacturing* (CAM). CAM is discussed in Chapter 12.

Tool and Fixture Engineering

Another responsibility of production engineering is for the specification and design of the tooling required in the production of the part. In metal and nonmetal cutting, the tooling includes *jigs* and *fixtures* to hold and position raw material and parts in machines while tools shape and finish the parts. Square-shaped vertical fixtures are visible in Figure 3–19 in front of the machining center. Parts to be machined are attached to all four sides of these fixtures, often called tomestones. In the forming area, the tooling often includes the design of dies (Figure 3–20) and molds to shape parts. The requests for new production machinery and other support equipment for the new product are made by production engineering.

In the typical manufacturing system, the tooling process begins after the part or product is completely designed. As a result, it is often impossible to design fixtures and tooling that allow the manufacture of a part under optimum conditions.

Figure 3–19   Production Work Cell Uses Multiple Pallets to Hold a Variety of Tooling for Many Different Parts.
(Courtesy of Cincinnati Milacron, Inc.)

Figure 3–20 Stamping Die.

However, in concurrent engineering the tooling is specified as a part of the product design process; as a result, the parts and production tooling are developed in parallel. The result is tooling and parts that fit like "a hand in a glove."

Work and Production Standards

To determine the manufacturing cost, the time required for the production of every part must be established. The standards group uses two methods to satisfy this requirement. In the oldest method, *direct time studies,* an operator is timed using a stopwatch as he or she goes through a production operation. After the job is timed, the percent of normal speed or the rate that the worker used in the production process is estimated between 80 and 120 percent. Using the time study data, an average production rate can be established for every step in the manufacturing process. Time studies are often used in purely manual operations.

The second and preferred method, *motion time measurement* (MTM), uses standard time data developed for basic work elements for manual tasks. All the individual time elements required to perform a new job are summed, and the result is a theoretical average time to perform the job working at an average rate. This process can be automated easily with a computer and MTM software.

Figure 3–21 illustrates the close relationship between the functions in production engineering. The planner codes the process into the process system, and the production planner and industrial engineer add supporting information. The part programmer uses the process plan information to develop the machine programs for use on automated production machines to create the part.

Plant Engineering

In some cases, the production requirements for a new product require a new manufacturing facility. *Plant engineering* addresses the design of a production facility. Many companies go through this process when the initial production facility is designed and built to meet the needs of a new product line. However, some manufacturers that have frequent model changes, such as those in the automotive industry, would have a large plant engineering group as part of production engineering to make major facility changes annually.

Analysis for Manufacturability and Assembly

The concept of *design for manufacturing and assembly* (DFMA), another responsibility usually embedded in production engineering, was introduced earlier in the synthesis filter stage of the design model. DFMA implies that the finished design is optimum for both the manufacturing processes required and the assembly techniques needed. DFMA is effective only if it is performed by production engineering as part of the design process using concurrent engineering techniques. DFMA concepts are addressed in greater detail in Chapter 5.

Manufacturing Cost Estimating

A crucial role for production engineering is the estimation of the product cost based on design drawing data and work and production standards information. Consider an example from a U.S.-based robot manufacturer. When a new model was designed, the complete set of design drawings and part documentation was delivered to production engineering. The 300 or more parts were distributed to the planners. Before manufacturing cost could be determined, the production process had to be established. With the routing and operations for every part established, the labor and overhead for each planned operation could be calculated. The cost for each part was then determined by summing the cost of individual operations. The total robot manufacturing and assembly cost was calculated by adding the cost of all the individual parts on the engineering bill of materials. The cost estimating process is much quicker now because of software packages that do a *cost roll-up* for an entire bill of materials with hundreds of purchased and manufactured parts. Establishing the product cost during the design process is critical for marketing. Therefore, integration of this process into the design activity through concurrent engineering concepts is essential to obtaining an order-winning criterion in the cost area.

Using a concurrent engineering model effectively to address these seven activities in production engineering improves the price, quality, and customer-response

Figure 3–21 Typical Process
Planning System.

(Courtesy of Chang/Wysk/Wang,
Computer-Aided Manufacturing, © 1991,
p. 409. Reprinted by permission of Prentice
Hall, Upper Saddle River, New Jersey.)

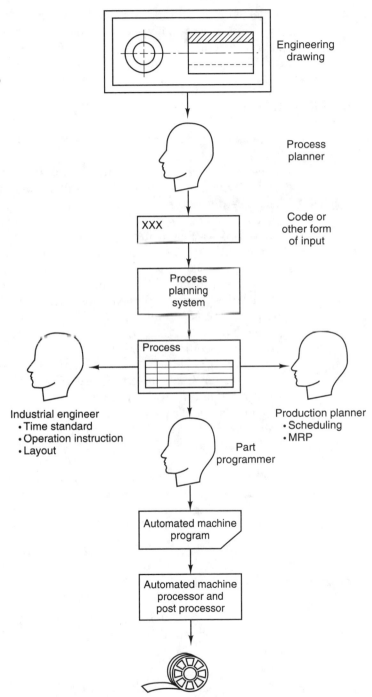

order-winning criteria. Therefore, production engineering is extremely important to the success of the enterprise and the development of a prosperous CIM implementation. Automation techniques frequently used in CIM to implement design and production engineering are described in the next two chapters.

3–6 SUMMARY

The development of new products and modifications to existing products begins in design and production engineering. This engineering group is the bridge between the external focus of sales, marketing, finance, and management and the internal focus present in production planning and manufacturing. Using form, fit, and function specifications from marketing, design engineering creates a design that satisfies the product requirements. The design process is modeled as an interactive process that includes five steps: conceptualization, synthesis, analysis, evaluation, and documentation. Conceptualization is divided into concept design and repetitive design, and synthesis is expanded to include design for manufacturing and design for assembly. A major problem exists with the conventional design model; the process is performed in isolation so that the design frequently cannot be produced economically in the shop. The conventional process reinforces the "great manufacturing divide" and does not focus on satisfying the order-winning criteria present in the market.

Concurrent engineering, a team process that brings all critical departments together from across the enterprise to the design activity, is required to remove the design isolation problem. Concurrent engineering, often called simultaneous engineering or early manufacturing involvement, is described as an iterative five-step process that is embedded into the design model.

Production engineering, frequently termed industrial engineering or manufacturing engineering in organizations, has six fundamental areas of responsibility: process planning, tool and fixture engineering, work and production standards, plant engineering, analysis for manufacturing and assembly, and manufacturing cost estimating. With a good understanding of the product design and documentation process, the next step is a study of the automation that supports the product design.

REFERENCES

ARNSDORF, D. R., *Technology Application Guide: CAD—Computer-Aided Design*. Ann Arbor, MI: Industrial Technology Institute, 1989.

CHANG, T. C., R. A. WYSK, and H. P. WANG, *Computer-Aided Manufacturing*. Upper Saddle River, NJ: Prentice Hall, 1991.

GRAHAM, G. A., *Automation Encyclopedia*. Dearborn, MI: Society of Manufacturing Engineers, 1988.

GROOVER, M. P., *Automation, Production Systems, and Computer Integrated Manufacturing*. Upper Saddle River, NJ: Prentice Hall, 1991.

MITCHELL, F. H., Jr., *CIM Systems: An Introduction to Computer-Integrated Manufacturing.* Upper Saddle River, NJ: Prentice Hall, 1991.

SHIGLEY, J. E., and L. D. MITCHELL, *Mechanical Engineering Design,* 4th ed., McGraw-Hill Book Company, New York, 1983.

SOBCZAK, T. V., *A Glossary of Terms for Computer-Integrated Manufacturing.* Dearborn, MI: CASA of SME, 1984.

QUESTIONS

1. Describe how product design fits into the organizational model for an enterprise.
2. What is the difference between a new design and an REA?
3. What are the responsibilities of design engineering?
4. What are the responsibilities of production engineering?
5. What are the responsibilities of the engineering release area?
6. Describe the design process model and describe briefly the function of each step in the process.
7. What is significant about the data and information that flow from the design process?
8. Why is the product design process critical for a price order-winning criterion?
9. Describe form, fit, and function.
10. Describe the difference between concept design and repetitive design.
11. What is parametric analysis?
12. Compare the interaction between conceptualization and synthesis in the design model.
13. Describe the two major types of analysis frequently used.
14. Define *concurrent engineering* in your own words.
15. What is the cradle-to-grave concept of product design?
16. What major flaw is present in the traditional design process?
17. Describe how concurrent engineering helps to eliminate the "great manufacturing divide" and to promote a congruence in the manufacturing strategy in the enterprise.
18. Describe the responsibilities present in production engineering.
19. Describe the function of routing sheets and operations sheets.
20. What is a shop packet?
21. What is the difference between direct time studies and motion time measurement?

PROJECTS

1. Using the list of companies developed in Project 1 in Chapter 1, identify the companies that use concurrent engineering in the design process.

2. Compare the product design model developed in this chapter with the design processes used by one company from the small, medium-sized, and large groups and describe how their processes differ from the design model in the text.

3. Compare the production engineering responsibilities listed in this chapter with those used in the companies selected in Project 2. Describe the differences that are present.

4. View the SME video "Simultaneous Engineering" and prepare a report that describes the concurrent engineering model and product design processes used by Motorola and IBM.

CASE STUDY: REPETITIVE DESIGN

A manufacturer of handheld power tools for the construction industry has a broad product line of tool sizes and types. In the initial development of each product line and in the yearly updates of the products, the process used for each power tool design was totally independent. Each new product was treated as a new design product and little reference was made to products designed earlier. Isolating each design was a result of the design department structure and a belief that a fresh start would produce new design ideas.

The broad product line had most of the characteristics of a repetitive design process, but the design specifications for current products with similar form, fit, and function characteristics were never reviewed. As a result, the company had to order and stock a total of 208 different washers to produce all the power tools. In many cases, the washers differed only in minor features such as surface finish.

Recognizing that repetitive design guidelines were required, the company analyzed every product design and grouped the washers into two categories:

- Washer designs that were common to more than one power tool product
- Washer designs that performed the same function as the common washers but were different in shape, size, material, or finish

After the analysis, the company used the repetitive design process to modify power tool designs around a small set of washers. It was able to build all the power tools with only eight different washers.

The difference in the cost of the washers may not be a significant part of the product cost. However, the dollars saved in managing the inventory of eight washers versus 208 is significant. In another study of repetitive design savings, NCR found the hidden costs enormously out of proportion to the part price. In one case, NCR estimated that the material, labor, and overhead costs associated with one small screw used in a point-of-sale terminal actually equaled $12,500 over the life of the product. These two examples illustrate that the use of repetitive design to minimize the number of component parts can pay big dividends.

Design Automation: CAD

The design department was one of the first areas in the enterprise to receive automation hardware and software. The technology most frequently implemented in the design area was called *CAD* and is defined as follows.

CAD is the application of computers and graphics software to aid or enhance the product design from conceptualization to documentation.

As the definition indicates, CAD technology supports all levels in the product design process. For example, *computer-aided drafting* (CAD) automates the drawing or product documentation process, while *computer-aided design* (CAD) is used to increase the productivity of the product designers. The parts, machines, and products illustrated in Figures 3–8 through 3–12 are examples of designs and documentation created with CAD software. The capabilities of the software for the CAD system to design a new product is quite different from the minimum resources required to document the product design. However, in every case, the user refers to the technology as CAD, a practice that leads to considerable confusion in defining the technology. Throughout the remainder of the book, the terms *computer-aided design* and *CAD* are used interchangeably when referring to the technology used to support all the stages in the design process. Since CAD is so widely used and has such a diverse application area, a full chapter is devoted to a description of CAD hardware and software systems. In this chapter we describe CAD technology fully, indicate how CAD systems fit into the CIM enterprise system, and provide a rationale for an investment in the technology.

4–1 INTRODUCTION TO CAD SYSTEMS

Computer-aided design is one of the oldest automation tools used to enhance productivity in manufacturing. A study of the evolution of CAD systems and the development of the technology over the past forty years illustrates the manufacturing changes that CAD stimulated.

CAD: A Historical Perspective

The historical time line in Figure 4–1 illustrates the significant developments for the CAD technology from its inception in the Sage Project at Massachusetts Institute of Technology (MIT) in 1963. For the Sage Project, aimed at the development of CRT (cathode ray tube) displays and computer operating systems, interactive computer graphics were initiated with a system called *Sketchpad.* The early CAD systems were basic graphic editors with few design symbols; by the end of the 1960s, however, CAD software could create two- and three-dimensional part drawings with the bounding edges represented by lines and curves. A new era in CAD started in the early 1970s with the introduction of *solid-part modeling.* With this advance in CAD technology, the three-dimensional part models included surfaces and the characteristics of a solid object to aid in analysis.

Throughout the 1960s and 1970s, CAD software was executed on mainframe or minicomputers, and the operators used graphics terminals with little internal computing power to support the modeling and drawing of parts. As CAD software increased in capability, mainframe computer manufacturers moved some of the computational power and control to create the graphic screen images out to each terminal. This decision reduced the computational overhead on the mainframe and permitted the central computer to support more CAD terminals. However, the use of mainframe systems for high-level CAD stations caused the cost per station for CAD to remain high. Figure 4–2 shows a typical configuration with a central computer connected to individual workstations and shared peripheral resources, such as printers and plotters.

An important event, the development of the microprocessor, occurred in the early 1970s. By the end of the decade, the stage was set for another paradigm shift in CAD, the era of PC (personal computer) -based CAD software. As the power of microprocessors increased throughout the 1980s, so did the capability of PC-based CAD software and systems. CAD part modeling and documentation software, with many of the functions found on the mainframe systems, was now available to individual designers on a stand-alone, relatively low-cost PC platform. The era of distributed computer-aided design was a reality, and the design process was changed forever.

While the PC-based systems were capable of performing many of the required design tasks, some operations required computing capability not present in the PC. A third CAD system platform, the reduced instruction set computer (RISC), was introduced in the early 1980s to meet this need. RISC-based systems, called *engineering workstations,* put the power of a small mainframe computer at every engineer's and draftsperson's desk. The CAD system, used to support the design process in large companies today, is a network of PCs, RISC-based engineering workstations, and mainframe computers. Figure 3–1 shows the configuration of a typical system. The size of the system is dictated by company size and project complexity, but all systems have similar operational capability. The capability includes (1) stand-alone PC- and RISC-based CAD workstations at each engineering and design drafting location, (2) the ability to share part data and product information with every station in the system, (3) access to part data files from

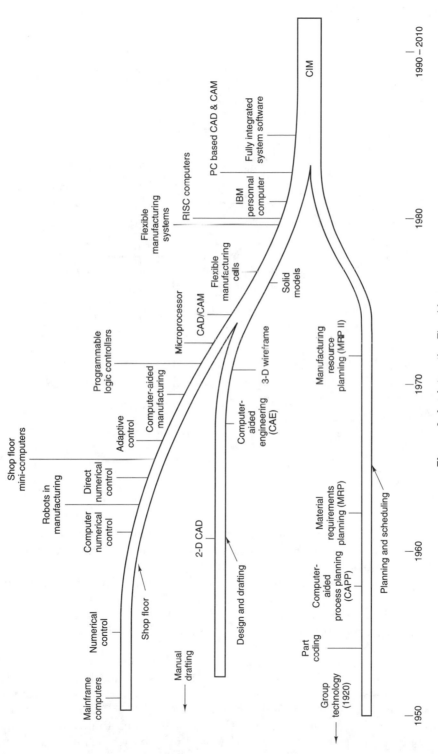

Figure 4–1   Automation Time Line.

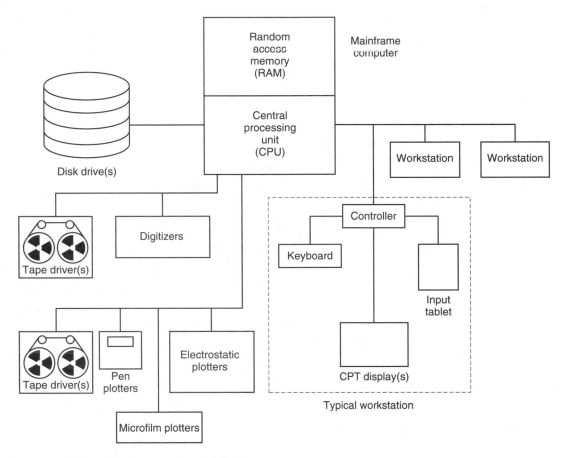

Figure 4–2   Mainframe Computer-Based CAD System.

the mainframe computers on the network, (4) shared peripheral resources such as printers and plotters, and (5) concurrent work on the same project from multiple workstations. In most new installations, the workstations perform all the basic CAD functions and use the mainframe for program storage, file service, and data storage.

The 1990s witnessed the maturing of the PC for CAD software delivery. The processor chips produced by Intel Corporation provided processing power that approached the low end of the RISC market, and the NT operating system from Microsoft Corporation became the de facto standard for PC industrial software development. As a result, many of the companies that marketed only RISC-platform CAD software started to offer an NT version of their software with most and in some cases all of the capability of the earlier RISC CAD versions. The lower entry price that these new powerful PCs provided helped to increase the number and sophistication of the CAD stations used by product design and documentation departments.

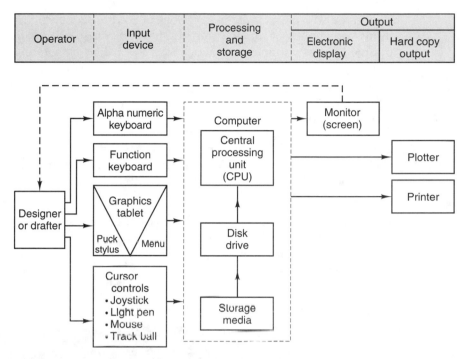

Figure 4–3   CAD System Block Diagram.

Basic CAD System

With the history of the CAD system understood, a description of the *basic systems components* is an appropriate first step to learn what a CAD system is and how it can be used. The basic system block diagram for CAD hardware is illustrated in Figure 4–3. The top bar indicates that the system is controlled by an *operator* who uses *input devices* to create an electronic product model in the *central processing unit* of the computer. The operator views the results of the product design or documentation on several different *output devices.*

A variety of CAD input devices is available to create lines and arcs in the CAD system. The devices frequently used include (1) a standard computer keyboard, (2) a command keypad, (3) a variable command input device, (4) graphics tablets, (5) a puck, (6) a mouse, and (7) a track ball. The function of each device is described below.

Keyboard.   The standard computer keyboard is the primary input device for alphanumerical data and text information entered into the CAD system. Some experienced CAD users prefer to enter system commands through this device as well.

Single and Variable Command Input Devices.   These devices are special command keypads or input devices that permit commonly used system commands to be entered with either a single keystroke or control sequence. Continuous variable inputs use groups of rotation controls. Input devices with rotational controls permit continuous variable control of command values by rotating the controls. For example, one control is usually assigned to the *zoom* command; as a result, the drawing scale is continuously varied by rotation of a knob. A command input keypad, used with CAD software such as CADAM, is illustrated in Figure 4–4. The 32-button keypad selects a command with a single keystroke and allows the operators to map or assign their most frequently used commands to the keypad keys. The Windows operating system permits CAD software vendors to build a graphical version of these input devices on the computer screen. The product designer uses a mouse pointing device to interact with the graphical display.

Graphics Tablet.   A flat digitizing surface offers another option for drawing and the selection of system commands. The tablet in Figure 4–5 is used to move the cursor to different parts of the monitor screen to create part geometry and to select drawing commands.

Figure 4–4   Command Keypad.

Figure 4–5   Digitizing Graphics Tablet with Puck.

Cursor Control.   The location and desired movement of the cursor on the monitor is controlled by a *joystick, light pen, puck, mouse,* or *track ball*. The puck, which is always used with a graphics tablet (Figure 4–5), has two or more buttons to initiate different system actions based on the location of the puck on the tablet when the button is pressed. The mouse and track ball are the most commonly used cursor control devices for PC CAD systems. The track ball provides control through movement of the ball with a finger. Both devices control the cursor location on the screen and have buttons to initiate system action.

The output devices listed in Figure 4–3 for a basic CAD system include a color computer *monitor, plotter,* and *printer.* During the creation of the product design, the operator views the drawing on a *monitor* or *screen.* The displays for most CAD systems are generally large-screen color monitors, with high picture resolution to show intricate detail and fine lines. When a paper or "hard" copy of the design is needed, a plotter is used for large sheet sizes and the printer is used for small drawings. The plotters come in a multitude of sizes, based on the largest drawing paper size that a plotter accepts. In addition to the standard pen plotters, a family of electrostatic plotters is available in either black-and-white or full-color models. The advantage offered by the electrostatic models is a shorter plotting or print time.

4–2 GENERAL SYSTEM OPERATION

With the key elements and components of CAD systems understood, the basic operation of the system needs to be addressed. The system is a marriage of hardware and software technologies, with each playing an equally important role. The hardware supports the electronic representation of objects, while the software allows the objects to be created with an infinite range of geometrical shapes and sizes.

CAD Hardware: Basic Operation

The CAD system components in the Figure 4–3 block diagram represent a minimum system configuration. The operation of that minimum system is covered in this section. The method used in the CAD system to develop, store, and display the geometrical part shapes is an excellent starting point. The picture or drawing created in CAD is represented in two very different ways in the computer. The format for saving the drawing data on magnetic (hard or floppy disk) or optical media uses *geometric* and *graphical* parameters for the lines, arcs, and curves in the object. However, when the drawing is loaded into the computer's active memory for display on the screen, the object is put into a *bitmap*. An example using the simple objects in Figure 4–6 illustrates the difference in the two techniques.

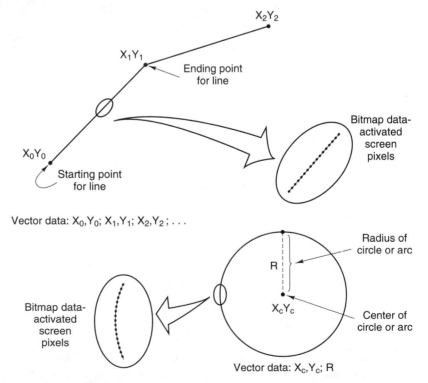

Figure 4–6   Storage of CAD Geometry.

The CAD drawing or object is displayed on the computer monitor by turning ON small dots, called *pixels,* on the inside of the monitor's screen. The object illustrated in the figure would have pixels turned ON every place a line or circle was present. The pixel is a single dot for monochrome (black and white) monitors or a group of three dots (red, green, and blue) for color monitors. The pixels are aligned in rows and columns with thousands of pixels in both the horizontal and vertical directions. The number of pixels in each direction is a measure of the *resolution* of the monitor, or the ability of the monitor to display fine lines and part detail. The larger the number of pixels, the greater the resolution and the more detailed the image. When an object is displayed, each pixel is assigned a memory location in the computer. The data stored in the memory location determine if the pixel is ON (visible) or OFF (usually black). Since each pixel on the screen has a corresponding memory bit or location, the data form is called a bitmap. Storing images in this fashion is memory intensive because most images require millions of pixels. Images saved on disks in the bitmap format are called bitmap or *raster* files.

The objects in Figure 4–6 could also be saved in memory using a *vector* data format. In the vector method, information to locate the starting and ending points of the line segments are saved in memory. For circles and arcs, the center and radius are saved. When a drawing saved in this format is retrieved from permanent storage, the geometric objects are *regenerated* by using the vector data to draw the lines and arcs into the computer's screen memory. The creation of the bitmap image in the screen memory area prepares the geometry for display on the screen. As objects become more complex with the addition of color or when a third dimension is added, the vector data associated with each point on a line increases. Therefore, in the vector format, the CAD database or computer file is just a series of numbers representing the vectors required to create the part. The length or size of the vector drawing file is determined by the number of line segments, arcs, and related geometric information present in the drawn object. Whereas vector files for drawings such as those illustrated in Figures 3–8 through 3–12 are usually less than 100,000 bytes, the bitmap image file for the same drawings would be megabytes in size.

The remainder of the hardware in the basic CAD system (Figure 4–3) supports the creation of the drawing and the production of hard copies for use in the enterprise. The cursor control and graphic tablet input devices, for example, are used to locate and mark the screen pixel that represents the ends of a line segment or the center of an arc. Moving the puck on the surface of the graphic tablet causes the cursor on the screen to move to the desired location or pixel for the start of a line. The other input devices in Figure 4–3 are used to enter drawing *commands* to indicate the type of geometry desired. For example, a line command indicates that the next point is the start of a line segment.

Hard-copy output devices create a paper or Mylar drawing using two different output formats. Laser printers, for example, create a bitmap image in the printer memory to generate the image for the paper. In contrast, pen plotters use the drawing data in vector format to regenerate the objects on velum or Mylar drawing medium.

CAD Software: Basic Operation

Creating CAD drawings requires the use of *command statements* selected from a menu with a mouse, puck, or track ball; entered from a keyboard; or inserted with a command input device. The command statements can be grouped into the following four general categories: *creation, editing, sizing and moving,* and *file manipulation.* The drawing screen for AutoCAD, one of the most frequently used PC CAD applications, is provided in Figure 4–7. The drawing screen in Figure 4–8 is from Pro/ENGINEER 2000i, one of the most frequently used CAD solutions on RISC-type CAD stations. The screen in Figure 4–8 is for the Pro/ENGINEER version that runs on a PC with the NT operating system.

The process of creating a drawing using a CAD system is different for every CAD application. However, the process usually starts with setting the default drawing parameters, such as units of measure, screen dimensional limits, and drawing grid. The development of the geometry is similar but not identical to the manual process. The primary difference is the productivity gains provided by specific draw commands, such as *Box,* which completely draws rectangles of any size, and *Flip,* which creates the second half of a symmetrical object. CAD software increases productivity during editing with commands such as *Trim,* which creates perfect intersections with just the click of the mouse or puck.

With the geometry created, the CAD software permits all part dimensions to be added quickly using either the accuracy of the CAD-drawn objects on the screen or manual input of dimensional values from the keyboard. The part dimensions are applied using basic American National Standards Institute (ANSI) standards or the more specific ANSI Y14.5M 1982 geometric dimensioning and tolerancing (GD&T) guidelines. In more sophisticated packages, the GD&T specifications are added to the drawing and checked for correct GD&T callouts using either the CAD software or third-party add-on software. In addition, the CAD software checks dimensional tolerances for consistency with the form, fit, and function requirements of the part.

With the drawing created and dimensioned, a permanent record of the electronic file is generated. The types of file formats supported by a CAD system vary widely but often include the file types illustrated in Figure 4–9. As the figure indicates, the CAD software offers an interface to vector, raster, and text data formats. In addition, CAD software translates drawing code to the broad range of standard file formats listed in the figure and is compatible with drawing files from RISC- and mainframe-based CAD software from different vendors. A description of the CAD and manufacturing data file format used for electronic data interchange (EDI) is provided in future sections. With the basic CAD system defined and the general operation understood, the broad scope of computer-based design is explored by classifying the systems. CAD systems are classified either by the hardware platform required for operation or by the type of geometric data used to define the part model.

4–3 CAD CLASSIFICATION: HARDWARE PLATFORMS

In the hardware platform classification process, CAD systems are divided into three groups: mainframe computers, engineering workstations or RISC systems,

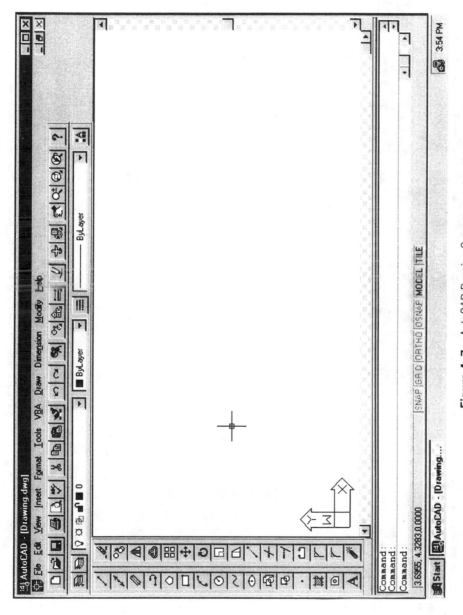

Figure 4–7 AutoCAD Drawing Screen.

111

Figure 4–8 Pro/ENGINEER Drawing Screen.

Figure 4–9 CAD Data Exchange.

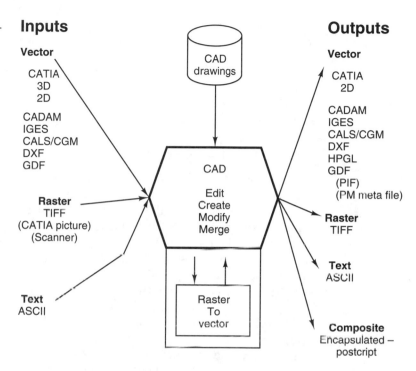

Inputs

Vector

CATIA
3D
2D

CADAM
IGES
CALS/CGM
DXF
GDF

Raster
TIFF
(CATIA picture)
(Scanner)

Text
ASCII

CAD
drawings

CAD

Edit
Create
Modify
Merge

Raster
To
vector

Outputs

Vector

CATIA
2D

CADAM
IGES
CALS/CGM
DXF
HPGL
GDF
(PIF)
(PM meta file)

Raster
TIFF

Text
ASCII

Composite
Encapsulated –
postcript

and microcomputers. Each group has unique characteristics that offer specific advantages and disadvantages in the implementation of CAD in computer-integrated manufacturing. A study of the characteristics of the different hardware platforms follows. Use the Websites listed in the Appendix to see the different types of CAD hardware.

Mainframe Computer Platforms

This group includes all the computers larger than engineering workstations used to run CAD software. Some of the hardware platform manufacturers include IBM, Compaq, Prime, Hitachi, and NEC. The CAD software, usually developed by third-party vendors, runs on the mainframe with multiple users working on terminals attached to the main computer. For example, large mainframe implementations in the automotive industry have 100 or more designers and draftspersons working on a wide range of product drawings and designs. While the computational capability at each terminal is significant in mainframe implementations, only a single image or copy of the CAD software and operating system exists on the mainframe. This feature is the most distinguishing characteristic of the mainframe CAD implementation and offers both advantages and disadvantages.

Advantages. Powerful design features are available at each workstation, multiple program execution is standard, a single database is shared by all operators, a good response is provided for large drawing files, support for large design and documentation projects is available, and an integrated software solution is offered for various manufacturing problems.

Disadvantages.   A large initial capital investment is required, cost per terminal is high for a small number of users, complex operating systems require dedicated system managers, the complex software is not suited for entry-level operators, and a single mainframe failure terminates all drawing productivity.

Engineering Workstation Platforms

Engineering workstations manufactured by companies such as IBM, Compaq, Hewlett-Packard, Sun Microsystems, and Silicon Graphics using various third-party CAD software are taking over most of the applications dominated by mainframes in the past. The feature that separates the workstation platform from other computers is the use of the reduced instruction set computer (RISC) chips for the central processor. These RISC chips used in the central processor are proprietary devices that are usually designed and produced by the company selling the RISC box. A second characteristic of the RISC workstation, *fast operation*, is a result of the RISC technology. The processing speed of both workstations and mainframes is measured in *millions of instructions per second* (Mips). RISC technology has been faster, with more Mips, than the Intel chips used in PCs. With new versions of chips from Intel, the PC microprocessors have been closing the gap, but RISC remains the fastest CAD platform. High-speed workstations are required when large complex vector images must be regenerated on the screen with subsecond response time. Each workstation requires an individual copy of or license for the full CAD software package; however, the workstations are frequently connected to a local area network (LAN) to share design files, product data, and peripheral resources such as printers and plotters. RISC-based machines offer many advantages and have some disadvantages.

Advantages.   RISC systems have a good price/performance ratio; design and analysis features are comparable to mainframe implementations; a lower initial capital investment is possible; many hardware vendors offer competitive RISC systems; the UNIX operating system, used on many of the vendors' platforms, is an industry standard; multiple program execution is standard; and failure of a computer affects only that workstation.

Disadvantages.   Multiple copies at the CAD software must be maintained, and the UNIX operation system is complex and less user friendly.

Microcomputer Platforms

The introduction of the personal computer (PC) by IBM in the early 1980s and the development of PC-based software such as AutoCAD changed the course of CAD applications. Starting with the 8080 microprocessor chips used in the first PCs, the Intel Corporation's family of microcomputer chips are the primary engine used for microcomputer platforms running PC CAD software. The operation and configuration of microcomputer CAD platforms is similar to the RISC implementation. The CAD software is loaded on each microcomputer in almost all applications. The computers are frequently connected to a network to share printers, plotters, and other system resources.

System speed is measured by (1) the type or number of microprocessors present in the computer, and (2) the frequency of the clock driving the microprocessor. Intel has been introducing a faster microprocessor about every twenty-four months, so the speed of computers and computer CAD application is continually increasing. The newer the chip, the greater the computational power and subsequent CAD software speed. The second factor in the response rate of the CAD software is the clock speed of the microprocessor. Computers using 800-megahertz system clocks or oscillators produce faster screen regeneration than systems using the same microprocessor with only a 500-megahertz clock.

As the power of the microcomputer CAD platforms increases, the systems take over markets served by low-end engineering workstations. This increase pushes the RISC-based systems to a higher level and further into the mainframe territory. Mainframe systems are hosting fewer and fewer CAD applications. As concurrent operating systems such as Windows 2000, OS/2, and UNIX become standard on the microcomputer platforms, and networks of RISC based systems start operating like a single large computer, the role of the mainframe CAD systems will be limited to special applications. PC-based CAD implementation offers both advantages and disadvantages to the small and medium-size company moving toward a CIM implementation.

Advantages.   A low initial capital investment and low incremental expansion cost are possible, a CAD operator can perform the system management tasks, operating systems (Windows 2000, OS/2, and Macintosh) are relatively easy to learn and use, and a large base of equipment and third-party software vendors produce competitive system costs.

Disadvantages.   Integration of the CAD drawing data into the total CIM database is more difficult, the drawing and object manipulation features are close but not equal to the capability of RISC CAD software, and production of complex or large projects with multiple designers is not well supported.

4–4 CAD CLASSIFICATION: SOFTWARE

The CAD software on all three hardware platforms is in a continuous state of change. For example, AutoDesk released over fourteen versions of the popular Auto CAD software in less than fifteen years. Each new release featured enhanced functioning plus a larger and more powerful set of drawing commands. In addition, new releases took advantage of the increased computing power of the latest Intel microprocessor in the PC hardware. While changes and continuous improvement in the CAD software on all platforms is unsettling, it results in a steady decrease in the price-to-performance ratio. This increased performance at less cost permits many industries to use CAD technology as a first step toward CIM.

CAD software falls into two broad categories, *2-D* and *3-D*, based on the number of dimensions visible in the finished geometry. CAD packages that represent objects in two dimensions are called 2-D software; while 3-D software permits the parts to be viewed with the three-dimensional planes—height, width, and depth—visible. In addition, the software at the 3-D level classifies the part geometry as a *wireframe, surface,* or *solid* model. Each of these classifications is described in the following sections. Use the Web sites for CAD vendors in Appendix 4–2 to view many examples of the major CAD applications.

CAD 2-D Wireframe Systems

A *wireframe* model is a CAD drawing in which the intersection of planes in the object are represented by lines and arcs. In a two-dimensional or 2-D wireframe model of a part, only two dimensions of the object are visible in a view. Figures 3–8 and 3–11 are examples of 2-D wireframes. Some 2-D CAD software permits a two-dimensional surface to be assigned an *extruded depth* value. The result is often called *2½-D* because it gives the impression of a third dimension when the object is displayed in an isometric view. The bracket in Figure 4–10 is an example of a part illustrated in 2½-D.

CAD software limited to 2-D or 2½-D has two distinctive characteristics: (1) the software is just an electronic emulation and extension of basic board drafting, and (2) stored geometric entities are limited to points, lines, and arcs. When 2-D wireframe models are saved, only two-dimensional data are attached to the location of each point, line, or arc.

Applications for 2-D software abound in two broad areas: basic drafting operations and part manufacturing. As a result, estimates from industry sources indicate that over 2 million 2-D CAD systems are in use, with the number growing 10 percent each year. Manual drafting has disappeared from most industry sectors because CAD is more productive. CAD wins the productive battle because parts change frequently and drawings are always modified. The power of CAD to enhance productivity becomes apparent when modifications of the drawings are required.

Two-dimensional part geometry, used in manufacturing applications, is the second major application for 2-D CAD. The part drawings required for every

Figure 4–10   Two-and-One-Half-Dimensional CAD Part.

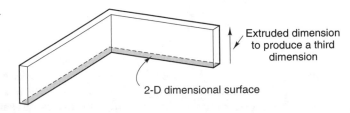

Extruded dimension to produce a third dimension

2-D dimensional surface

manufactured object are good examples of 2-D CAD applications in manufacturing. In addition, a large number of 2-D CAD files are still used to generate computer programs for automated cutting tools called *computer numerical control* (CNC) machines.

In the past, CAD software applications for the microcomputer were usually restricted to 2-D applications. However, many programs that started as only 2-D, like Auto CAD, have added 3-D capability to current versions. In addition, many CAD applications that were designed for 3-D work on UNIX-RISC workstations have been rewritten to run on the new powerful PC workstations. Many of the engineering workstation and mainframe CAD software solutions can be used in a 2-D mode to support basic drafting requirements.

CAD 3-D Wireframe Systems

The 3-*D wireframe* system is an extension of the 2-D concept. The primary difference between 2-D and 3-D wireframe CAD software is the association of a third dimension, usually a Z data value, with each point, line, and arc. The 3-D model can be viewed from any angle by rotating the object's geometry around the X, Y, or Z axis. Additional characteristics of the 3-D model include: (1) a data base for 3-D part geometry that is not significantly larger than the 2-D model, (2) an easy-to-use command set with primitives (points, lines, and arcs) and construction techniques similar to the 2-D application, (3) a data set that can be used to build a surface model, and (4) a data set that can be used to generate process programs for automatic cutting machines.

The creation of a 3-D wireframe of a part can start with two or three views of the object in 2-D. The operator selects edges of surfaces in the 2-D views, and the 3-D software creates a 3-D wireframe. The plastic pencil sharpener of Figure 3–8 is illustrated in Figure 4–11 as a 3-D wireframe using AutoCAD software. A second approach to creating a 3-D wireframe uses the drawing commands to create a 3-D wireframe directly without the aid of 2-D views. Building a part geometry requires the application of 3-D lines, arcs, and curves. Notice how difficult it is to visualize the shape of the object when all the lines in the object are present. The same drawing, with *hidden lines* removed, is illustrated in Figure 4–12; the shape of the object is now easy to see. This example illustrates that the availability of automatic hidden-line removal in 3-D wireframe CAD software is crucial for a successful application. In summary, the application areas best suited for 3-D wireframe CAD software include:

- The drawing of objects with only *planar* surfaces. If the object has a *sculptured* surface, the location of edges is difficult unless the part is viewed from specific angles.
- The drawing of parts with geometric and surface features that make it easy to distinguish open spaces in the part from solid material.
- The drawing of early conceptual designs where rapid creation of the overall part model without many details is desired.

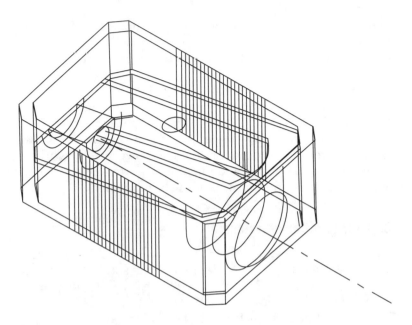

Figure 4–11   Three-Dimensional Wireframe of Pencil Sharpener.

■ The drawing of large models where the use of solid models would create a database too large for system resources.

Examples of 3-D wireframe CAD software appear on all three types of computer platforms. The microcomputer-based solutions have fewer productivity-enhancing features than the RISC-based systems. AutoCAD is an example of microcomputer software with 3-D wireframe capability; Pro/ENGINE is an example of software that runs on both the microcomputer and RISC platforms.

Figure 4–12   Three-Dimensional Drawing with Hidden Lines Removed.

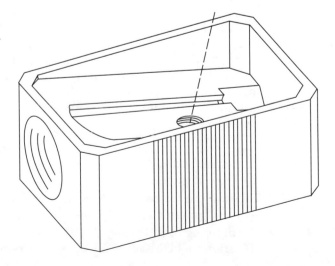

Figure 4–13   Ruled Surface.

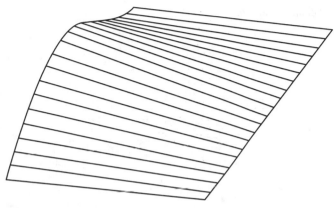

Surface Model Software Systems

Objects created with *surface* model systems combine the points, lines, and arcs from the 3-D wireframes with surfaces. The surfaces are defined mathematically in three ways:

1. *Ruled surface.* A *ruled surface* (Figure 4–13) is a surface formed by connecting two lines or arcs with straight lines.
2. *Surface of revolution.* A *surface of revolution* (Figure 4–14) is a surface formed by rotating a 2-D drawing of the cross-section of the object around the X, Y, or Z axis.
3. *Sculptured surface or free-form surface.* A *sculptured surface,* like the fillet connecting the aircraft wing to the fuselage in Figure 4–15, is formed by approximating the complex surface shape with a *surface patch.* The technique used most frequently to form surface patches is based on *Bezier* and *B-spline* curve-generation mathematics. The nonuniform rational B-spline (NURBS) technique is used most frequently for complex surface patch generation.

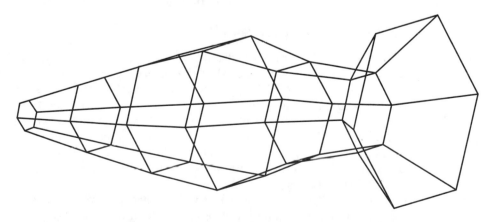

Figure 4–14   Surface of Revolution.

Figure 4–15   Sculptured Surface.

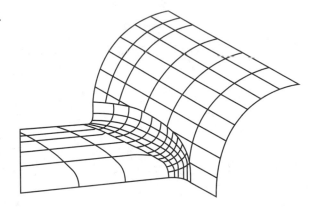

A full description of B-spline operation is provided in Appendix 4–1. Other surface types used by some CAD systems include *C-poles, S-poles,* and *Coons patches.*

Other characteristics of surface model systems include (1) the need for more powerful computer platforms to handle the calculations; (2) a larger database for the geometric model, due to the information required to create the surface; and (3) a steeper learning curve because analogous operations in manual drafting do not exist. The surfaces of the model are frequently shaded and hidden lines are removed to highlight features of the complex object to make it easier to visualize. In general, surface modeling products exist as an adjunct to either wireframe or solid modeling products.

The major application for surface modeling systems is the design and drawing of objects formed from a collection of irregular surfaces. Examples include automotive body design, television cabinets, liquid soap bottles and containers, and appliances. All RISC-based engineering workstation CAD software packages and some CAD packages on microcomputers offer surface modeling options.

Solid-Model Software Systems

The solid model is a mathematically complete and unambiguous representation of part geometry. The solid model of the object has all the properties of the actual part, including physical (size, mass, and material), mechanical (strength and elasticity), electrical (resistance, capacitance, and inductance), and thermal (conductivity and expansion coefficient). Therefore, the part model has all the characteristics of the real-world object. As a result, a test performed on the solid-part model provides information on the performance of the actual part. As a result of the benefits of solid models, industry data indicates that 60 percent of the roughly 2 million 2-D CAD stations currently used will be moving to 3-D tools, mostly solid modeling software. The computer-aided engineering (CAE) applications described in Chapter 5 illustrate how the properties of solid models are tested in simulated systems. An example of a CAD solid model is provided in Figure 4–16.

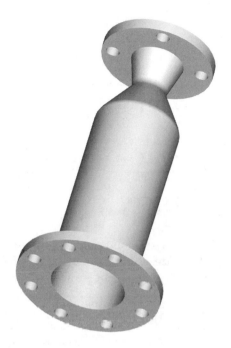

Figure 4–16   Solid Model of a Rocket Engine.

Two techniques, *constructive solid geometry* (CSG) and *boundary representation* (B-rep), are used by a majority of the software vendors to create solid-model CAD images. The product designer can use the two techniques individually on a part or in a hybrid or mixed mode where the benefits of each technique are available. A third solid-modeling technique, *sweeping,* is used for certain types of geometry. Less frequently used techniques include *pure primitives, spatial occupancy,* and *cell decomposition.* The operation of each type of the most frequently used solid modelers is described below.

Constructive Solid Geometry.   The CSG software, sometimes called explicit manipulations, builds a solid geometry from a primitive shape library. The library includes many basic object shapes including *blocks, cubes, cylinders, cones, spheres, wedges,* and *toruses.* The shapes are combined using the *union* (un), *intersection,* and *difference* (dif) Boolean operators. The example shown in Figure 4–17 illustrates this process. Note that the two full boxes are combined using the union operator to create the base of the part. The cylinder and block use the same union operator to create the pin on the part. Finally, the difference operator subtracts the wedge from the base for the complete solid part model. The CSG tree structure and compact code (Figure 4–18) produce an efficient database for the solid object. The CSG technique is the quickest and easiest method to produce solid-part models. However, many shapes, automotive body designs, for example, cannot be created by combining shapes with a CSG modeler.

Figure 4–17 CSG Solid Modeler.

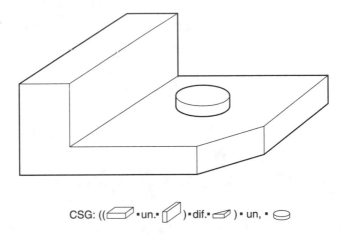

CSG: ((⬭·un.· ▱)·dif.· ⬭) · un, · ⬭

Boundary Representation. The B-rep solid modelers, sometimes called para-metric modelers, represent objects by describing the bounding faces. The edges, curves, and points in the faces are then defined. The same object modeled with the CSG techniques is used in Figure 4–19 to illustrate the B-rep approach. The faces or surfaces are usually defined using Bezier, B-spline, or NURBS surface modeling techniques. (NURBS are described in Appendix 4–1.) As a result, the B-rep solid modeler can represent any object regardless of surface contour complexity. How-ever, it is difficult and complex to create a B-rep solid model, and the vector file for the completed model is usually quite large.

Sweeping. The third type of solid modeler is used for cutter path simulation. The two types of sweeping actions, translation and rotation, are good for certain types of geometry. The sweeping action required to produce a rectangle and cylin-der are illustrated in Figure 4–20.

Figure 4–18 CSG Tree and Code.

CGS: Code

Chamfer = Wedge .. at ..
Pin = Cylinder .. at ..
Box_2 = Block .. at ..
Box_1 = Block .. at ..
Obj_1 = Box_1 un. Box_2
Obj_2 = Obj_1 un. Cylinder
Part = Obj_2 dif. wedge

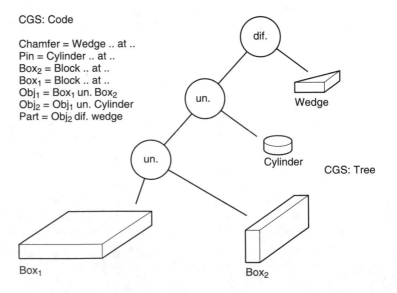

CGS: Tree

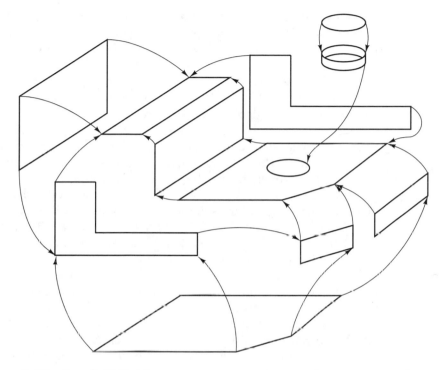

Figure 4–19   B-rep Solid Model.

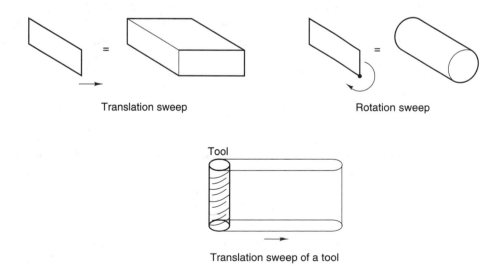

Translation sweep

Rotation sweep

Tool

Translation sweep of a tool

Figure 4–20   Sweeping Solid Modeler.

Solid modeling software offers several alternative methods for creating and manipulating solid geometry, including *explicit, parametric, variational,* and *features-based* modeling. Parametric analysis, as it applies to design, was introduced in the last chapter. When applied to solids modeling, parametric analysis is associated with one- and two-way *associativity.* Associativity means that the relationship is established between the solid model of a part and the 2-D working drawings prepared for the production. If a dimension is changed on the solid model, all of the 2-D shop drawings are updated automatically. The opposite is also true: A change in some of the dimensions in the 2-D drawing would be reflected in the solid representation of the part. This reverse direction is important because changes in the solid model would normally have to be made using the CSG or the B-rep functions in the software. Using associativity, the drivers controlling model shape can be changed on the 2-D drawing, and the solid model responds accordingly.

The *variational* characteristic is a subset of the parametric model. When variational capability is present, the user can add dimensional and geometric constraints without adhering to an exact sequential definition for the geometry and constraints. *Feature-based* modeling is another subset of parametric analysis that permits the use of familiar command language to build part geometry. It is used most frequently for holes, bosses, and rounds. In practice, an operator can use feature-based commands to build a countersunk hole by specifying the geometry desired and then providing appropriate radii and depth. Feature-based design currently focuses primarily on design requirements and does not support manufacturing-related operations.

The hybrid system that combines the CSG and B-rep techniques offers the best overall solid modeling solution. For objects without contour surfaces, the CSG works well; when a contour surface is encountered, the B-rep portion of the modeler is used. Another type of hybrid system combines the surface creation techniques found in surface modeling with solid modeling products. Solid modeling software provides many benefits, including calculation of full mass properties, true cross-sections, interference checking between models of parts, photorealistic graphics, and seamless exchange of data to downstream applications such as finite element analysis, generation of numerical control code, and stereolithography.

Solid model software is required when the drawing database for the part must include the real-world characteristics of the object. Examples that require solid models include:

- Operational testing and simulation of parts and assemblies using the physical, mechanical, electrical, and thermodynamic properties present in the part database
- Process planning of manufacturing operations using computer software

Most RISC-based engineering workstation and some microcomputer CAD packages offer solid modeling options. For example, the Pro/ENGINEER software from Parametric Technology Corp. has a solid modeler module for both RISC and PC platforms.

4–5 APPLICATION OF CAD TO MANUFACTURING SYSTEMS

A study of over 300 firms conducted by the Industrial Technology Institute in Ann Arbor, Michigan, indicated that CAD is used most often in three application areas: Concept and Repetitive Design, Drafting, and New PDM and the Internet.

Concept and Repetitive Design

CAD software is used for a large percentage of the design function in manufacturing, including the design of the product plus the design of all the systems required to support the production process, such as fixtures for machining, assembly, or quality checking; gauges; and material-handling pallets. CAD and the design functions form a natural union because the part geometry is used frequently in the production process. For example, the curve surface data created on the CAD system during the design of a shampoo bottle can be sent to the mold manufacturer through electronic data interchange (EDI) for use in machining the injection molds needed to produce the bottles. The design of the plastic injection mold for the bottle is less expensive and faster because the surface geometry is directly available from the CAD database. The design data captured in CAD are reused often by many other departments and systems in the manufacturing process.

Drafting

The second major application for CAD is in the creation of all the working drawings required for product manufacturing. In most organizations CAD was first introduced in this area as a replacement for manual drafting. In addition, CAD is used to create the documents that will be referenced during design verification and ordering of raw material and component parts. Manufacturing support departments require CAD as well; for example, facility planning, maintenance, and security make use of CAD to develop visual presentation of data that describe their operation.

New PDM and the Internet

The deployment of a common product database in the CIM enterprise has frequently been called *product data management* (PDM). The new PDM is *product development management* and addresses the seamless sharing of CAD and other product/engineering data across the enterprise. Creating product drawings is one of the first activities in the manufacturing sequence. In this third CAD application area, the product details contained in the CAD drawings are shared with departments across the enterprise through a common gateway using product data management (PDM) principles and software. The process includes (1) creating the product with CAD, (2) converting the CAD part geometry and attributes file to the format required by other departments, and (3) saving the different versions of the drawing files in the CIM data repository. For example, Figure 4-9 indicates the file format conversions that some CAD systems support for creating enterprise data. The enterprise units that frequently use part geometry and specifications created in CAD are computer numerical machines (CNC) in manufacturing, manufacturing resource planning (MRP II) software systems in production planning and control, and text/graphics documentation software systems in marketing and other front office departments. In addition, most of the computer-aided engineering (CAE) software described in Chapter 5 starts with CAD geometry and attributes.

Integrating product data within the extended enterprise will liberate an organization's overall design and manufacturing process. Studies indicate that, for every CAD user in a manufacturing enterprise, nine additional nondesign personnel review, approve, and use the same product data. Enlarging the circle of CAD data users is hampered by the many different computer systems and software applications that must interface with the CAD product data. The solution taken by many CAD product leaders is to use the Internet and intranet as the common interface for all PDM files. Companies like Parametric Technology Corp., IBM/Dassault Systems, SDRC, and Unigraphics have all released CAD support products that give the corporate users with a standard Web browser a window into critical product data. The Internet permits data generated on UNIX-type CAD workstations to be viewed on PCs with the Windows 98 or Windows 2000 operating systems.

The level and intensity of CAD used is not evenly divided across the design, drafting, and database development application areas because the activity level is dictated by the nature of the products being designed and produced. However, most enterprises implementing CIM are heavy CAD users in at least two of the three application areas just described.

Selecting CAD Hardware and Software for an Enterprise

Five criteria are frequently used to define the type(s) of CAD system(s) a company will use in an integrated solution. The selection process for a CIM-based CAD system becomes clear after each of these five criteria is examined. Every criterion affects both the hardware platform choice and the software solution.

Type and Complexity of Geometry and Drawings.   A simple cross-section view on the working drawing of a complex geometry such as an aircraft wing rib can be represented easily using a 2-D CAD system. However, a NURBS-based surface or solid modeler is necessary to draw the rib (Figure 4–15) as a pictorial. Factors that affect complexity are:

- *Degree of axisymmetry.* Objects that are symmetric or the same about one or more axes can be represented with less sophisticated CAD software.
- *Types of curved surfaces.* If the curves are circular or spherical, the object is less complex and only 2-D software is required. However, if conic or quadric surface data are present, the geometry is complex and the software must match.
- *Interior complexity.* An object with detailed interior surfaces represented with a 3-D wireframe is hopelessly confusing. Such a part would require software with surface or solid modeling capability.
- *Parts in the assembly.* Objects with a large number of parts require analysis of fit, tolerance, and interference; therefore, the demand on the CAD software increases.

Concept Versus Repetitive Design.   The type of design practiced by a company is a major factor in the selection of a CAD system. Repetitive design needs are often satisfied on a 2-D system, while concept design generally demands a surface or solid modeler with CSG and B-rep capability. In addition, the CAD software supporting the repetitive design function should have parametric design capability.

Size and Complexity of the Product.   The size and complexity of the finished product affects the type of CAD system needed because big projects require several designers who each work on separate parts of the same design. One of the key considerations in this criterion is the system response time. In a study conducted by IBM, an improvement in the response time on a CAD system from ½ second to ¼ second doubled the number of commands performed on the system and thus the productivity. Examples of complex projects include commercial aircraft, oil refineries, office buildings, bridges, cars, and microprocessors.

Enterprise Data Interfaces.   A major CIM axiom states, "Create data once and use it many times." CAD is affected heavily by this precept because CAD is where most of the product geometry data originate. The CAD hardware and software must support file formats that permit the part geometry and specification data attached to the drawing to be used by (1) marketing for product brochures and manuals; (2) manufacturing engineering for CNC code generation, engineering change orders, routing, process planning, and quality analysis; and (3) production planning and control for product structure and part specification.

External Data Interfaces.   Electronic data interchange (EDI), a technology that transfers data from one computer to another, is critical for vendors working with large companies. Two factors in EDI affect the CAD selection: (1) the type of EDI interface between the contractor's and vendor's computers, and (2) the transferrability of the part geometry data file between the CAD software used by the contractor and vendor. Vendor certification with large companies or the federal government often dictates the EDI standard and CAD software the supplier must use.

Figure 4–21 indicates how the need for the different types of CAD software varies in the different manufacturing systems and design processes introduced in Chapter 1. A similar chart (Figure 4–22) indicates how the required number of computer stations varies for the different areas. The final comparison chart (Figure 4–23) indicates the type of CAD software recommended for different types of drawings and manufacturing interfaces. Take a few minutes and study these charts. Keep in mind that these are general recommendations and comparisons; in practice, there are always exceptions that increase or decrease the requirements for the CAD hardware and software used in a specific application. Note that the boxes in Figure 4–21 indicate the level of need for CAD technology. A filled box indicates CAD would be highly necessary for product success. In contrast, a clear box indicates little need for CAD and a partially filled box indicates a CAD need falling between the two extremes.

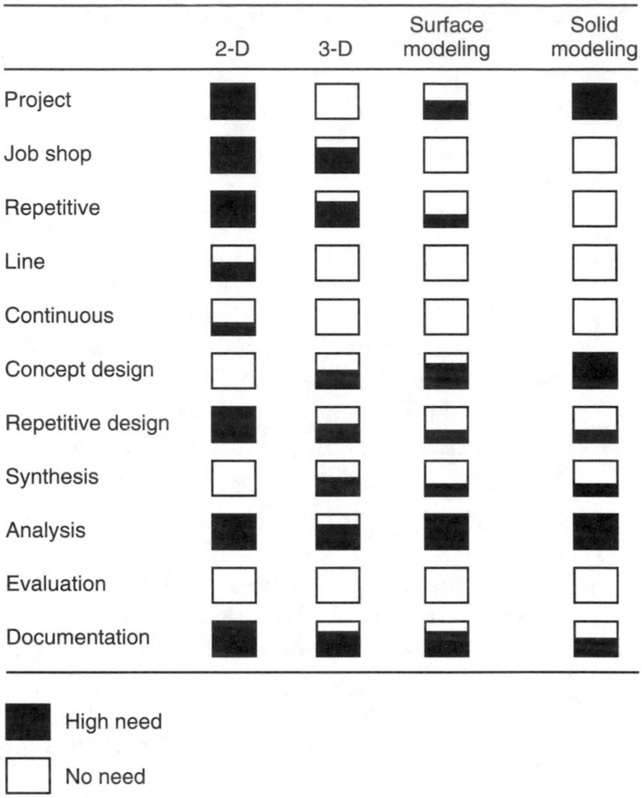

Figure 4–21 Need for CAD Stations in Various Types of Manufacturing and the Product Design Processes.

	2-D	3-D	Surface modeling	Solid modeling
Project				
Job shop				
Repetitive				
Line				
Continuous				
Concept design				
Repetitive design				
Synthesis				
Analysis				
Evaluation				
Documentation				

■ High need

□ No need

CAD Software Selection Guidelines

Data collected by Anderson Consulting indicate that engineers and designers rarely devote more than fifteen hours per week (Figure 4–24) to engineering layout; as a result, they are part-time CAD users. CAD systems in general are designed for full-time users, especially the surface and solid model software required for many design applications. To ensure that the CAD software selected will be used efficiently, the following selection/operational guidelines should be followed:

1. Select the CAD system and software options carefully to ensure the best fit available between the users and the CAD tools.
2. Customize the CAD software for each user or application area so that it meets specific needs. For example, a designer could be provided with macros that take care of the tedious and repetitive operations.
3. Select CAD software with a *design automation language* or one that has *feature-based design* capability. In each case, the designer specifies only product features or specifications and the design rules take over. For example, specifying the horsepower, length, and shaft rotation speed in revolutions per minute produces a coupler design with the shaft diameter, flange size, bolt size, and number of bolts required.

Figure 4–22 Number of CAD Stations in Various Types of Manufacturing and the Product Design Processes.

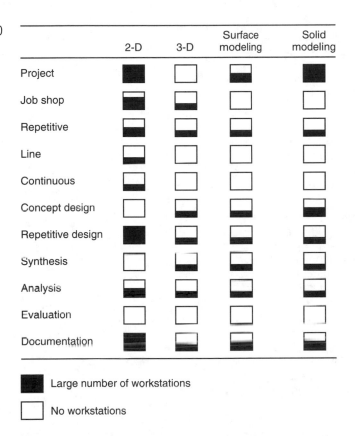

	2-D	3-D	Surface modeling	Solid modeling
Project	■	□	▬	■
Job shop	▬	▬	□	□
Repetitive	▬	▬	▬	▬
Line	▬	□	□	□
Continuous	□	□	□	□
Concept design	□	▬	▬	▬
Repetitive design	■	▬	▬	▬
Synthesis	□	▬	▬	▬
Analysis	▬	▬	▬	▬
Evaluation	□	□	□	□
Documentation	■	▬	▬	▬

■ Large number of workstations

□ No workstations

4. Even with well-matched systems and users, customized support programs, and easier-to-use software, the CAD systems will remain complex. Therefore, regular education for all users is critical.

4–6 SUMMARY

Computer-aided design (CAD) is used at all levels in the design process, with the heaviest use in the design documentation area. The technology was started at MIT in 1963 during the Sage Project when the Sketchpad system was developed. CAD software capability increased significantly in the 1970s with the introduction of 3-D capability. The introduction of solid modeling software, powerful engineering workstations, CAD hardware platforms, and powerful PC-based CAD software caused the use of CAD applications to increase exponentially in the 1980s.

The basic CAD system includes the items listed in the block diagram in Figure 4–3. Two components, a hardware platform and CAD software, form the basic CAD production station. The hardware platform includes a basic computer system with additional peripherals to expedite the entry of part geometry data. The computer systems used to produce the CAD database fall into three

Part, drawing, and interface characteristics	Wireframe		Surface and solid models			
	2-D	3-D	Ruled/revolution	UBS	NUBS	NURBS
Working drawings	•					
Assembly drawings	•			•	•	•
Pictorial drawings						
Normal surfaces		•	•			
Inclined surfaces		•	•	•		
Oblique surfaces		•	•	•		
Quadric surfaces						•
Simple sections			•	•	•	
Conic sections						•
Technical illustrations		•	•	•	•	•
Welding/drafting	•					
Surface development drafting	•	•	•	•	•	•
Architectural and structural drafting	•	•	•			
Map drafting	•	•				
Electrical and electronic drafting	•					
Aerospace drafting	•	•		•	•	•
CAM interface		•		•	•	•
DTP interface	•	•	•	•	•	•
MPC interface	•	•				
CMM interface		•				
CAE interface		•		•	•	•

Figure 4–23   Type of CAD Software for Modeling the Range of Part Geometries and Manufacturing Interfaces.

categories: mainframes, RISC-based engineering workstations, and microcomputers. The variation in CAD software is dictated by the type of computer platform chosen for the application. The capability of the software to handle complex part geometry increases with hardware power. For example, basic or entry-level CAD software running on a small microcomputer is restricted to 2-D drawing applications. In contrast, Pro/ENGINEER software running on a RISC/UNIX workstation could produce a surface or solid model of a complex shape such as the skin of a commercial aircraft. CAD software is divided into the following classifications: 2-D, 3-D, surface model, and solid modeling software.

A study of CAD applications indicates that the software is used in three industry application areas: design (concept and repetitive), drafting, and supporting product data management in a common CIM data base. The selection of CAD hardware and software systems to support these three application areas requires analysis of five criteria: (1) the type and complexity of part geometry and drawings, (2) the concept versus repetitive design, (3) the size and complexity of the product, (4) the enterprise data interface, and (5) the external data interface. Using

Figure 4–24 Time Study of a Design Engineer's Week.

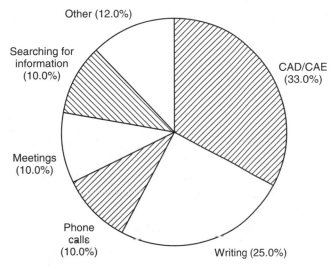

Other (12.0%)

Searching for information (10.0%)

CAD/CAE (33.0%)

Meetings (10.0%)

Phone calls (10.0%)

Writing (25.0%)

these criteria, the CAD hardware and software can be matched to the needs of the CIM system, the department, the product, and the user.

The power of CAD systems offers tremendous productivity opportunities; however, using powerful graphics software is often difficult. The workplace must support the part-time CAD users at the upper end of the design process through better training and better design aids. The software, especially programs at the high end, must become more intuitive.

REFERENCES

ARNSDORF, D. R., *Technology Application Guide: CAD—Computer-Aided Design.* Ann Arbor, MI: Industrial Technology Institute, 1989.

CHALMERS, C. E., "Windows and the Web: Leveraging CAD Across the Enterprise," *Integrated Manufacturing Solutions*, March 1999, pp. 26–31.

CHANG, T. C., R. A. WYSK, and H. P. WANG, *Computer-Aided Manufacturing.* Upper Saddle River, NJ: Prentice Hall, 1991.

GRAHAM, G. A., *Automation Encyclopedia.* Dearborn, MI: Society of Manufacturing Engineers, 1988.

KUTTNER, B. C., and M. A. LACHANCE, "Nurbs: CAD by the Numbers," *Actionline*, January 1991, pp. 20–23.

MITCHELL, F. H., Jr., *CIM Systems: An Introduction to Computer-Integrated Manufacturing.* Upper Saddle River, NJ: Prentice Hall, 1991.

SOBCZAK, T. V., *A Glossary of Terms for Computer-Integrated Manufacturing.* Dearborn, MI: CASA of SME, 1984.

VERSPRILLE, K., "Which CAD is Right for You," *Integrated Manufacturing Solutions*, March 1999, pp. 1–9.

QUESTIONS

1. Define CAD.
2. Describe the significant events in the development of CAD technology.
3. Describe the significance of the microcomputer in the growth of CAD applications.
4. List the basic components associated with a CAD system.
5. Describe a bitmap and define the term *pixel.*
6. How is the resolution of a drawing defined?
7. How do vector data images compare to raster images?
8. What are the four general categories into which CAD software commands are grouped?
9. Describe some of the productivity gains provided by CAD applications.
10. Describe the classification of CAD systems based on hardware platforms.
11. Describe the positive features of RISC-based CAD hardware platforms.
12. Describe the positive features of microcomputer-based CAD hardware platforms.
13. Compare and contrast the operation and capability of entry-level RISC systems with high-end microcomputer CAD systems.
14. Describe the CAD classification process based on the software.
15. Describe the positive features of 2-D wireframe CAD software.
16. Compare and contrast 3-D wireframe CAD software with software in the 2-D classification.
17. What application areas are best suited for 3-D software?
18. Describe the three methods used to define surfaces in surface modeling software.
19. What are the primary differences between geometry represented by a solid model and a surface model?
20. Describe the two commonly used techniques for creating a solid model.
21. Describe the three major functions for which CAD is used.
22. Describe the five criteria used in the selection of a CAD system.

PROBLEMS

Use microcomputer CAD software available in your local college or university for the following problems.

1. Make a list of the CAD commands used for the creation of part geometry.
2. Make a list of the CAD commands used to edit or change part geometry.
3. Make a list of the CAD commands used to size, move, and copy part geometry.
4. Use the CAD system to draw lines, boxes, rectangles, circles, arcs, octagons, ellipses, and triangles, and to create text. Experiment with the different techniques available for the construction of the basic geometric shapes.

5. Use the CAD system edit commands to move, copy, flip, mirror, rotate, and scale the objects created in Problem 4.

PROJECTS

1. Using the list of companies developed in Project 1 in Chapter 1, list the type of CAD hardware and software used in each company.
2. Use the CAD selection criteria and the data in Figures 4–21 to 4–23 to analyze the companies in Project 1 that do not currently use CAD or do not use it effectively. Based on the analysis, list the CAD platform and software features that would increase productivity.
3. Select three companies with the largest CAD installations, one from each of the small, medium-size, and large groups, and describe how CAD is used across the enterprises.
4. Develop a table that illustrates the types of computer platforms supported by the CAD vendors in Appendix 4–3 at the end of the chapter. Indicate software that runs on Microsoft NT/2000, Unix, or both.
5. Develop a table that illustrates the major features of CAD software provided by the CAD vendors in Appendix 4–2 at the end of the chapter. The minimum features should include 2-D, 3-D, wireframe, surface, and solid modeling; associativity; and parametric, variational, and features-based modeling.
6. Describe the product data management (PDM) options provided by some of the vendors listed in Appendix 4–2 at the end of the chapter.
7. Use the Unigraphics Web site to determine what *virtual product development* and *predictive engineering* include.
8. Select a PC and RISC-based CAD solution from the CAD vendors in Appendix 4-2. Then use the minimum machine specifications from the CAD vendor Website to build a computer from the computer vendors at the end of the chapter that could be used for the CAD software.

APPENDIX 4–1: B-SPLINES TO NURBS

The surface of a forming die for a car fender includes curves that cannot be represented by circular arcs. The CAD system uses line segments called *B-splines* (BS) to draw the complex contours of the surface. For more complex surfaces, such as conic sections, the system needs *nonuniform rational B-spline curves* (NURBS) to develop the surface. As a result, a basic understanding of BS and NURBS and their origin is essential.

This process started with the need to represent curves mathematically in the computer system. Circles and arcs of circles are relatively easy to present using the equation for a circle. All other types of curves require polynomial equations, such as the quadratic $At^2 + Bt + C$, to describe the shape of the curve. The more complex the curves, the larger the power or degree of the polynomial equation. To avoid higher-order polynomials, the complex curve is divided into segments, with each segment represented by a much simpler polynomial. Five low-degree polynomials are used as an alternative to one equation of high degree. When these segments, called *splines*, are linked together smoothly and efficiently, the curve becomes a *B-spline*.

The *knots* along the curve mark the connections between splines and determine the B-spline's shape and the underlying properties. When splines of equal length are used, the knots are spaced uniformly along the B-spline and a curve like that in Figure 4–25a can be represented. The result is called a *uniform B-spline* (UBS). Representing curves with a shape like that described in Figure 4–25b requires nonuniformly spaced knots, and the curve becomes a *nonuniform B-spline* (NUBS). As these B-spline features are added to the CAD software, the system becomes capable of representing curves with greater complexity. Some curves have shapes that cannot be represented with B-splines formed from linking single polynomials, regardless of the spacing on the knots. In these applications the B-splines are created by forming a ratio of two polynomials so that *rational* polynomials are constructed. As a result, the new curve-generating software, called *nonuniform rational B-splines* (NURBS), can model the most complex curves and surfaces.

Figure 4–25   UBS and NUBS.

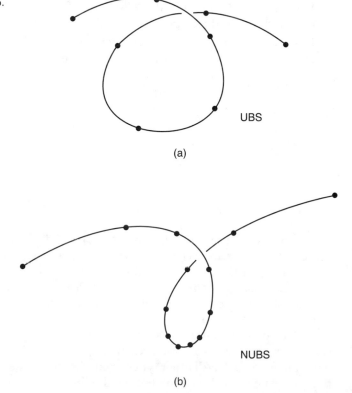

UBS

(a)

NUBS

(b)

APPENDIX 4–2: WEB SITES FOR CAD VENDORS

All the major CAD vendors have extensive Web sites that describe their products and offer other useful data for selecting CAD software. Visit the following Web sites:

CAD Vendors	Web Address	Comments
Applicon	www.applicon.com	Select *Bravo manufacturing* to view CNC products.
Autodesk	www.autodesk.com	Creators of the popular AutoCAD software—select *Products* and then *Demos* to view the list of demos available.
Baystate Technologies	www.cadkey.com	Select *Gallery* to see images of designs, case studies, and virtual html images.
Bentley Systems	www.bentley.com	Select *Products* to view software options and *Bentley Gallery* to see case studies.
CoCreate	www.cocreate.com	Select *Search* and type in *case studies.*
Dassault Systems	www.catia.ibm.com	Select *Catia* and then select *Gallery* to view design examples.
Microcadam	www.microcadam.com	Select *Gallery* to view sample designs and case studies.
Parametric Technology	www.ptc.com	Creator of the popular ProENGINEER CAD/CAE software— select *Products* to view all of the CAD-related products.
SDRC	www.sdrc.com	Creator of the popular I-Deas CAD/CAE software—select *Products and Services* to view the array of CAD/CAM software available.
Solid Edge	www.Solid-Edge.com	Midrange CAD software— select *Solid-Edge V7, Success Stories,* and *Hall of Fame Designs* to view products, sample models, and case studies.

SolidWorks	www.solidworks.com	Select *Partners,* then select *SolidWorks Gold Logo Products* to view the CAE support; select *products* and *gallery* to view product data and images/ case studies.
Unigraphics Solutions	www.Unigraphics.com	Select *Product Overview, Images,* and *Customer Successes* to view full product lines, sample models, and case studies.
Visionary Design Systems	www.ironcad.com	Select the down arrow on the selection box to link to *gallery* and *product info* to view solid images and product data.

APPENDIX 4–3: WEB SITES FOR COMPUTER SYSTEMS

All the major computer vendors have extensive Web sites that describe their products and offer other useful data for selecting CAD computer systems. Visit the following Web sites:

CAD Vendors	Web Addresses	Comments
Acer Computers	www.acer.com	Select *Products* to view systems.
Compaq Computers	www.compaq.com	Select *Products* to view systems.
Dell Computer	www.dell.com	Select *Business Systems* to view systems.
Fujitsu Computers	www.fujitsu.com	Select *Computers* to view systems.
Gateway Computers	www.gateway.com	Select *Corporations* to view systems.
Hewlett Packard	www.hp.com	Select *Business* to view systems.
Hitachi	www.hitachi.com	Select *Information Systems* to view systems.
IBM	www.ibm.com	Select *Business* to view systems.
Motorola	www.motorola.com	Select *Products* to view systems.

NEC	www.nec.com	Select *Computers* to view systems.
SGI	www.sgi.com	Select *Products* to view systems.
Sun Microsystems	www.sun.com	Select *Products* to view systems.

Design Automation: Computer-Aided Engineering

Over the last twenty years the term *computer-aided engineering* (CAE) has had several meanings. To some, it implied software that worked with CAD files to analyze geometric designs; others considered CAE as the umbrella covering all computer software used in design and manufacturing. More recently, the term *CAx* has been used to indicate that many operations in manufacturing are computer aided; you would supply the applicable capital letter/initial for the *x*. As a result of the past confusion, the study of *computer-aided engineering* should start with a definition.

> *CAE is the analysis and evaluation of the engineering design using computer-based techniques to calculate product operational, functional, and manufacturing parameters too complex for classical methods.*

Study of the form, fit, and function characteristics of products is covered by the terms *operational* and *functional* in the definition, while an examination of the match between the design requirements and the production capability is included in the phrase "manufacturing parameters." The expression "too complex for classical methods" indicates that the process performed by CAE software could be performed manually, but the quantitative difficulty, number of the computations, or time required for analysis would be prohibitive.

CAE fits into the *design process* (Figure 3–14) at the synthesis, analysis, and evaluation levels and is also consistent with the concurrent engineering (CE) principles described earlier. At the synthesis level, the primary CAE activity is focused on manufacturability using design for manufacturing and assembly principles. The output from the CAE operation at the analysis and evaluation levels is used by the CE team to determine the quality of the product design. Based on this CAE data, the product design is cycled through the top four steps in the design process until an optimum solution is generated.

Computer-aided engineering provides productivity tools to aid the production engineering area as well. Software to support *group technology* (GT), *computer-aided process planning* (CAPP), and *computer-aided manufacturing* (CAM) are grouped under the broad heading of CAE. As a result of these applications and the use of CAE in the *design process*, CAE is required in all five manufacturing system classifications and in the four production strategies described in Chapter 2. The level and variety of CAE software used, however, depends on the amount of concept and repetitive design practiced in the company and the type of manufacturing systems present. In the sections that follow, we describe the type of CAE software commonly used throughout the *design process* and in the production engineering area.

5–1 DESIGN FOR MANUFACTURING AND ASSEMBLY

The synthesis stage in the *design process* enriches the product by adding basic geometric detail and reshapes the product by applying *design for manufacturing and assembly* (DFMA) guidelines and constraints. The DFMA process answers the question: Is the design optimum for manufacturing and assembly? *DFMA* is defined as follows:

> *DFMA is any procedure or design process that considers the production factors from the beginning of the product design.*

The definition states that every design activity, from conceptualization to evaluation, must focus on generation of a design that meets market expectations and can be manufactured successfully.

Brief History of DFMA

The DFMA effort started with two separate thrusts: *producibility engineering* and *design for assembly* (DFA). Producibility engineering was an alternative to the older manufacturing reviews, which were typically held after the design was completed. Producibility guidelines focused on the producibility of individual parts rather than the total product with all parts assembled. Therefore, the major thrust was to produce simpler parts that were individually easier to manufacture. Producibility engineering guaranteed easily manufactured parts; as a result, the part count on the product often increased and caused a more complicated product structure. In many cases the total product was more difficult and costly to manufacture and assemble.

The underlying principle of *design for assembly* was to reduce the number of parts by eliminating or combining them to achieve a simplified product structure. While the assembly efficiency was improved, the total cost of the product was not minimized because using the best manufacturing processes was not a high priority.

In 1977, design for manufacturing and assembly (DFMA) started as an extension of the design for assembly (DFA) process. The two methods (design for manufacturing [DFM] and DFA) must interact because the key to successful DFM is product simplification through DFA. DFMA is a holistic approach to design analysis because both the manufacture and assembly of the finished product are considered simultaneously. This consideration of manufacturing and assembly during design results in lower total product costs.

Justification of DFMA

The use of DFMA early in the design process is necessary because 70 percent of the product costs are set by the time the product leaves the initial design team. The product costs are fixed because important decisions about material, processes, and assembly requirements are settled early in the design. Any effort to reduce cost after the design stage can only influence the remaining 30 percent of product cost. In addition to cost issues, 80 percent of quality problems are often a result of a poor design.

The greatest effect that DFMA has is in the reduction of total part count in a product. Reduction in part count leads to a reduction in total product cost and an increase in reliability of the product. The interface between separate parts is the major source of product failure and poor quality; therefore, a reduction in the part count results in fewer interfaces between parts and a better product. Also, with fewer parts there is a reduction in dies, molds, stamping presses, assembly operations, inventory, material handling, and paperwork.

DFMA: The Manual Process

The DFMA process is both *manual* and *computer based.* From the manual perspective, DFMA provides a step-by-step procedure to query the designer about part function, material limitations, and part access during assembly. The software used for DFMA calculates assembly time, product cost, and a benchmark theoretical minimum number of parts. In addition, a manufacturing element in the software assesses various material options (steel versus plastic) and manufacturing processes (machining versus die casting) for the most cost-efficient production of parts.

DFMA starts with effective design for assembly because assembly still remains a large part of the manufacturing effort and cost. When products are designed to make assembly easier and faster, costs are reduced and higher quality products result. Ten design rules or guidelines, listed in Appendix 5–1, are used by designers to create products with good assembly characteristics. Applying the guidelines to a production part will help explain how the rules are used. Study the spindle/housing assembly introduced in Chapter 3 and repeated in Figure 5–1 so that you are familiar with the parts and assembly. Note that the U-shaped sheet

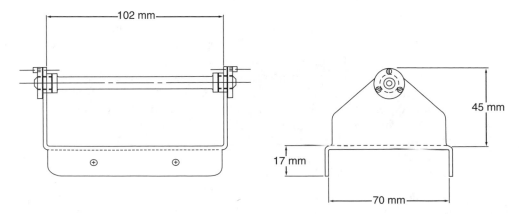

Figure 5–1   Spindle/Housing Assembly.

Source: Boothroyd/Dewhurst, "Product Design for Manufacturing and Assembly," © April 1988, Manufacturing Engineering Magazine, *p. 42. Reprint by permission of the Society of Manufacturing Engineers, Dearborn, MI.*

metal bracket supports a stainless steel spindle. Two nylon bushings are inserted into holes in the sheet metal bracket and the spindle turns inside the bushings. Each bushing is held in place with three threaded fasteners screwed into tapped holes in the bracket.

Exercise 5–1

Apply the ten design guidelines in Appendix 5–1 to the spindle/housing assembly shown in Figure 5–1 and determine how many of the rules were violated.

Solution

- *Rule 1 (violated).* The total part count is 10, with only one moving part in the assembly. Any time that two mating parts, such as the bracket and the bushings, do not move relative to each other, every effort should be made to combine them into a single part.
- *Rule 2 (unknown).* With the information provided, it is not possible to determine if some of the surface area could be eliminated to reduce surface processing and finishing requirements.
- *Rule 3 (violated).* The part has a natural base, the sheet metal bracket, on which the assembly is constructed. The rule is violated, however, because the parts are assembled from the sides and not the top.
- *Rule 4 (satisfied).* Access to the parts during assembly is adequate, and standard tools could be used.
- *Rule 5 (violated).* There are no grooves or aligning guides for the bushings on the bracket to aid in aligning the holes for the screws.
- *Rule 6 (satisfied).* The parts are all symmetric.
- *Rule 7 (violated).* The screw size does not have adequate surface area for easy gripping.
- *Rule 8 (violated).* Six separate fasteners are used in the assembly.
- *Rule 9 (violated).* Self-locking features were not used in this design.
- *Rule 10 (violated).* This component is probably not standard.

The result of the analysis is that *seven* rules are violated, *two* are satisfied, and *one* is unknown. The value of using the ten design guidelines is that problem areas (all violations) are identified and attention is focused on satisfying the guideline in question. As a result of this analysis, the spindle/housing design could be improved in several areas.

Another approach used to analyze the quality of a design is illustrated in Figure 5–2. The assembly method scoring chart assigns a score from 0 to 100 points based on ease of assembly. Study the chart until you understand the general layout and the way a score is calculated. Note that there are five assembly motions or approaches (left side) from *assembly stack from top* to *assembly from bottom*. Each of these has two

Best ←

Approach \ Connection	Weld	Solder	Stake	Adhesive	Pin	Nut	Tape	Screw	Snap ring	Snap fit	Nothing
Assembly stack from top ■ (without hold-down)	10	20	30	40	50	60	70	80	90	100	
Assembly stack from top ● (with hold-down)		10	20	30	40	50	60	70	80	90	
Assembly from side ■					30	40	50	60	70	80	
Assembly from side ●					20	30	40	50	60	70	
Assembly from bias ■						20	30	40	50	60	
Assembly from bias ●						15	20	30	40	50	
Rotated parts ■						10	10	20	30	40	
Rotated parts ●						5	5	10	20	30	
Assembly from bottom ■								5	10	20	
Assembly from bottom ●										10	
	Special tool or equipment required			Small tool required							Nothing

Good →

Legend:
- ■ Without hold-down
- ● With hold-down

Fastening or assembly method

Comments: Assign points to the open boxes; if the part you are placing in the assembly falls in one of the gray marked boxes, no points would be received. The upper right hand corner box would be assigned the highest points, and, as you go to the left or down, the points would decrease. After you evaluate your assembly, you add up the total points and divide the sum by the number of parts in your assembly; this gives you a design score.

$$\text{Example: Design score} = \frac{\text{total points in boxes}}{\text{number of parts in assembly}} = \frac{750 \text{ points}}{10 \text{ parts}} = 75\%$$

Figure 5–2 Assembly Method Scoring Chart.
Source: Courtesy Xerox Corporation.

categories, with and without a hold-down, indicated by a black square or circle. In addition, the connection (across the bottom) can vary from *weld* to *snap fit*. The best score (100) is for an assembly from the top using a snap fit. The shaded areas indicate a zero score for that operation. The application of the chart is illustrated in the following exercise.

Exercise 5–2

Calculate the assembly score using the assembly method scoring chart shown in Figure 5–2 for the spindle/housing assembly shown in Figure 5–1. Assume that the assembly starts with the insertion of the base into a holding fixture with a hold-down required. Use the following procedure for computing the assembly design score from the chart:

1. Start with the base of the assembly; next, assume that the base is placed in an assembly fixture with a snap fit from the top. Determine if hold-downs are required and enter either a 90 or a 100 in the score sheet.
2. Select the next logical part component to be added to the assembly and locate the appropriate *approach* row and *connection* column to assemble that part. The score in the box where the row and column intersect is entered on the score sheet for that part.
3. Repeat step 2 until every part, component, and subassembly have been added and the assembly is complete.
4. Add up the part assembly scores to get a total score for the assembly.
5. Compute the assembly score by dividing the chart total by the total number of parts.

Solution

The following chart indicates the assembly score for the spindle/housing bracket in Figure 5–1.

Assembly part	Score	Comment
Base	90	
Spindle	60	From side, held by screws
Right bushing	80	From side, held by snap-in
Left bushing	80	From side, held by snap-in
Screw 1	60	From side, screw
Screw 2	60	From side, screw
Screw 3	60	From side, screw
Screw 4	60	From side, screw
Screw 5	60	From side, screw
Screw 6	60	From side, screw
Total	670	

The assembly score is 670/10, or 67. The score of 67 would indicate a below-average design, which is what the analysis in Exercise 5–1 indicated. The score could vary depending on how the chart is interpreted. Despite this possible variation, the assembly method scoring chart provides a good analysis tool for comparison of alternative designs.

DFMA: Computer Support

CAE software in the DFMA area aids the designer by calculating cost for alternative design solutions. With DFMA software, the designer enters the specifications for the part design and the software provides a quantitative analysis of the alternative designs. An alternative solution for the spindle/housing assembly is illustrated in Figure 5–3. The requirement for bushings was eliminated by making the entire bracket or housing from nylon, so the part count is reduced from ten to two.

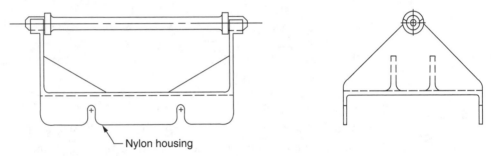

Nylon housing

Figure 5–3   Alternative Spindle/Housing Assembly.
Source: Boothroyd/Dewhurst, "Product Design for Manufacturing and Assembly," © April 1988, Manufacturing Engineering Magazine, p. 42. Reprint by permission of the Society of Manufacturing Engineers, Dearborn, MI.

DFMA software from Boothroyd Dewhurst, Inc. was used to analyze the two spindle/housing assembly designs. Figure 5–4 shows the cost analysis of the two designs provided by the DFMA software. Note that the cost includes tooling, material, manufacturing, and assembly. The DFMA software provides additional analysis of injection molding cost, machining cost, and material alternatives.

In summary, DFMA encourages designers to break with old concepts in design and produce less costly products. Frequently, DFMA designs produce more complicated individual components but result in a simpler product structure and a lower total production cost.

5–2 COMPUTER-AIDED ENGINEERING ANALYSIS

Testing the design at the analysis step in the *design process* requires a broad array of CAE software. The software selected is a function of the type of test desired. Generally, software is available on all computer platforms: main-frame, RISC

(a) Design using sheet metal housing cost

	Assembly	Material	Manufacturing	Tooling
Housing	0.02	1.74	1.56[a]	7,830[b]
Bush (2)	0.09	0.01	0.06[c]	9,030[d]
Screw (6)	0.35	0.72	—	—
Spindle	0.04	0.26	1.29	—
Total	0.05	2.73	2.91	16,860

[a] Includes $1.35 for drilling and tapping screw holes.
[b] Three separate die sets for blanking, punching, and bending.
[c] Molded bushings have three-cored holes for screw clearance.
[d] Ten-cavity mold for least-cost manufacture.

(b) Design using injection-molded housing cost

	Assembly	Material	Manufacturing	Tooling
Housing	0.02	0.14	0.24	10,050[a]
Spindle	0.02	0.26	1.29	—
Total	0.04	0.04	1.53	10,050

[a] Two-cavity mold for least-cost manufacture.
Note: A Comparison of the Two Spindle/Housing Assembly Designs Shows Significant Cost Reductions as a Benefit of DFA.

Figure 5–4 DFMA Cost Analysis.
Source: Boothroyd/Dewhurst, "Product Design for Manufacturing and Assembly," © April 1988, Manufacturing Engineering Magazine, p. 44. Reprint by permission of the Society of Manufacturing Engineers, Dearborn, MI.

workstations, and microcomputers. CAE and CAD are closely linked. In most applications, the data and information, used as input for the CAE software, are in the form of a drawing of the product created in CAD. The part geometry file produced in CAD is used by the CAE software to get the data needed for analysis.

The applications of CAE at the analysis step fall into two broad categories: *finite-element analysis* and *mass property analysis*. Software in each of these categories is described in the following sections.

Finite-Element Design Analysis

The most frequently used CAE application is *finite-element analysis* (FEA), defined as follows.

> *FEA is a numerical program technique for analyzing and studying the functional performance of a structure or circuit by dividing the object into a number of small building blocks, called finite elements.*

Most FEA applications fall into two categories, *mechanical systems* and *electronic circuits*. The FEA process for the mechanical system begins with the creation of the geometric model of a part or structure using CAD software. The 3-D CAD model is divided into a finite number of small pieces (elements) that are connected to each other at points called nodes. Figure 5–5 illustrates the FEA mesh and nodes on a piston of an internal combustion engine. The FEA software has mathematical equations that describe how these nodes respond when an external stimulus or force is applied. To analyze the static strength of a bridge, for example, a load is applied to the FEA model of the bridge that is created from a CAD drawing of the structure. The FEA equations predict how each of the finite elements throughout the structure will respond. The material, composition, and other variables that describe the bridge are represented in the equations. The simultaneous solution of the equations representing each finite element produce the overall response of an entire part or structure to the applied force. Both graphical and numerical

Figure 5–5   Finite-Element Analysis.

solutions are provided; for example, Figure 5–6 illustrates the temperature gradient across an object when a heat transfer FEA is performed.

FEA software covers a wide range of applications that include:

- *Static analysis.* Static analysis software determines the deflections, strains, and stresses in a structure under a set of fixed loads. Typical structures include aircraft, bridges, buildings, cars, dams, and machine parts.

- *Transient dynamic analysis.* Transient dynamic analysis software calculates the deflection and stress under changing load conditions using the natural response time and frequency for the structure. Typical structures would be the same as those above.

- *Natural frequency analysis.* Natural frequency analysis software computes the stress on a structure caused by vibration at the natural frequency of the structure. For example, the destructive power of low-frequency vibrations created by earthquakes can be applied to structural models of buildings and bridges to determine the limits for catastrophic failure.

Figure 5–6   FEA for Thermal Analysis.

- *Heat transfer analysis.* Heat transfer analysis software determines the temperature distribution plus steady-state and transient heat transfer in a structure when thermal loads are applied and boundary conditions are known. Heat transfer analysis is used on end products and production processes. For example, the efficiency of automotive cooling system designs for removing heat from engine components is tested with heat transfer FEA. In the production area, mold-filling analysis using a heat transfer FEA model is used to test plastic injection mold design for optimum plastic flow, filling speed, and cooling parameters.

- *Motion analysis.* Motion analysis software, sometimes called kinematic analysis, computes the geometric properties (displacement, velocity, and acceleration) required for a mechanical mechanism to produce a desired motion. Few products designed today do not have moving parts, and motion FEA allows the designer to put the parts of the solid model into motion. Under these simulated motion conditions, parameters such as limits of motion; interference; and the geometric properties of displacement, velocity, and acceleration on any part can be analyzed.

- *Fluid analysis.* Fluid analysis software determines the flow, diffusion, dispersion, and consolidation characteristics of a fluid under different controlled conditions. Design of piping systems with complex turns for large volumes of fluid requires a model of the fluid flow. Fluid FEA allows the optimum piping system design for the type of fluid carried.

Use the Web sites provided in Appendix 5–2 to find current examples of each of these applications.

The key to successful FEA is the selection of the *mesh* that is composed of elements connected at nodes. The size and location of the elements is critical for generation of useful results. Note in Figures 5–5 and 5–6 that the size of the elements varies. The elements are generally smaller at points on the geometry where the effects of the analysis are most pronounced. For example, in Figure 5–6, the elements are smallest at the hole in the side of the cylinder. Selecting appropriate elements is performed automatically by FEA software with mesh generation features, but manual adjustment of the generated mesh is often necessary for optimum results. In the past, special training was necessary to operate the FEA software systems and to interpret the FEA output data. For some analyses, FEA specialists had to perform the analysis work. However, the ease of use of current FEA software permits design engineers to perform the analysis on their CAD designs.

FEA software is available for all three computer hardware classifications. The ANSYS software is a widely recognized, large-scale, general-purpose finite-element analysis program for engineering. The original versions of the ANSYS program were oriented toward batch processing for mainframe computers and emphasized structural and heat transfer analyses. The program has been extended to include other physical phenomena (magnetic fields and fluid flow), nonlinear effects, on-line documentation, an extended graphics library, solid modeling, light-source shading, and many other features. SuperSAP is a popular FEA package for the microcomputer.

Mass Property Analysis

Mass property analysis, a CAE function used frequently in the design process, is invoked through a command included in most CAD software. The mass property command signals the CAD system to calculate and return numerical values that describe properties of the drawing geometry selected. In the most basic application, the *area* of a 2-D CAD shape or the *volume* of a 3-D solid object is calculated and displayed. More complex mass property analysis produces the *mass, bounding box, centroid, moments of inertia, products of inertia, radii of gyration,* and *principal moments* with *X-Y-Z directions about centroids.* These complex parameters are important for analysis of part geometry that moves and rotates in the final application. Detailed explanations of these parameters are beyond the scope of this book, however.

An example from a company that produces extruded aluminum rails illustrates how important basic mass property data are for some manufacturers. A typical cross-section of a rail used for building commercial aluminum windows is pictured in Figure 5–7, and the CAD drawing cross-section used to produce the extrusion die for production of the aluminum part is shown in Figure 5–8. CAD could not be justified based on drawing the simple outline required for the extrusions; however, the extrusion area is important because it determines the amount of metal used and the weight per linear foot of rail. The cross-section of the extrusion is drawn with 2-D CAD software, and the area is calculated using the mass properties feature in CAD. As a result, the cross-section for the rail and die are designed to the desired weight standards, and the cost of the extrusion is accurately established. A secondary benefit is derived from electronic transfers of the CAD data file. The CAD file of the cross-section is sent from the computer of the rail manufacturer by modem over phone lines to the computer of the company that produces extrusion dies. The die manufacturer processes the CAD shape file for

Figure 5–7   Extruded Aluminum Rail.

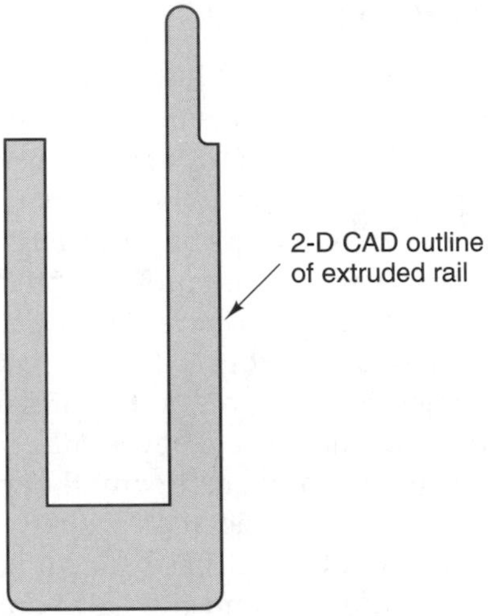

Figure 5–8 Die Drawing for
Extruded Aluminum Rail.

2-D CAD outline
of extruded rail

the extrusion and sends the converted file to a wire electronic discharge machine (EDM) on the shop floor to cut the dies. The single CAD database shared by the two companies guarantees that the die will produce the exact desired extrusion.

Mass property analysis of CAD solid models is also critical for companies that design plastic injection molded or metal die cast parts. DFMA software, used to evaluate trade-offs between assemblies in Figures 5–1 and 5–3, requires data from mass property analysis. The base for the spindle/housing assembly shown in Figure 5–3 is a nylon part produced on a plastic injection molding machine. Accurate values for the critical DFMA input parameters, *part volume* and *projected area,* are difficult to obtain without mass properties software. The DFMA software input screen is illustrated in Figure 5–9. In this example, the part was drawn using 3-D solid CAD software, and the mass properties feature provided the area and volume data. The output data from the DFMA analysis of the injection molded housing (Figure 5–10) provide additional mass property data, *part weight,* and a host of cost data.

Mass property analysis is available on computer hardware in all three categories from mainframe to microcomputers, usually as a command in the CAD software. Mass property analysis is important to every type of manufacturer where product design is performed.

Other CAE Design and Analysis Software

The exponentially expanding software market has many more examples of special-purpose programs for analysis of CAE designs. Two additional types of analysis software used frequently included *circuit analysis* and *assembly interference, fit, and tolerance.*

Software in the CAE circuit analysis area supports a broad range of design activities across the electronic industry. In general, the software performs numerical analysis on electronic circuits to determine electrical performance and worst-case

Computer Input Screen for Nylon Housing

ESTIMATION OF INJECTION MOLDING COSTS FOR: HOUSING   THERMOPLASTIC: 6/6 NYLON

Dimensional Data		Part Complexity	
Part volume = 25.51 CM^3		Outer surface or cavity (0–5)?	2
Projected area = 71.25 CM^3 L = 95 MM W = 75 MM D = 57 MM		Inner surface or core (0–5)?	0
Thickness maximum = 2.5 MM Average = 2MM			
Quality and Appearance		Mold Complexity	
		Standard two-plate mold?	Y
Tolerance factor (0–5)?	3	Three-plate mold?	N
Appearance factor (0–5)?	1	Multiplate stacked mold?	N
Colored resin?	N	Hot runner system?	N
Textured surface?	N	Number of side cores or pulls?	0
		Number of unscrewing devices?	0

At any time press: <H>elp, <V>olume, <A>rea, or <C>omplexity calculator.

Figure 5–9 DFMA Input Screen.

Source: Boothroyd/Dewhurst, "Product Design for Manufacturing and Assembly," © April 1988, Manufacturing Engineering Magazine, p. 44. Reprint by permission of the Society of Manufacturing Engineers, Dearborn, MI.

Results for Injection Molding of Nylon Housing

ESTIMATED INJECTION MODLING COSTS FOR: HOUSING   THERMOPLASTIC: 6/6 NYLON

Total Production Volume (Thousands)	Number of Cavities	Number of Total Mold Base Costs ($) Cavities	Cavity/Core Manufacturing Costs ($)	Total Mold Cost ($)	Mold Cost per Part (Cents)
100	2	3589	6462	10,051	10.1

Machine Size (kN)	Machine Rate ($/hr)	Cycle Time (Seconds)	Manufacturing Cost per Part (Cents)
1600	72	20.3	23.9

Part Volume (cm³)	Part Weight (Grams)	Polymer Cost ($/kg)	Polymer Cost per Part (Cents)
26	29	4.69	13.5

Total part cost (cents) = 47.5

Select required option:

1. Screen edit
2. Show mold cost/cycle elements
3. Print results and responses
4. Change basic cost data
5. Change responses/polymer
6. Exit

Figure 5–10 DFMA Mold Analysis for Alternative Design.

Source: Boothroyd/Dewhurst, "Product Design for Manufacturing and Assembly," © April 1988, Manufacturing Engineering Magazine, p. 45. Reprint by permission of the Society of Manufacturing Engineers, Dearborn, MI.

conditions. No circuit types are excluded; discrete, integrated, and hybrid circuit analyses are available. In addition, a host of design software is available to support integrated-circuit chip design, surface-mount technology design, and printed circuit board design. Circuit analysis software such as ECAP and SPICE is used for analysis of electronic and power circuits.

The assembly interference, fit, and tolerance software focuses on design and analysis of mechanical parts and assemblies. A software package called Valisys runs on mainframes and RISC workstations using CIM part files created with CATIA CAD software. Valisys features are worth describing because they illustrate the type of cross-discipline data generation and collection required in a CIM environment. In addition, a review of the software emphasizes the critical relationship between product design, manufacturing, and quality. The Valisys software has five separate modules.

Design Verification. Starting from the initial design, the Valisys software from Tecnomatix Technologies (see Website in Appendix 5–2) checks the dimensioning and tolerancing information against the geometric dimensioning and tolerancing (GD&T) standard ANSI Y14.5. In addition, the software uses the existing part geometry to create *soft gauges*. Hard gauges are made by manufacturing to check the quality of a manufactured part. For example, if the size and separation distance between two holes is critical for an assembly, a metal gauge like the mating part is produced. This hard gauge is just a metal plate with two pins that can be inserted into the two holes to test the manufactured parts. The soft gauge is a 3-D wireframe model of the hard gauge and is created by Valisys as part of the CATIA CAD geometry file. When a part is manufactured, the critical features of the finished part are measured on a coordinate measuring machine (CMM). These dimensional data are checked electronically by the soft gauge for proper mating part fit. The CAE Valisys soft gauge technique provides better quality monitoring with less investment in hard tooling.

Tolerance Analysis. The tolerance analysis function in Valisys analyzes the fit between mating parts under worst-case tolerance conditions. Production tolerances on the part are set at the maximum value because this design feature of the software considers part fit as part of the design. In addition, optimal design geometry and drafting text for clearance and threaded holes for fasteners are generated automatically.

Quality Engineering. The quality engineering function in Valisys addresses inspection of the manufactured part using a coordinate measuring or numerical control (NC) machine. Inspection paths and an inspection program for the measuring device are generated from the CAD geometry by Valisys to collect critical dimensional data on the manufactured part. An online graphical model of the "as-built" part is created with the data collected. A comparison of the "as-built" model with the soft gauge indicates whether the part passes or fails inspection. In addition, recommendations for rework are provided for parts that fail inspection.

Inspection.   The inspection function controls the measuring machines and manages the acceptance and rejection of parts. Recommendations for process improvement are generated from the data collected during the quality measurement phase.

Process Control.   The process control module in Valisys collects quality data and creates control charts required for *statistical process control* (SPC). The data collected in this module are used to determine the accuracy and repeatability of the manufacturing process, to determine how closely the "as-built" part matches the design specifications, and to anticipate trends in the machine process so that corrections are made before out-of-tolerance parts are produced.

5–3 COMPUTER-AIDED ENGINEERING EVALUATION

The design analysis process provides ample data on the various design alternatives. The examination of those data to determine the degree of match between the actual design and the initial design goals and specifications is one part of the evaluation process. Examination of the data is performed by every member of the concurrent engineering team and recommended changes in the design are made. The reiterative nature of the *design process* makes it difficult to separate CAE activities in the analysis and evaluation functions. The important point is that CAE software to analyze and evaluate design quality is available. One CAE activity traditionally performed at the evaluation stage of the design is *prototyping*.

Prototyping

Building a prototype of a design is an age-old practice. The prototype is an original model of the design built to evaluate operational features before the start of full production of the product. The style of the prototype is dictated by the tests that are planned. For example, the automotive industry builds small-scale (about 1 foot long) models of new cars from solid metal or wood for wind-tunnel tests of the aerodynamics of the body. The same industry builds full-sized fully working models of new engine designs and subjects them to operational tests.

The tools used for standard prototyping are conventional production machines. Frequently, prototype parts are machined from nonferrous metal or plastic; however, with the use of more complex plastic injection molded parts in products, the prototype process becomes difficult. Machining the complex shapes of injection molded parts is difficult, expensive, and time consuming. While prototyping a design is still a critical evaluation process, the requirement to cut lead time to market requires faster prototyping techniques. Several different technologies, called *rapid prototyping,* are reducing the time required to develop prototype parts.

Rapid Prototyping

Rapid prototyping, a technique used to build a sample of a new design quickly, is a reliable tool in the evaluation process. These systems electronically divide a

3-D CAD model of a part into thin horizontal cross-sections and then transform the design, layer by layer, into a physical model or prototype.

Rapid prototyping systems are driven by RISC workstations or large micro-computer platforms. Starting with a CAD 3-D fully closed surface or solid model file, the CAD software converts the geometry into a file format compatible with the rapid prototype system. For example, the stereolithography system from 3D Systems, Inc. uses an STL file format. The conversion software for the STL file format is available for all the major CAD software. Visit the Web sites provided in Appendix 5–3 to see examples of the following rapid prototyping systems. Some of the rapid prototyping systems currently in use are discussed below.

Stereolithography.   The process employs a tank of liquid photosensitive poly-mer with a vertically controlled table in the polymer and a servo-controlled laser focused on the surface of the polymer (Figure 5–11). The table is positioned with its top just below the level of the liquid in the polymer tank. The computer in the system reads the STL CAD file for the prototype part and cuts the part into cross-sections from top to bottom. The cross-sections are typically from 0.0015 to 0.005 inch thick. Thinner slices are possible to produce a smoother model, but the processing and fabrication time is increased significantly. The computer system stores the cross-section data as SLI files and merges them to create files for controlling the laser and table elevator mechanism. The laser traces the area of the bottom cross-section on the thin liquid layer of polymer on the table; the laser light causes the polymer to harden. With the first cross-section of the part created on the table, the table elevator is lowered by the cross-section thickness by the computer

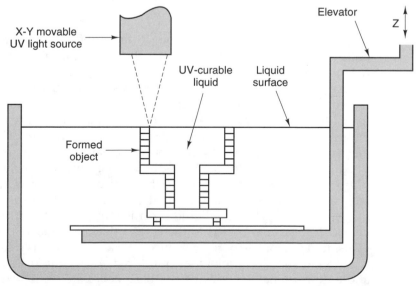

Figure 5–11   3-D Systems Stereolithography.

so that another thin layer of polymer covers the bottom cross-section. The laser then solidifies the next cross-section. This process continues until the entire part is created from liquid polymer. Processing time is a function of cross-section thickness and the size of the area traced. On average, the system processes about an inch of thickness per hour for a small part. After the part is produced in the system, it must be cleaned thoroughly and cured in an oven to develop the full strength of the material. Accuracy varies between 0.1 and 0.5 percent of overall dimension from small to large parts. Currently stereolithography is the most accurate process. Figures 4–11 and 4–12 show 3-D AutoCAD drawings of a plastic pencil sharpener used to produce the prototype in Figure 5–12. Study the drawings and compare the original pencil sharpener in Figure 5–12 with the models produced through sterolithography.

Figure 5–12 Original Pencil Sharpener and Part Produced Through Stereolithography.

Solid Ground Curing. Solid ground curing (SGC) is a prototyping process for building models that are solid, which eliminates curling, warping, support structures, and any need for final curing. The process is similar to stereolithography but has the following variation. The process starts with a CAD file and renders the object as a stack of slices. The slice is printed on a glass photo-mask using an electrostatic process similar to laser printing. The part of the slice that represents solid material on the part remains transparent. Next a thin layer of photoreactive polymer is spread evenly across the working surface below the mask. An ultraviolet floodlight is projected through the photo-mask onto the newly spread layer of liquid polymer. The resin exposed to the light polymerizes and hardens, while the resin not illuminated remains a liquid and is removed with a vacuum. The void in the part created when the liquid resin is removed is filled with liquid wax and chilled. The layer is milled to the correct thickness and the layering process repeats until a complete part is produced. Finally, the wax is removed by melting or rinsing and a finished prototype is produced. A layer thickness of .004 to .006 inches and a dimensional accuracy of +/− 0.02 inches are combined with the ability to build up to 100 layers per hour.

Selective Laser Sintering.   Selective laser sintering (SLS) employs a high-energy laser to fuse or sinter powder into a solid object (Figure 5–13). Using a technique similar to stereolithography, the SLS laser traces the shape of each cross-section, fusing the thin layer of powder. A mechanical roller then spreads more powder across the top of the finished layer, and the laser traces the next cross-section. Currently, the SLS systems can work with three materials: wax (for investment castings), polycarbonate, and poly (vinyl chloride). In the future, ABS, nylon, and some ceramic materials may be added to the list. Accuracy varies between 0.00 to 0.015 inches depending on the size of the part. Layer thickness varies from 0.003 to 0.02 inches.

SLS has the advantage of using a wider variety of materials with better mechanical properties than photopolymers and at lower cost. SLS works better on some parts with complex internal shapes but does not hold tolerances as tightly as does stereolithography. Dimensional variation of plus or minus 5 mils per inch is normal for these systems. The SLS surface finish of PVC parts is quite smooth, but other materials exhibit a laminated appearance.

Figure 5–13 DTM Corporation's Selective Laser Sintering.

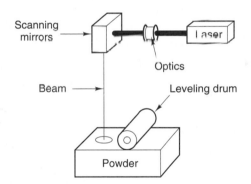

Three-Dimensional Printing (3DP).   *Three-dimensional printing* (3DP) was developed at The Massachusetts Institute of Technology and is licensed for use by several companies. The process utilizes a thin layer of powdered material spread one layer at a time. The powder can be any material, including ceramics, metals, polymers, and composites. Adhesive is then applied to an area that represents the cross-section of a part or the mold for a part. The adhesive is dispensed in droplets through a device similar to an inkjet printer head.

Three-dimensional printing functions by building parts in layers in a process similar to SLS. Each layer begins with a thin distribution of powder spread over the surface of a powder bed. The binder material is selectively applied to form the object. A piston that supports the powder bed and the part-in-progress lowers so that the next powder layer can be spread. This layer-by-layer process repeats until the part is complete. The formed part is placed in an oven to fuse the bond material together, leaving the fabricated part. The advantages of this process include high geometric flexibility; no supports required for overhangs, undercuts, and internal volumes; and any material that can be produced in a powder form will work.

A variation of 3DP, used by Soligen Corporation, is called *direct shell production casting* (DSPC). In this variation, the powder is ceramic and the part produced is a casting. Using this technique, castings for complex parts are produced much more quickly than the traditional model building process would permit. In addition to the speed advantage provided by DSPC, the process also permits parts too complex for standard model-making techniques to be manufactured efficiently and quickly.

Fused Deposition Modeling. Fused deposition modeling (FDM), illustrated in Figure 5–14, builds up each cross-section by moving a thin extruded "wire" of plastic or wax just above the part location and heating it to its melting point. Again, the part is built one cross-section at a time. The process is fast and employs relatively cheap materials. Like stereolithography, fixtures to support internal part

Figure 5–14 Stratasys's Fused Deposition Modeling.

Source: Courtesy Stratasys, Inc., Eden Prairie, MN.

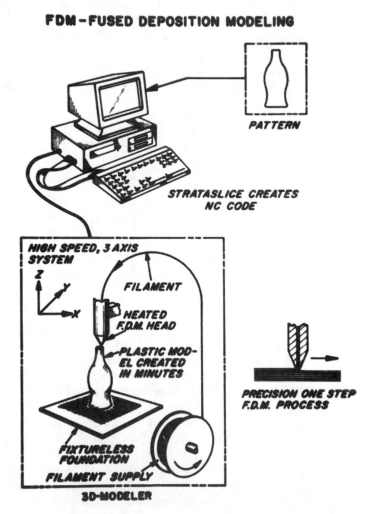

geometries are needed; however the process is much faster than stereolithography. Some geometries produced by FDM have a grainy appearance and materials are limited to investment casting wax and a "nylon-like" material. Figure 5–15 shows prototypes built with FDM technology. A system developed by Stratasys, Inc. to produce FDM models is illustrated in Figure 5–16. Note the CAD workstation with the solid model of the part visible on the monitor.

Laminated Object Manufacturing.   In the LOM system, parts are built up from sections cut from thin sheets of stock. The stock may be paper, plastic, or polyester composite. As the part is built, each sheet (0.002 to 0.02 inches thick) is glued with adhesive to the already constructed part, then trimmed with a laser. LOM is five to ten times faster than other rapid prototyping processes because the laser beam traces only the outline of each cross-section, not the entire area. In addition, production of very large parts is possible with an accuracy of ±0.005 inches over the dimensions of the machine.

Figure 5–15   Parts Produced Through Fused Deposition Modeling.
Source: Courtesy Stratasys, Inc., Eden Prairie, MN

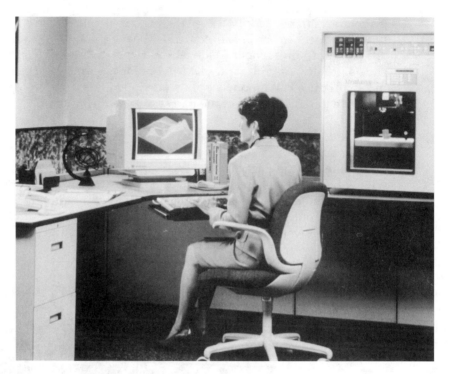

Figure 5–16   Rapid Prototyping System for Fused Deposition Modeling.
Source: Courtesy Stratasys, Inc., Eden Prairie, MN.

5–4 GROUP TECHNOLOGY

Group technology (GT) is not an automation strategy associated with either the de-sign or the production engineering area, but the implementation of some form of GT is often necessary to achieve the order-winning criterion described in Chapter 1. For example, GT is a critical first step for computer-aided process planning and many of the production engineering activities described in the next section.

Defining Group Technology

Group technology is defined as follows.

> *GT is a manufacturing philosophy that justifies small and medium-size batch production by capitalizing on design and/or manufacturing similarities among component parts.*

Under a group technology implementation, dedicated production cells are created in which families of parts grouped by a selection code are produced on a set of production machines selected for the part group. For example, study the parts shown in Figure 5–17 and list all features common to all the parts. At first glance the parts shown in the figure appear to have no common features. However,

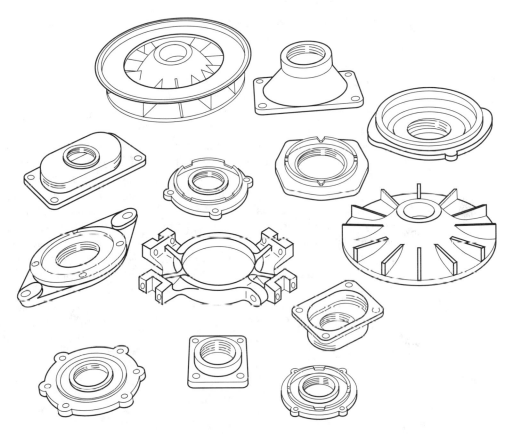

Figure 5–17   Family of Parts for Group Technology.

with further study the common features and machining operations become apparent: (1) similar in size and shape, (2) raw material is a casting, (3) all require internal hole boring in a single direction, (4) all require face milling, and (5) most have drilled holes in a single direction. Common features that are not obvious from Figure 5–17 include required dimensional tolerance, type of material, and surface finish demands. As a result of the similarity in process operations, a single production cell could be built to machine the part family shown in Figure 5–17.

GT offers a structured method of classifying parts based on geometry and production characteristics. With GT coding on all parts and components, a company can sort all manufactured parts into part families suitable to a single production cell. Conversion to GT and cell manufacturing supports the following order-winning criteria: shorter lead and setup times, reduced work-in-process and finished-goods inventories, and less material handling. In addition, GT helps simplify production planning and control and is necessary for the successful implementation of other production engineering software applications.

Coding and Classification of Parts

Coding is a systematic process of establishing an alphanumeric value for parts based on selected part features. *Classification* is the grouping of parts based on code values. Coding and classification in GT are highly interactive because the coding system must be designed to produce classified groups with the correct combination of common features. Relevant part features are used to place parts into groups using either the *hierarchical, chain,* or *hybrid* code structures.

The hierarchical code structure, called a *monocode,* is based on the biological classification system established by Linnaeus. This type of coding, often called a *tree structure,* divides all parts of the total population into distinct subgroups of about equal size. Study the example shown in Figure 5–18, where the total population is all cylindrical parts. The group of parts is initially subdivided into two groups based on the ratio of the length divided by the diameter of the parts. The values on the conditions for the initial branch (> 1.5 and < 1.5) are selected so that each of the subgroups is about equal in size. Additional subgroups are selected based on the presence of gears and the type of machining performed. The number of digits in the code is determined by the number of levels in the tree. Another characteristic of the code numbers at each level is that only the least significant digit is different. Study the tree structure shown in Figure 5–18 until these last two concepts are clear. The advantage of the hierarchical structure is that a few code

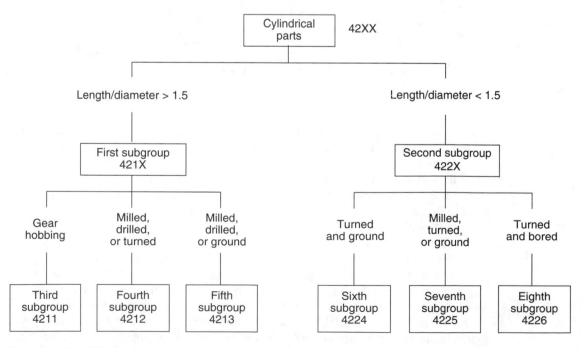

Figure 5–18 GT Monocode.

numbers can represent a large amount of information. The singular disadvantage is the complexity associated with defining all the branches.

The chain structure, called a *polycode,* is created from a code table or matrix like the example shown in Figure 5–19. Study the example matrix; note that the type of part feature and digit position are defined by the left vertical columns. The numerical value placed in the digit position is determined by the feature descriptions across each row.

Digit	Class of feature	Possible values of digits			
		1	2	3	4
1	External shape	Shape$_1$	Shape$_2$	Shape$_3$	—
2	Internal shape	None	Shape$_1$	—	—
3	Number of holes	0	1–2	3–5	5–8
4	Type of holes	Axial	Cross	Axial and cross	
5	Flats	External	Internal	Both	
6	Gear teeth	Spur	Helical		

Figure 5–19   GT Polycode Structure.

Exercise 5–3
A part with the code 311412 has a 2 in position one, 1 in position two, 4 in position three, . . . and a 3 in position six. Determine the features of the part using the polycode chart shown in Figure 5–19.

Solution
The features described by the code are no gear teeth (3 in position six), external flats (1 in position five), all axial holes (1 in position four), 5 to 8 holes (4 in position three), no internal shape or cutout (1 in position two), external shape in the shape 2 category (2 in position one).

The major advantages of polycodes are that they are compact and easy to use and develop. The primary disadvantage is that, for comparable code size, a polycode lacks the detail present in a hierarchical structure.

A hybrid code captures the best features of the hierarchical and polycode structures. One of the best examples of a hybrid code is the Optiz code and classification system (Figure 5–20), developed in 1970. Note that the code starts and ends with a polycode and has a hierarchical code in the middle. Industry currently uses over 90 GT coding systems; the code selected is dictated by the type of product, production system, and total mix of all parts and components. With the code and classification system selected, the development of a GT production cell follows.

GT Production Cells

Batch manufacturing has traditionally taken place in a functional layout where similar machines are grouped together (for example, all mills grouped together, all

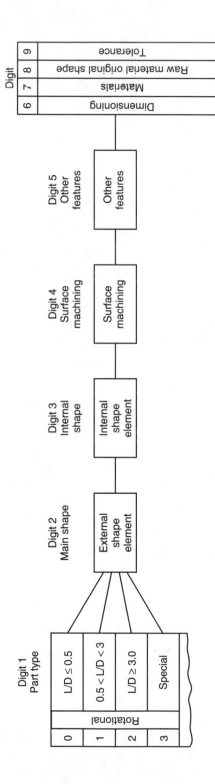

Figure 5–20 Adoption of an Optiz Type Code that Combines Monocode and Polycode Structures. The Code Can Be Expanded for Nonrotational Type Parts.

lathes grouped together, etc.) in the production facility. In the batch production process, parts are routed through these various work centers according to a specified sequence of operations. For example, the layout of a typical job shop is illustrated in Figure 5–21. Note that all similar machines are located in the same geographic area on the shop floor. Parts are routed to the machine for the specific operation required by the design. A partial list of the inefficiencies include high in-process inventory, a large number of parts handlers, longer lead times, and longer setup time. The application of GT part families to batch production requires a physical rearrangement of the production facility. With GT applied to the production part mix in Figure 5–2, a family of parts is created and three linear production flow lines with cells are developed (Figure 5–22) to handle the same production. Instead of organizing production around machine similarity, "groups" of different machines are identified based on their ability to produce families of parts.

Another method for identifying part families and the associated groups of machine tools is called *production flow analysis* (PFA). The primary grouping and classification data used in this process are the operation sequences and machine routings used for the parts under study. Parts with identical or similar routings are grouped together into families. From these family groups, logical machining cells and a GT layout are produced.

Regardless of the method used to code, classify, and eventually to reorganize the shop floor, the primary benefit from the GT exercise is better organization, identification, and understanding of the products in manufacturing.

5–5 PRODUCTION ENGINEERING STRATEGIES

The production engineering function, highlighted in the organizational model in Figure 3–2, is closely linked to design. In the preceding chapters, we have described the impact on the design function as a result of the computer and software revolution. The degree of change in the design function because of the development of computers is paralleled in production engineering. The term *computer-aided manufacturing* (CAM) was coined in the 1960s when early mainframe and special-purpose

Figure 5–21 Traditional Process or Job Shop Layout.

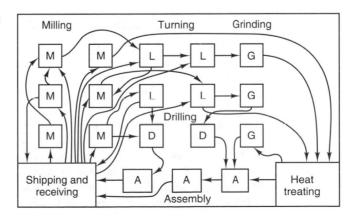

Figure 5–22   GT Layout.

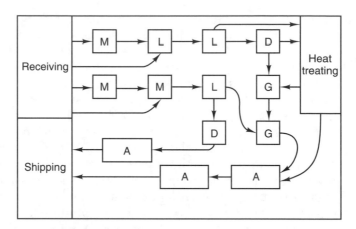

computers were interfaced to production machines to automate some of the production engineering and manufacturing functions. The introduction of computer automation into the production engineering areas continues to increase.

Production engineering has the responsibility for the development of a plan for the manufacture of goods and services for the enterprise. The elements in the plan, listed in Chapter 3, include process planning, production machine programming, tool engineering, work and production standards, plant engineering, analysis for manufacturing and assembly, and manufacturing cost estimating. The analysis for manufacturing and assembly was described in Section 5–1. A description of the automation strategy associated with the rest of the elements in the manufacturing plan is provided in the following sections.

Computer-Aided Process Planning

Manual process planning includes the creation of all paperwork necessary to direct the flow of raw materials and parts through production and assembly. The process planner determines the sequence (Figure 5–23) and machines that will transform the raw material into a finished part. For example, in a typical job shop, the planner studies the drawing of the part, selects data from the machinability data handbook, checks the tooling and fixtures available, selects raw material stock, and then selects the metal-cutting operations available on the shop floor that are necessary to produce the part. The plan for the spindle in Figure 5–1 would start with a cutoff saw operation (Figure 3–17) to cut a 90-mm length from a ¾-inch stainless steel bar. Following the deburring operation, the planner would route the material to a turning center or lathe (Figure 3–17) and the contour would be cut. Following the lathe operation, the spindle would be routed to a centerless grinder to finish the bearing surfaces on each end. For a simple part like the spindle, the planning process is clear. However, for more complex parts, the order of operations depends on the planner's knowledge and experience.

For example, four planners were asked to plan a part; study the results from the four planners illustrated in Figure 5–23. The order of operations in each is

		Process planner		
	One	Two	Three	Four
1	Machine first face	Hole drilled in two steps: a. 20 mm dia b. 38 mm dia	Outside surface– 70 mm dia– turned	Hole drilled to finish in two steps: a. 30 mm dia b. 40 mm dia
2	Hole finished in three steps: a. Drill 10 mm b. Drill 38 mm c. Bore 40 mm	Machine first face	Hole drilled to finish in one step with drill of 40 mm dia	Outside surface– 70 mm dia– turned
3	Outside surface– 70 mm dia– turned	Cutoff	Machine first face	Machine first face
4	Cutoff	Machine second face	Cutoff	Cutoff
5	Machine second face	Outside surface– 70 mm dia– turned	Machine second face	Machine second face
6		Hole finished to 40 mm dia by boring		

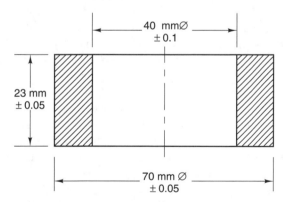

Figure 5–23   Comparison of Process Planning by Four Different People.
Source: Schaffer, "Implementing CIM," © American Machinist, August 1981, p. 82, Reprint by permission of American Machinist, a Penton publication.

different, and the method used to produce the 40-mm hole varies. Two of the planners decided they could get the required finish for the 40-mm hole from a drill, and two required a boring operation. The two calling for a boring operation were more experienced and correct. Consistent and correct planning requires both knowledge of the manufacturing processes and experience. Two automation techniques, *variant* and *generative* process planning, to improve the processing planning operation in the integrated environment are embodied in *computer-aided process planning* (CAPP).

The CAPP variant approach uses a library of manually prepared process plans in a database and a retrieval system to match components on new parts to existing process plans of similar components. When the process plan is valid for a family of components, it is called a *standard* plan and is stored in the enterprise database with the family key number for identification. The retrieval method and the logic of the variant system are predicated on the grouping of parts into families, as in group technology. In most situations, the standard plan must be modified to some extent before the plan can be used with the new component parts. After the preparatory stage (Figure 5–24), where the families of standard process plans are developed and saved in the variant database, the system is ready for production components. For a new part, the flow indicated in Figure 5–25 would be used for the variant process. A new production component is given a family code and then passed through a part-family search routine to find the family to which the component belongs. The standard plan for that family of components is retrieved and a human planner makes the adjustments necessary for the new component. Figure 5–25 illustrates the variant process planning procedure. Study Figures 5–24 and 5–25 until you understand the variant process. The primary advantage provided by the variant technique is the reduction of process planning time by almost 50 percent.

The CAPP *generative* technique for the creation of process plans is both more difficult to develop and more highly automated. In general, the generative process planning system creates plans for new components without referring to existing plans or with the assistance of a human planner. Generative CAPP utilizes a process information knowledge base that includes the decision logic used by expert human planners. Frequently, the heuristic planning knowledge is captured in artificial intelligence (AI) software, called *expert systems.* The AI software is designed to store and imitate the human decision-making process. The generalized system in Figure 5–26 illustrates the generative CAPP operation. A part drawing is received from design and needs a process plan created for manufacturing. The first step is to convert the design specifications into an input format compatible with the CAPP automation. The three techniques frequently used include code, a descriptive language, and CAD. In each technique, the complete design specification for the part is converted into a format compatible with the decision engine in the CAPP software. The decision logic portion of the CAPP system uses manufacturing database information, such as production machine capability, tooling, fixtures, and time standards, and the design specifications to arrive at an operational process plan. The three most commonly used decision logic algorithms are listed in Figure 5–26.

Leading CAPP software captures the knowledge and experience that has been developed on the shop floor, creates detailed process plans with accurate time standards, and then communicates this information to material requirements planning (MRP) and enterprise requirements planning (ERP) databases. METCAPP is an example of CAPP software that provides these types of benefits in the planning

Figure 5–24   Developing a Variant
Process Planning System.

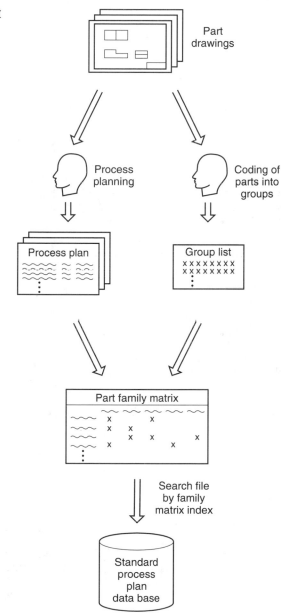

process. METCAPP was developed by the Institute of Advanced Manufacturing
Sciences in Cincinnati, Ohio, for machining operations. METCAPP uses cutting
speeds and feeds from the industry-standard Machining Data Handbook, which
makes accurate cut time and cost calculations possible. METCAPP uses solid part
models to extract design features. The software separates part features into the
appropriate setups and machining operations, then generates a complete process

Figure 5–25 Variant Process Planning System.

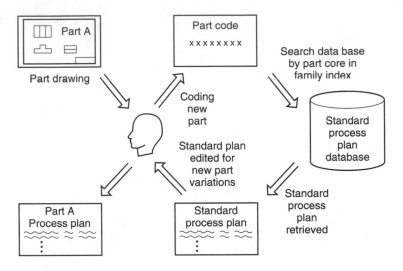

Figure 5–26 Generative Process Planning System.

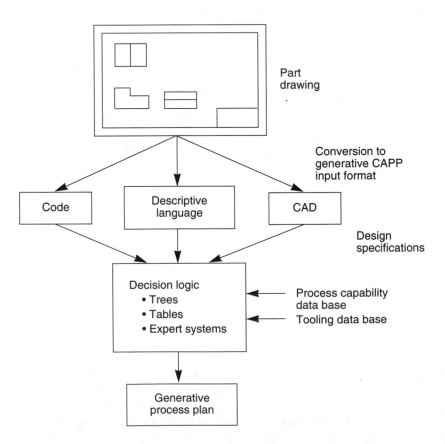

plan for the part production. Using the solid part model from the CAD software, METCAPP imports the model and automatically generates a full-blown process plan with detailed work instructions for every step in the machining process. METCAPP is supplied with a database of 12,000 time standards, machining systems rules for forty-two common features, data on 18,000 cutting tools, and specifications for over 100 machine tool models. The interface between CAPP and the enterprise is illustrated in Figure 5–27. Note how the CAPP software forms the interface between design and the shop floor, thus providing critical integration of the product data from initial design to final part production.

The advantages provided by generative CAPP include:

- Process plans are created rapidly and consistently.
- Totally new process plans are created as quickly as plans similar to existing components.
- An interface to MRP and ERP software is possible.
- Documentation requirements of ISO or QS 9000 standards are completed and best practices are promoted, even without experienced workers.

Computer-Aided Manufacturing

The term *computer-aided manufacturing* (CAM) is used to describe a wide range of automation technologies. The reason for the confusion over the definition of CAM is explained by a brief review of the history of shop floor automation. The emerging technologies chart in Figure 4–1 indicates that *numerical control* (NC) was the start of the shop floor automation evolution. As U.S. industry emerged from World War II, the NC technology and computers began to shape automation on the shop floor. The next significant jump in productivity was the development of *direct numerical control*

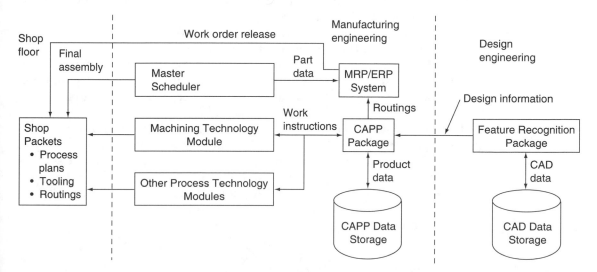

Figure 5–27   CAPP in the Enterprise System.

(DNC) machine tools. Entire shop floors of CNC or NC machines were connected directly to a large mainframe computer that acted as a central program repository and control center. Later, NC tape-programmed production machines were replaced with *computer numerical control* (CNC) equipment that integrated a computer-type controller into the production machine. The CNC production equipment could now store and execute part programs without program tapes or a DNC connection to the mainframe computer and with less operator intervention. In the 1960s, other production systems, especially in the process control industries, incorporated computers into production processes. In the 1970s, computer hardware shrunk in size, dropped in price, and increased in capability; as a result, "smart" production machines with powerful computers were available to every size industry. As a result of the rapid development of computer-driven shop floor automation, the term *CAM* no longer applies just to automated machine tools. To understand the broad new role that CAM plays in a full integrated enterprise, a clear definition is required.

> *CAM is the effective use of computer technology in the planning, management, and control of production for the enterprise.*

One of the major CAM applications, used by the discrete-part manufacturing industry, is the production of finished parts with information extracted directly from design drawing data. In this application, often called *CAD/CAM,* the part geometry created with CAD in the design engineering area is used with CAM software to create machine code capable of machining the part on almost any CNC machine. The block diagram in Figure 5–28 illustrates the robust data interface between CAD drawing files and the CAM files required to machine the part. Study the diagram until you are familiar with all the names in the boxes.

The drawing of a machined part is created using conventional CAD software and drawing techniques; no special commands or controls are embedded into the graphic file. However, it is customary to put all the geometry of the part on a separate drawing layer with dimensions, notes, text, and other nongeometric information on other layers. This arrangement permits the part geometry, critical for the CAM process, to be stripped away from all the other drawing information by turning off all layers except the one with geometry. With the part geometry captured from the CAD drawing file, the geometry information is transferred to the CAM workstation.

The file format used to transfer the part geometry to the CAM workstation depends on several factors. Some CAM software has internal CAD capability to perform limited drawing of part geometry, and some of the CAD software systems have CAM capability as an option. For example, most of the microcomputer based CAM software systems, such as SmartCAM and MasterCAM, have basic 2-D and 3-D CAD capability. In the RISC computer area, software, like CATIA, offers true CAD/CAM, as the list of application modules indicates:

- *3-D design:* a 3-D graphics modeler
- *Drafting:* a basic drafting system
- *Advanced surfaces:* a sculptured surface modeler

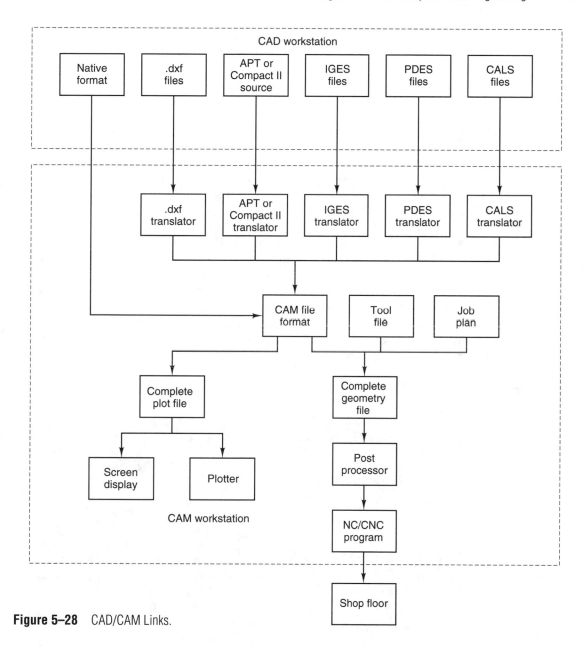

Figure 5–28   CAD/CAM Links.

- *Solid geometry:* a solid geometry modeler
- *Building design:* an architectural design modeler
- *Library:* a library for custom symbols and objects
- *Numerical control:* an NC part programmer generator
- *Robotics:* a robot cell simulator and programmer

Most of the other CAD vendors offering workstation software support full CAM capability. Many of the CAD vendors who started in the UNIX environment have migrated their software to the PC and NT operating systems. As a result, the integration at CAD and CAM at the PC platform level has made great strides. If CAD and CAM are integrated into the same software system, the CAD file in its *native* format can be transferred to the CAM application without the need for a format translation. Locate the native type of data transfer in Figure 5–28.

Another popular CAD file format used to translate drawing files between different brands of microcomputer CAD software is the .dxf (drawing interchange file) file format. Developed by the authors of AutoCAD, the .dxf format is accepted by most CAM software. For example, the AutoCAD drawing of the switch rotor displayed on the screen in Figure 5–29 was used with SmartCAM software to produce the plastic rotor pictured in Figure 5–30. The .dxf format is used most frequently with 2-D part geometry models. The 3-D features in the plastic rotor in Figure 5–31 were created by adding depth dimensions to the rotor geometry in the CAM software after the 2-D file was transferred using the .dxf format. The SmartCAM screen in Figure 5–32 shows an isometric view of the switch rotor after the Z-axis dimensions were added. The .dxf format permits part geometry created on a microcomputer-based CAD software system to be sent to CAM software to prepare NC or CNC programs to cut the part. Inside the CAM software, a .dxf translator changes the .dxf file to a format that is native to the CAM system. Locate the .dxf file transfer in Figure 5–28.

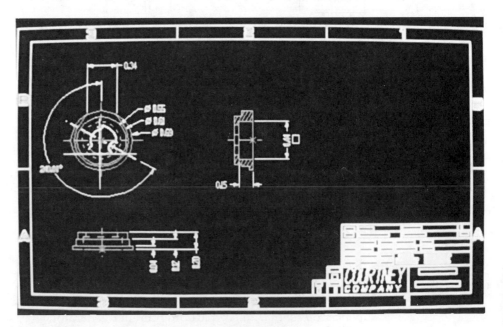

Figure 5–29   AutoCAD Drawing of Part for CAM System.

Figure 5–30   Milled Rotor.

Figure 5–31   SmartCAM
Two-Dimensional Representation of Rotor Part.

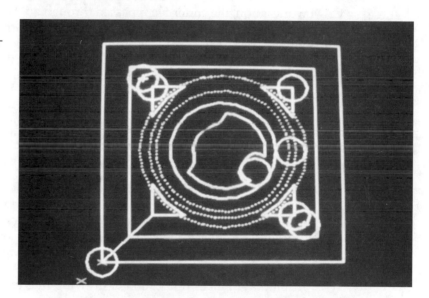

Figure 5–32   SmartCAM Three-
Dimensional Representation of
Rotor Part.

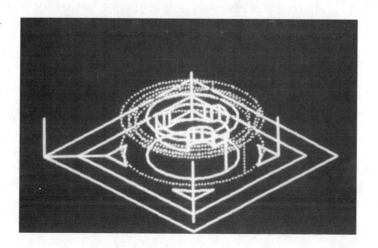

The *automatically programmed tool* (APT) file format developed in 1956 at MIT and COMPACT II developed by Manufacturing Data Systems Inc. in Michigan are two popular part-programming languages in the United States. While .dxf is the common file format for transfer to CAM on microcomputer software, APT is used more frequently on mainframe and RISC-based CAD/CAM solutions. APT and COMPACT II are NC programming languages with a format similar to the Fortran language. A sample APT program to mill a pocket is presented in Figure 5–33. After the program in Figure 5–33 is complete, a generalized solution in terms of a series of cutter location points, called a *CL* file in APT, is created. The cutter file is then passed through a postprocessor, resident in the host computer or a CAM workstation, to create the NC program code. If the cutter file is sent to the CAM workstation, the cutter file is passed through an APT/COMPACT II processor to convert the file into a format compatible with the CAM software in the workstation. The CIMpro software from Intercim Corp. is an example of CAM workstation software that supports transfer of APT CL files and also offers a programming language called Intercim APT. Other NC programming languages in use include:

- *ADAPT and AUTOSPOT:* an IBM part programming language
- *UNIAPT:* small computer version of APT
- *MAPT (Micro-APT):* a microcomputer version of APT

Before continuing, study Figure 5–28 and locate the APT file transfer path.

The *initial graphics exchange specification* (IGES), jointly developed by industry and the National Institute for Standards and Technology in the 1970s, is the most widely used file format for CAD data exchange for mainframe and RISC-based CAD software systems. For example, the CATIA software described earlier supports the IGES standard. The complex drawing files usually associated with this type of computer makes translation of drawing files between different software vendors difficult without several errors. While far from perfect, the IGES standard is the best supported common format for 3-D CAD models at the present time. When IGES is used to bring part geometry into the CAM software, an IGES translator is required. The translator converts the standard IGES file format into a format compatible with the CAM software. Locate the IGES file exchange in Figure 5–28.

The last two file formats listed, the *product data exchange standard* (PDES) and *computer-aided acquisition and logistics support* (CALS), are more recent efforts to develop robust standard data part formats that extend beyond part geometry. Some projects have been undertaken to use these standards in the CAD/CAM interface, but widespread use in this application has not occurred. The CALS standard was adopted for use in U.S. government purchasing applications and is used by companies working on government contracts. Again, a translator is necessary in the CAM system to convert these file formats to one acceptable to the CAM software. Locate these last two formats in Figure 5–28.

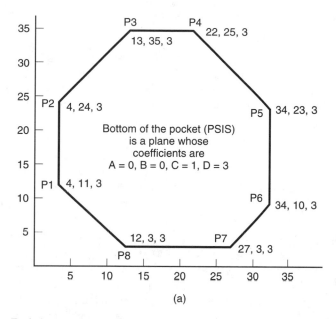

(a)

Pocket part program

```
1   REMARK POCKET POLYGON COLLAPSE DEMONSTRATION TEST
2   NOPOST $ $ NO POSTPROCESSING FOR THIS TEST CASE
3   CLPRNT $ $ PRINT CUTTER CENTER DATA
4   TOLER/.001 $ $ TOLERANCE BAND
5   $ $ POINTS DEPINING POCKET PERIMETER
6   P1 = POINT/4, 11, 3 $ $ STARTING POINT OF POCKET DEFINITION
7   P2 = POINT/4, 24, 3
8   P3 = POINT/13, 35, 3
9   P4 = POINT/22, 35, 3
10  P5 = POINT/34, 23, 3
11  P6 = POINT/34, 10, 3
12  P7 = POINT/27, 3, 3
13  P8 = POINT/12, 3, 3 $ $ ENDING POINT OF POCKET DEFINITION
14  H = .01 $ $ SCALLOP HEIGHT MAXIMUM
15  D = .38 $ $ CONSTANT CUTTER DIAMETER
16  CR = .19 $ $ CUTTER CORNER RADIUS
17  D2 = SQRTF ((D * B) - B ** 2) $ $ A BALL END MILL EFFECTIVE CUTTER RADIUS
18  CV = (4 * D2)/D $ $ A MEASURE OF POCKETING CUT OFFSET
19  CUTTER/D, CR $ $ BALL END MILL
20  FEDRAT/50 $ $ MODAL FEED RATE
21  FROM/0, 0, $ $ STARTING CUTTER POSITION
22  GO TO/20, 20, 5 $ $ MOVE CUTTER TOWARD AND OVER CENTER OF POCKET
23  PSIS/(PLANE/0, 0, 1, 3) $ $ BOTTOM PLANE OF POCKET
24  POCKET/D2, CV, CV, 3, 10, 10, 1, 1, P1, P2, P3, P4, P5, P6, P7, P8, $ $ STATEMENT
25  GO DLTA/0, 0, 2 $ $ CLEARANCE POSITION OF CUTTER
26  GO TO/0, 0, $ $ END CUTTER POSITION
27  FINI $ $ END OF PART PROGRAM
```

(b)

Figure 5–33   Part and APT Program.

Source: Chang/Wysk/Wang, Computer-Aided Manufacturing, © *1991, p. 275.*
Reprinted by permission of Prentice Hall, Upper Saddle River, New Jersey.

177

The technology just described used the CAD/CAM interface to move the part geometry required by the production equipment to the CAM *file format box* in Figure 5–28. This transferred part geometry file is used for two functions: (1) display of the cutting file for evaluation, and (2) preparation of a program to make the part on the production machine. In the first situation, the part geometry file is converted to a *plot file* that can be displayed on a computer screen or plotted on a paper plotter. Check the process shown in Figure 5–28 for the screen and plotter output. The computer screen from a SmartCAM program shown in Figure 5–34 indicates the tool paths to produce the part pictured in Figure 5–30. Screen outputs are used to check the NC program code before it is sent to the production machine. Note that a single cutter size is used and all the cutter motion is represented. The example used in the last several figures started with a 2-D CAD file. CAM packages also support 3-D either through a .dxf file transfer or by creating the 3-D graphics directly in the CAM software. Figure 5–35 shows a 3-D mold created in SmartCAM.

The second function, preparation of the machine program, has several additional steps. To prepare a machine program, the part geometry file is merged with a *tool file* and a *job plan*. The tool file has a list of available tooling with the appropriate tool offsets; the job plan includes a recommended sequence of operations for a specific type of production machine. The completed part geometry file is presented to a *postprocessor* that converts the file to a sequence of machine codes. Each

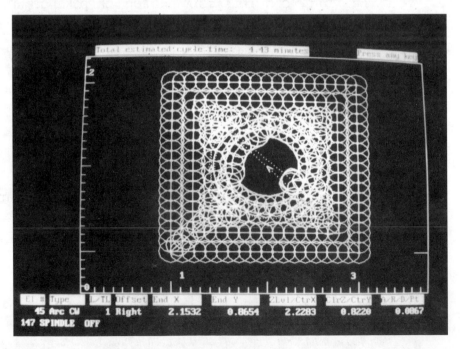

Figure 5–34   SmartCAM Screen of Rotor Part and Cutter Path.

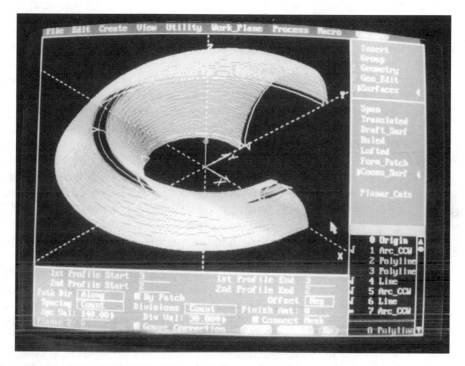

Figure 5–35   Mold Cavity Created Using 3-D Graphics in SmartCAM Software.

production machine vendor has machine-specific postprocessor software, so machine code is generated to cut the part on a specific type of CNC production machine. The completed production program is sent to the machine on the shop floor over a local area network, through a direct link to the machine or on permanent storage media. In the past the permanent storage medium was paper or Mylar tape, but current systems use magnetic storage devices.

Programming discrete-part production machines is the most common application of CAM technology on the shop floor. The machines that are programmed to produce a product from CAD geometry include most of the metal-cutting machines: mills, lathes, turning centers, punch presses, and nibblers; and boring, electronic discharge milling (EDM), and laser cutting machines. In addition, CAM includes a host of other computer-controlled applications on the shop floor, too broad a range to list and describe. The linking of shop floor and continuous process production areas to the enterprise computer system is easy to accomplish. Through this link, a two-way data exchange occurs between the machines and process and the departments controlling the production. The control of the many different systems and processes on the shop floor is possible only because of the rich selection of CAM software available to manufacturing.

Production and Process Modeling

The analytical models of parts, products, and manufacturing systems use mathematical equations that uniquely describe the behavior of the manufactured parts and production systems under study. The part or system is analyzed over a range of operating conditions by forcing the part or system variables in the equations to vary between their limits. When the variables reach worst-case conditions, the simulation provides a look at the part or manufacturing system under the most stressful conditions. The stress imposed by the worst-case environment may cause a catastrophic failure in the part, product, or manufacturing system. For example, consider the design of the aluminum alloy struts for the landing gear of a commercial jet aircraft. In the best-case scenario, the full weight of the aircraft must be supported by the struts when the plane is sitting at rest. However, the struts must be sized to handle the worst-case stress conditions that occur during landing and braking. The lightest design that can handle the worst-case condition with some safety margin is required. Subjecting an analytical computer model of the strut to the most vigorous landing conditions results in an optimum design and a significant reduction in engineering changes when the strut is sent to manufacturing. The techniques described earlier in the finite-element analysis section are additional examples of analytical models. Analytical models are applied to production problems as well.

Figure 5–36 illustrates an example of an analytical computer model used to analyze the design of a manufacturing cell including an industrial robot. The 3-D wireframe models of the cell and robot are tested using kinematic motion analysis to determine the optimum position for the robot and associated cell machines and hardware. For some types of parts and for some elements of the production system, the equations required for the model are well defined and understood. However, the previous example illustrates two important points for analytical models: (1) the use of models is a cost-effective and fast method of determining the limits of a design, and (2) the results obtained from the analytical model are valid only to the degree that the equations in the model truly describe the part, product, or manufacturing system. The last point is important to consider because manufacturing is a highly complex system of machines and operations; as a result, developing the equations to describe the operation accurately is often difficult. When a computer model is not possible, *simulations* offer a reliable alternative to analytical modeling of the more complex parts and manufacturing systems.

Production and Process Simulation

The construction of a physical model to analyze product and process behavior is called *prototyping*. To avoid the high cost of building this physical model, many manufacturers use computer-generated models called simulations to study the system. In manufacturing, *computer simulation* is defined as follows.

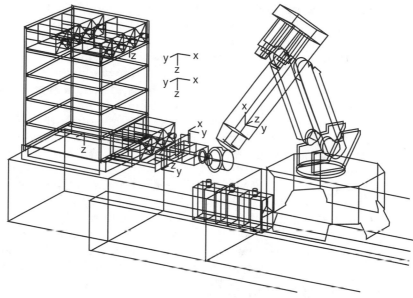

Figure 5–36   Screen Output for Work Cell Simulation.

Source: James A. Rehg, Introduction to Robotics in CIM Systems, Fourth Edition, © 2000, p. 118. Reprinted by permission of Prentice Hall, Upper Saddle River, New Jersey.

> *Computer simulation is the development of a theoretical or graphical model of a process or production system to evaluate behavior under varied conditions and in changing environments.*

Two computer simulation techniques are used to test manufacturing systems: discrete-event and continuous.

Discrete-event simulations use symbols to represent objects and resources, such as parts and machines, in the construction of a manufacturing model. The interaction between symbols is defined mathematically or with a logical relationship. Since the mathematical expression is only for a part of the manufacturing system, the complexity of the problem is reduced. The computer simulates the operation of the modeled process by capturing the operational data for objects and resources over time. The time for those intervals is established in the simulation by the computer clock. For example, if a machining work cell produces a finished part every five minutes, the output from the cell occurs at discrete intervals. The operational state of the objects and resources in the manufacturing system are saved in a series of variables, called *state variables*. The manufacturing system is modeled by establishing the relationships between the symbols, defined in the simulation program, and an event calendar with the times for discrete events. Examples of events include job_arrival, begin_operation, end_of_operation, machine_breakdown, and machine_repair. For example, if the event calendar triggers a job_arrival event, production is initiated with the relationship between the part and the machine determining when the end_of_operation event is triggered. The time between the two events is the production time. Parameters embedded into the simulation software permit the production time to be affected by tool wear, setup problems, and other events, such as machine_breakdown. As the simulation executes, the statistical data associated with the objects and resources are captured in the state variables. Analysis of the state variable helps solve manufacturing problems plaguing the shop floor or identifies problems before the production system is built. For example, if the queue time for parts at machines is saved as a state variable, the queue variables with large values identify bottlenecks in the flow through the production system. In addition, many of the new manufacturing simulation programs provide a graphic overview of the production facility. The production flow is illustrated in color on the computer monitor, with problems highlighted for easier recognition.

Discrete-event simulation models use two programming methods: general-purpose computer languages, such as C and Pascal, or languages designed specifically for simulations. Commonly used simulation languages include:

- General Purpose Simulation Systems (GPSS)
- General Activity Simulation Program (GASP)
- Simulation Language for Alternative Modeling (SLAM)
- Research Queueing Package (RESQ2)

The simulation languages do not require the programming proficiency needed for languages such as C and Pascal; however, the special-purpose languages often fit one class of simulation better than others. Therefore, as the simulations become complex and sophisticated, the user is forced to adopt more powerful general-purpose languages.

Continuous processes require a different type of simulation strategy because the state of the system changes continuously over time. Examples of continuous systems include production of synthetic fiber, rubber, and many petroleum-based products. The continuous systems are modeled using the mathematical and logical relationship between production components described in the last section. In addition, the model includes one or more differential mathematical equations that describe the rate of change in the state variables with respect to time. This last element in the model makes the continuous process different from the discrete-event technique. State variables are used to capture the critical production statistical data with the values plotted in real time or recorded at a specified sampling rate. In the past, all complex continuous process simulations were performed on analog computers, but now most of the simulations are executed on high-speed digital machines.

The advantages of manufacturing simulation include:

- The optimum solutions for the manufacturing layout and production flow are identified.
- Alternatives (new products) and changes (quantity and mix of products) in manufacturing are evaluated rapidly.

- Production problems associated with product flow and material movement are identified.
- Manufacturing performance under various production rates is easily studied.
- Decisions that can change the manufacturing environment or product flow are analyzed before implementation.

The applications of discrete-event and continuous manufacturing simulations will increase as the price-to-performance ratio of computer platforms continues to fall.

Maintenance Automation

Machine and plant maintenance are major cost centers in manufacturing. The dollars spent in this area add no direct value to the products produced in the enterprise; however, if maintenance is ignored, a successful CIM implementation would be impossible to sustain. Automation in the maintenance area can include both hardware and software. The hardware includes better tools that are designed to shorten the time required to do plant and machine maintenance. Software designed to improve plant and machine maintenance has two components: (1) faster and more accurate identification of malfunctions in production hardware and manufacturing systems and (2) better management of the maintenance operations.

Artificial intelligence techniques in the form of *expert systems* permit the development of maintenance software to assist maintenance personnel in identifying machine problems more quickly and easily. An expert system is a computer program containing a series of rules that mimic the logical processes of an expert. For example, an expert system written for troubleshooting would use the same logic that a skilled troubleshooter with years of experience would follow in finding a faulty circuit in a production machine. When less experienced technicians use this expert system software, they are helped by the experienced troubleshooter's logical set of measurements to identify the bad component or part. Expert systems or similar software is often embedded into the computer that controls a machine so that some internal faults can be diagnosed by the machine and reported to maintenance. In some cases, faulty machines are connected to the phone line through a modem, and a computer in an off-site location, sometimes another state, checks the machine and identifies the problem area.

The other area of maintenance automation focuses on improving the efficiency of the management of the maintenance operation. Software in this area performs functions that include scheduling of all planned and preventive maintenance, building databases of equipment and production machine part numbers, statically tracking machine maintenance cost, tracking the use of supplies and spare parts in maintenance inventory, supporting bar code input of machine numbers for annual capital equipment inventory checks, and producing a host of reports that help determine how maintenance operations can improve.

Production Cost Analysis

Establishing accurate product costs is critical in all five of the manufacturing systems (project, job shop, repetitive, line, and continuous) because price is an order-winning or order-qualifying criterion for many products. The accuracy of the cost estimate is especially vital in the first two manufacturing systems, project and job shop, because every job is a new product or design. In addition, a mistake in estimating fixed or variable cost for the product or job is not buffered by long production runs and large repeat orders. It is an equally important factor in the four production strategies, especially in the assemble-to-order and make-to-stock area. When customer lead time is so short that products must be available from stock or assembled from completed subassemblies, the work-in-process and finished goods inventory costs rise. The first step in controlling these inventory costs is an accurate determination of product cost in all phases of manufacturing. The information provided in this section focuses on hardware and software use in the analysis and evaluation phase of the product design model to estimate product cost.

The ambiguity present in the CIM model becomes more apparent with increased understanding of the technologies present. The benefits of specific types of hardware and software are a function of how effectively the organization uses the results across all enterprise departments. The integration of enterprise systems makes the lines that separate departmental functions and responsibilities less clear. For example, the design for manufacturing and assembly (DFMA) analysis described earlier used computer software to force a good design for assembly and manufacturing. The product cost information generated from the DFMA software is critical for accurate cost estimating. Study Figures 5–4, 5–9, and 5–10 again and note the wealth of cost data present. The DFMA analysis of machining operations on the spindle/housing assembly (Figure 5–37) also provides cost data in the metal-working area. In the past, large manufacturing operations would have estimators working separately from the design area. The advent of concurrent engineering and the integrating effect of the CIM software force a desegregation of many enterprise areas. For example, the department responsible for loading the production manufacturing and control software with manufacturing cost data must work closely with engineering to obtain reliable values.

Cost estimating in the job shop is especially critical for the following reasons:

- The number of quotes for new parts is high, with some shops quoting over 5000 different parts a year.
- The response time on quotes must be fast, in some cases within twenty-four hours.
- The direct labor cost is often a significant part of the total manufactured part cost.
- Price is often the order-winning criterion.

The shortage of experienced estimators and the inconsistencies present in the manual system create many errors. Products that introduce consistency into estimating in

Worksheet from Machining Software

Spindle   Material: stainless steel— ferritic free-machining   Machine: manual turret lathe

Operation number	Operation <R>ough or <F>inish	<S>teel <D>isposable <C>arbide <G>rind	Setup time (hours)	Load/unload time, etc. (seconds)	Number of operations	Set tool, engage cut, etc. (seconds)	Volume (in.^{3})	Area (in.^{3})	Machine Time (seconds)	Operation cost(s)
1 Face	F	C	1.41	29	1	9	—	0.15	0.66	0.30
2 Cylindrical turning	R	C	0.22	—	1	9	0.03	0.28	2	0.08
3 Cylindrical turning	F	C	—	—	1	9	—	0.28	0.50	0.07
4 Cylindrical turning	R	C	—	—	1	9	0.28	2.98	16	0.19
5 Cylindrical turning	F	C	—	—	1	9	—	2.98	4	0.10
Batch 10,000 Totals			3.48	58		81	0.35	7.50	27	1.29

Cost/part($):   Material = 0.26   Setup = 0.01   Operations = 1.29   Total = 1.56

(Move indicator to required row/column/page using keypad functions)
Press <INS>ERT. <DEL>ETE, <C>HANGE, < M<M>ATERIAL, <H>ELP, OR <O>K

Figure 5–37   DFMA Cost Analysis for Spindle Machining.
Source: Boothroyd/Dewhurst, "Product Design for Manufacturing and Assembly," © April 1988, Manufacturing Engineering Magazine, *p. 46. Reprint by permission of the Society of Manufacturing Engineers, Dearborn, MI.*

the job shop setting are available. The systems automate the estimation process for metal-cutting operations by integrating a computer database with standard production times and costs with a data-entry device to get basic part shapes into the computer. On one of the systems, the human operator traces the outline of the part on a drawing and interacts with the software by answering questions about the part and production process. The software searches through extensive databases of production machines and a broad range of standards to extract manufacturing costs tailored to the specific machines and direct labor standards in the job shop. The geometric analysis and mass properties commands present in CAD software also provide data important for accurate estimates of material and production cost.

5–6 DESIGN AND PRODUCTION ENGINEERING NETWORK

The CAD, CAE, and production engineering automation, described in Chapters 4 and 5, places two demands on the enterprise infrastructure: (1) easy, accurate, and instantaneous movement of part geometry files and product data between departments in the enterprise; and (2) a single, common database for all enterprise information, part files, and product data. To satisfy the last two conditions, the enterprise must have the automation systems and computers in all department areas linked to common data storage through an *information* and *data network.* Planning for electronic data communications in the enterprise must extend beyond internal divisions. The competitive nature of world markets demands data links with external vendors, equipment suppliers, and technology service bureaus. The most frequently used technology for external networks is electronic data interchange (EDI). The rapid development of the Internet makes links between companies fast, inexpensive, and reliable, resulting in significant changes in business models in many product sectors.

The Basic Enterprise Network

Figure 5–38 describes a typical enterprise network; be sure you are familiar with the layout and terms before continuing. Make frequent reference to the figure while the general operation of the enterprise network configuration is described. The best place to start is by defining an *enterprise network.*

> *An enterprise network is a nonpublic communications system that supports communications and the exchange of information and data among various devices connected to the network over distances from several feet to thousands of miles.*

The enterprise-wide network is usually not a single long network; instead, the total network is several smaller networks connected to a main network called the *backbone.* Each of the separate networks, called *local area networks* (LANs), has an *area controller* or *network server* to handle the local administration of network protocol and to provide users with network services. Networks that have similar protocols and operational software are connected together with a computer called a *bridge, repeater,* or *router.* Some suppliers of network hardware differentiate among the functions of bridges, repeaters, and routers. However, for this overview, those three devices are considered as devices to link LANs with similar operation software. Therefore, the network backbone uses devices to tie the system together and to allow any user on any LAN to get access to the *mainframe central database* or data stored on any area controller or server in the system. In some true peer-to-peer network systems, common data on all computers in the network is accessible from any network station. In addition, the network is not limited to the confines of the enterprise because of the *gateway, remote bridge,* and *modem* capability shown. The gateway is used to connect networks with different protocols or operating software. The Internet and electronic data interchange (EDI) networks would be interfaced to the enterprise with gateway computers. The remote bridges connect networks with similar protocols and operational software that are separated by large geographic distances. For example, a company could have corporate offices in South Carolina and plants in New York. The networks in each location would be tied together through remote bridge computers connected to high-speed digital data lines called *T1 carrier lines.* The separation between the facilities would be transparent to the users; for communications and data exchange, it would appear

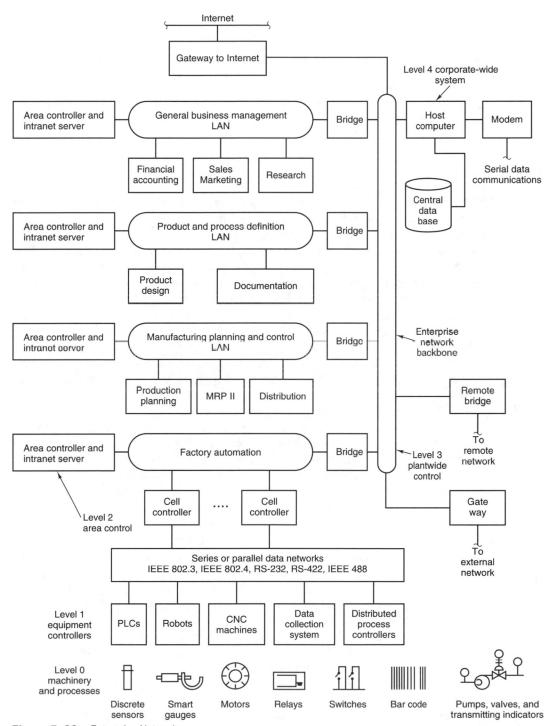

Figure 5–38 Enterprise Network.

that the distant network was in the same office area. The final external interface illustrated in Figure 5–38 is a *serial data communications* interface that uses a *modem* to convert computer data into a standard serial data stream for transmission over standard phone lines. The most popular standards for connecting between the host and the modem are the *RS-232* and *RS-422* interface definitions.

5–7 SUMMARY

The term *computer-aided engineering* has had many definitions since its inception in the early 1970s, but today it clearly means the analysis and evaluation of engineering design using computer-based software to calculate product parameters too complex for manual calculations. CAE is used in the design process from synthesis to evaluation and supports concurrent engineering efforts directly. CAE provides productivity tools to aid the production engineering area in group technology (GT), computer-aided process planning (CAPP), and computer-aided manufacturing (CAM).

The design for manufacturing and assembly (DFMA) process considers the production factors from the beginning of the design and is a major CAE technology capable of significant saving in production costs. At the analysis phase in the product design model, CAE uses the geometry file from CAD as the model for the analysis. The applications at this stage of the design include finite-element analysis and mass property analysis. Other CAE analysis tasks include software for circuit analysis and assembly interference, fit, and tolerance.

The CAE activity at the evaluation stage in the product design model includes examination of the data collected through all of the analysis processes and the building of prototypes of the design. The newest prototyping technique, called rapid prototyping, quickly constructs full-size models of products directly from the geometry stored in the CAD model. The two rapid prototyping techniques used most frequently are stereolithography and fused deposition modeling.

Group technology (GT), the grouping of parts based on design or manufacturing factors, is a technique used to support automation of the production engineering area. In GT, parts are coded and classified using one of the three following techniques: hierarchical, chain, or hybrid. With parts grouped by similar codes, creation of production cells to produce similar parts efficiently is possible. The major production engineering strategies used to increase productivity include computer-aided process planning (CAPP), computer-aided manufacturing (CAM), production modeling, maintenance automation, and automated product cost analysis. Of these technologies, CAM is the most widely used, with software implementations on all three computer platform types.

The enterprise must have the automation systems and computers in all department areas linked to common data storage through an information and data network. The CIM enterprise requires a central management facility to handle common product data and communication across all the internal and external entities in the organization. Electronic data interchange (EDI) extends the network beyond the boundaries of the enterprise to external vendors, equipment suppliers, and technology service bureaus. Specifically, the product and process segment described in Chapters 3 through 5 must be linked through a network to achieve the productivity required for survival.

REFERENCES

BOOTHROYD, G., and P. DEWHURST, "Product Design for Manufacture and Assembly," *Manufacturing Engineering,* April 1988, pp. 42–46.

CHANG, T. C., R. A. WYSK, and H. P. WANG, *Computer-Aided Manufacturing.* Upper Saddle River, NJ: Prentice Hall, 1991.

GETTLEMAN, K. M., "Stereolithography: Fast Model Making," *Modern Machine Shop,* November 1989.

GRAHAM, G. A., *Automation Encyclopedia.* Dearborn, MI: Society of Manufacturing Engineers, 1988.

GROOVER, M. P., *Automation, Production Systems, and Computer-Integrated Manufacturing.* Upper Saddle River, NJ: Prentice Hall, 1987.

MITCHELL, F. H., Jr., *CIM Systems: An Introduction to Computer-Integrated Manufacturing.* Upper Saddle River, NJ: Prentice Hall, 1991.

SOBCZAK, T. V., *A Glossary of Terms for Computer Integrated Manufacturing.* Dearborn, MI: CASA of SME, 1984.

WELTER, T. R., "Designing for Manufacture and Assembly." *Industry Week,* September 1989, pp. 79–82.

WOHLERS, T. T., "Make Fiction Fact Fast." *Manufacturing Engineering,* March 1991, pp. 44–49.

QUESTIONS

1. Define *CAE* in your own words.
2. Describe how CAE fits into the product design model developed in Chapter 3.
3. Define *DFMA* in your own words.
4. Why is the combination of DFMA more effective than the use of DFM or DFA separately?
5. What is the major justification for using DFMA?
6. What is the difference between FEA and mass property analysis?
7. Define *FEA* in your own words.
8. What two categories of FEA software are used most frequently?
9. In mechanical applications of FEA, describe the process used to analyze a part for structural integrity.
10. Describe three of the six FEA software application areas outlined in the chapter.
11. Describe the part of the FEA process most critical for valid analysis data.

12. Describe mass property analysis and list three typical parameters provided.
13. Describe the Valisys software and list the type of features that are useful to the analysis and evaluation process.
14. Describe the differences between prototyping and rapid prototyping.
15. Compare and contrast the operation of the different types of rapid prototyping systems.
16. Define *group technology* in your own words.
17. Describe the hierarchical type code used to classify GT parts.
18. Describe the chain code used to classify GT parts.
19. What are the advantages and disadvantages of each of the GT code types?
20. Compare GT production cells with the machine production layout used in a conventional job shop. What advantages does GT offer over the conventional setup?
21. Describe the differences between the two techniques used to develop routings in CAPP.
22. What are expert systems and how can they improve the CAPP process?
23. Define CAM in your own words.
24. Describe the process used to manufacture a part on a CNC machine using CAM technology.
25. Compare the DXF, IGES, PDES, and CALs file transfer protocols and describe the differences.
26. Describe the difference between process modeling and process simulation.
27. Compare the two simulation techniques: discrete-event and continuous.
28. How can expert systems help in maintenance automation?
29. Why is cost estimating especially critical in a job shop manufacturing system?

PROBLEMS

1. List the major order-winning criteria, presented in Chapter 1. Next to each criterion, list the CAE technologies presented in this chapter that support that order-winning criterion.
2. Select a group of three to five similar products that are easily disassembled and determine how many of the ten DFA guidelines listed in Appendix 5–1 are violated. (The click-type ball-point pen is an example of a product group to analyze.)
3. Select a moderately complex consumer product or a subassembly of a complex product. Apply the assembly method scoring chart shown in Figure 5–2 to the product and determine the total score.
4. Using the product from Problem 3, apply the ten DFA guidelines and determine how many were violated. Using these results, suggest design improvements that would reduce the number of violations. Apply the assembly method

scoring chart shown in Figure 5–2 to the improved design and compare this score with the original score from Problem 3.

5. Select a group of products or parts and design a hierarchical code that could be used to group objects into common categories. (Use the parts from Problem 2, if possible.)

6. The chain code for a part is 121322. Use the polycode structure in Figure 5–19 to determine and list the features present in the part.

PROJECTS

1. Using the list of companies developed in Project 1 in Chapter 1, create a matrix that compares the CAE technology in use at each of the companies.

2. Using the three companies selected in Project 3 in Chapter 4, determine what CAE technology could be installed that would improve design efficiency and productivity. List any changes that would be required, including changes in the design process and CAD system.

3. Select one company from the list in Project 1 that includes assembly as part of the manufacturing process. Apply the ten DFA guidelines listed in Appendix 5–1 and the assembly method scoring chart (Figure 5–2) to one of the assemblies. From the results, suggest changes that would improve the assembly process and product.

4. Using CAD software available at the college, draw several regular and irregularly shaped 2-D or 3-D objects. Use the mass property analysis command for the CAD software to generate mass property data for the objects.

5. From the list in Project 1, select three companies that route parts on the shop floor. Describe the technique used by the process planners to route the parts to production machines. Using copies of sample routing sheets from each company, compare the various routing processes. Determine if CAPP would be useful for these companies.

6. From the list in Project 1, select all the companies that use CNC machines, and identify the type of CAM software and system used.

7. Select three companies from the list in Project 1 and prepare a report on how they generate product cost data.

8. Select several companies from the list in Project 1 that have computer networks installed. Prepare a report describing how the networks are used and what departments are served by the network system.

9. View the SME video "Design for Manufacturing" and prepare a report that describes the benefits of DFMA at one of the following companies: Storage Technology, Caterpillar, Xerox, or IBM.

10. View the SME video "Simulation" and prepare a report that describes the benefits of manufacturing simulation at one of the following companies: Rohr Industries, Intel, or General Electric.

11. Using data from the CAD Web sites listed in Appendix 4–2, write a paper that describes how the CAD software from one of the CAD vendors integrates with the CAE functions covered in the chapter.

12. Identify two CAD companies listed in Appendix 4–3 that have an integrated CAM solution and describe the CAM functions offered.

13. Develop a list of application software for CAD, CAM, and CAE using the Web sites listed in Appendixes 4–2 and 5–2 that would support the design and manufacturing engineering departments of a small, medium-size, and large company.

14. Develop tables that compare the software capability provided by vendors for the following integration software: computer-aided manufacturing (CAM), finite-element analysis (FEA), and product data management (PDM).

15. Compare the operation and features of the HMS CAPP software listed in Appendix 5–2 with the METCAPP software described in the chapter.

APPENDIX 5–1: TEN GUIDELINES FOR DESIGN FOR ASSEMBLY*

1. *Minimize the number of parts.* Combine or eliminate parts whenever possible. Combine parts in an assembly that do not move relative to each other unless there is strong justification otherwise.

2. *Minimize assembly surfaces.* Reduced surface processing results from reduced surface areas.

3. *Design for top-down assembly.* Establish a base part on which the assembly is built, and provide for assembly in layers from above the base. Insertion of parts from above takes advantage of gravity and usually results in less costly tooling and fewer clamps and fixtures. Use subassemblies to avoid violation of this rule.

4. *Improve assembly access.* Design for easy access, unobstructed vision, and adequate clearance for standard tooling. If possible, allow parts to be added to the assembly in layers.

5. *Maximize part compliance.* Design with adequate grooves and guide surfaces for mating parts. Use the ANSI Y14.5M 1982 standard "Geometric Dimensions and Tolerancing" to ensure compliance between mated parts after processing.

6. *Maximize part symmetry.* Symmetrical parts are the easiest to orient and handle. Where symmetry cannot be included, design in obvious asymmetry or alignment features.

7. *Optimize part handling.* For easier handling, design parts that are rigid rather than flexible, have adequate surfaces for mechanical gripping, and have barriers to prevent tangling, nesting, or interlocking.

*Adapted from an article "Designing for Manufacture and Assembly" by Welter (1989).

8. *Avoid separate fasteners.* First, fasteners should be eliminated by using snap-fits and other strategies. Second, the number of separate fasteners should be reduced to a minimum. Third, the fasteners used should be standardized to reduce variation and ensure availability.

9. *Provide parts with integral self-locking features.* Use tabs, indentations, or projections on mating parts to identify and maintain orientation through final assembly.

10. *Focus on modular design.* When parts have a common function, use a standard component or module; when parts must be interchangeable, use a common or standard interface.

APPENDIX 5–2: WEB SITES FOR CAE VENDORS

All the major CAE vendors have extensive Web sites that describe their products and offer other useful data for selecting CAE software. Visit the following Web sites:

CAE Vendors	Web Address	Comments
Algor	www.algor.com	A broad line of FEA, analysis, and simulation software. Select *Software Products* and *Customer Application Stories* to view products and case studies.
Ansys	www.ansys.com	Specialist in finite element analysis plus thermal, computational fluid dynamics, magnetics, and electrical-field analysis.
Boothroyd Dewhurst	www.dfma.com/software/index. html	Design for assembly and manufacturing software
Cosmos	www.cosmosm.com	FEA solutions. Select *Success Stories* for a list of case studies.
Enterprise Software Products division of SDRC	www.femap.com	Specialist in finite element analysis (FEA).
MatrixOne	www.matrix-one.com	Specializes in management of product data, documents, and configuration management tools. Select *Customer Success* to view a list of case studies.

MCS	www.mcsaz.com	CAD/CAM software. Select *Software* and *Success Stories* to view solutions.
Metaphase	www.metaphasetech.com	Specializes in management of product data, documents, configuration management tools, and Web-based user interfaces.
MSC Software	www.macsch.com	Simulation software for analysis. Select *Products* and *Success* to view simulation products and case studies.
SDRC	www.buildingblocksinc.com/smartmill.html	Distributors of SmartCAM software and CAM software.
Smart Solutions Ltd.	www.smarteam.com	Specializes in product data management.
Structural Research and Analysis Corp.	www.srac.com	FEA solutions. Select *Success Stories* for a list of case studies.
Tecnomatix Technologies	www.valisys.com	Virtual quality control and analysis software.
Teksoft	www.teksoft.com	CAM software. Select *Camworks* to view on-screen CAM examples and to download demos.

APPENDIX 5–3: WEB SITES FOR RAPID PROTOTYPING VENDORS

Rapid Prototyping Vendors	Web Address	Comments
Cubital	www.cubital.com	Solid ground curing (SGC) equipment.
Helisys	www.helisys.com/	Laminate object modeling (LOM). Select *Products* and *How People Use Us* to view products and case studies.
MIT	me.mit.edu/groups/tdp/index.html	Description of 3D printing (3DP) from the Massachusetts Institute of Technology.

Soligen	www.soligen.com	Direct shell production casting (DSPC), a form of fused depositon modeling and 3D printing.
Stratasys	www.stratasys.com	Fused deposition modeling (FDM). Select *Products* and *Customer Solutions* to view products and case studies.
3D Systems	www.3dsystems.com	Stereolithography Select *Products* and *3D Systems@ Work* to view products and an excellent list of industry cases.

Managing the Enterprise Resources

After you have completed Part 3, it will be clear to you that:

- The systems used to manage and control the production operations are part of CIM.
- The fundamentals of manufacturing planning and control are consistent with the fundamentals of CIM.
- Computer-based systems for material and capacity management are at the heart of CIM systems.
- Modern methods and techniques such as just-in-time manufacturing and lean production systems have a place in the CIM environment.
- Analysis of current enterprise operations and the elimination of unnecessary non-value-added operations must be completed early in the CIM implementation process.
- A new generation of systems integrates the core elements of material requirements planning and manufacturing resource planning that support true enterprise-wide systems for operational control.
- The continuing emphasis on serving the customer is leading to new systems for electronic business and customer resource management.

Introduction to Production/ Operations Planning

In this chapter, we introduce the production and operations planning process necessary for a successful CIM implementation. Charles Kettering, a prominent industrialist, once said: "I expect to spend the rest of my life in the future, so I want to be reasonably sure what kind of future it is going to be. That is my reason for planning." Enterprises spend their life in a rapidly changing future; as a result, they share the same need for planning as that of Mr. Kettering.

In the past, the term *production* was often associated with factories, machines, and assembly lines. The scope of the management function was narrowly focused on problems related to manufacturing, with an emphasis on the procedures and skills necessary to run a factory. Since the early 1970s, however, evolutionary changes have occurred in enterprises across the world; as a result, production methods and techniques are now applied to a wide range of activity outside of manufacturing. For example, service areas such as health care, recreation, banking, finance, hotel management, retail sales, education, and transportation use management techniques derived from experiences in production management. In addition, many manufacturing organizations apply production management techniques to service functions such as information management and distribution. As a result, the traditional production management view was expanded and the term *operations management* coined to describe better the activities performed in both manufacturing and service organizations.

In the next three chapters, we address the operation and automation needs of the operations management area. The operations management activities utilize systems and shared knowledge to provide information needed for customer support, the definition of products/processes, and manufacturing operations. Refer to the SME enterprise wheel shown in Figure 6–1. The manufacturing infrastructure is illustrated in Figure 6–2. The planning functions have formal interfaces with both the design and production departments and informal relationships with most of the enterprise. The operations management functions are a critical part of the CIM implementation.

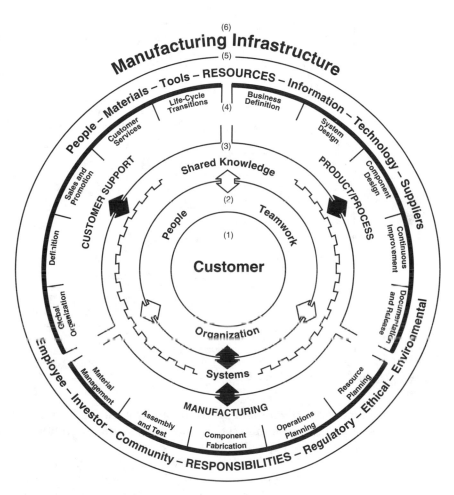

Figure 6–1   Manufacturing Planning and Control on the SME Enterprise Wheel.
(Courtesy of the Society of Manufacturing Engineers, Deerborn, Michigan, 48121, Copyright 1993. Third Edition.)

6–1 OPERATIONS MANAGEMENT

The operations management area has responsibility for the administration of enterprise systems used to *create goods* or *provide services.* The health care industry serves as an example for each area. The activities performed by hospital management, a service sector operation, includes running the hospital, managing in- and outpatient medical services, providing food service, supervising and training of staff, and housekeeping. Now compare this service side of health care to a manufacturer of health care equipment. The factory management must design new products, redesign current models, test designs, order raw material, determine product mix and quantity to produce, schedule the production machines, maintain

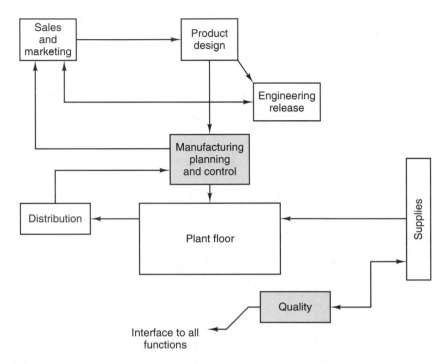

Figure 6–2 MPC in the Organizational Structure.

production hardware and software, and adjust fixed and variable resources to meet changes in the market. The manufacturing operation has a *tangible output* that can be seen and touched.

This book focuses on manufacturing operations and the management of a *product-oriented* enterprise employing CIM concepts. However, changes in the global economy and enterprises over the last thirty years require the application of CIM principles to areas outside of manufacturing operations. The principles that provide an *order-winning* advantage to the manufacturing sector work in the service area as well. Clearly, the *computer-integrated enterprise* (CIE) needs an *operations management* emphasis for the service and manufacturing segments that must coexist. While the next three chapters spotlight the management concepts required for a product-oriented operation, the principles are also applicable across other types of CIEs.

6–2 MANUFACTURING PLANNING AND CONTROL

All planning has a time horizon that dictates the number of days, months, or years into the future described by the plan. In general, enterprise planning is divided into the three levels illustrated in Figure 6–3.

The *strategic plan* is generally long range, one year to many years, and focuses, for example, on future capacity, products, and production plant locations. Planning

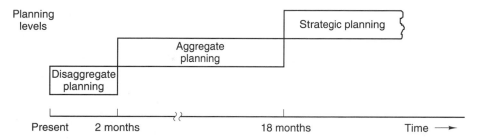

Figure 6–3   Enterprise Planning Levels.

for a new building or a production line takes many months. A strategic plan that covers two to four years should help identify long-range capacity needs while there is enough time to plan and implement them. It is fairly common for companies to consider a strategic view that looks ahead three, five, or even as many as ten years into the future. The strategic planning process is performed at the highest management level in the enterprise and the plan is usually updated annually. Tracking the progress being made toward the planned goals and objectives is an ongoing effort.

The *aggregate plan* has an intermediate-length time horizon of two to eighteen months, with an emphasis on planning levels of employment, output, inventories, back orders, and subcontractors. The shortest time horizon, *disaggregate planning*, provides short-range planning with detailed plans that include machine loading, part routing, job sequencing, lot sizes, safety stock, and specific model order quantities. The disaggregate and aggregate planning processes have broad enterprise participation and are a major part of the manufacturing planning and control (MPC) model illustrated in Figure 6–4. Study the figure until you are familiar with all the terms. An overview of each area in the MPC model is provided in the following sections, and a detailed study of some sections is provided in the following two chapters.

Introduction to Aggregate Planning

Aggregate planning with a time horizon from two to eighteen months is a "big picture" view of the production plan and includes all the functions in the top block in Figure 6–4.

> *The goal of aggregate planning is the generation of a production plan that utilizes the enterprise resources efficiently to meet customer demand.*

Planners working at the aggregate level consider how output rates, inventory levels, and back orders affect the variable and fixed resources of the enterprise. Variable resources, such as full-time and temporary employees and subcontractors, are usually expressed in total labor hours per period. The fixed resources—facilities and machines—are often expressed as capacity in total machine hours per period. At this level, similar products or models with common characteristics are combined

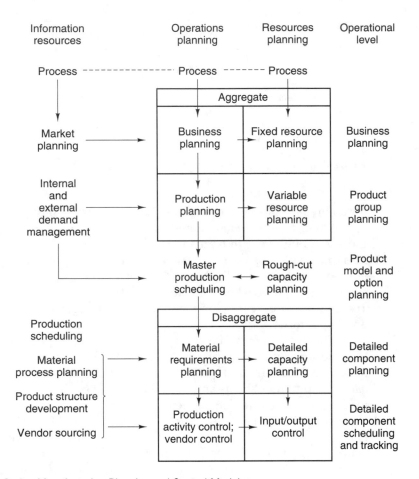

Figure 6–4   Manufacturing Planning and Control Model.

Source: Adapted from a chart from Dan Steele, University of South Carolina.

into product groups for planning purposes. For example, in the automotive indus-try, the aggregate number of cars or trucks planned would not specify two-door ver-sus four-door models, four-cylinder versus six-cylinder engines, or other options that would not affect the results of the aggregate plan. With an aggregate plan es-tablished, three logical conditions relating the forecasted customer demand and enterprise production capacity are possible: (1) demand and capacity are about equal, (2) demand exceeds capacity, and (3) capacity exceeds demand. When the first condition is present, the planners' activity shifts to matching existing capacity to forecasted demand in the most efficient manner. The last two conditions require additional action before allocating production resources. For example, additional product promotion may be necessary to bring the lagging demand up to the current level of capacity. If demand exceeds capacity, however, expanding facilities or con-tracting with other manufacturers may be necessary. In the service sector, a similar process occurs, with the demand for services balanced against capacity limitations present. Aggregate planning forces management to address demand/capacity issues early in the production/service cycle so that the most cost-effective capacity solution is applied to meet customer demand.

Introduction to Disaggregate Planning

The term *disaggregate* means *to separate into component parts.* At this planning level (Figure 6–4), a feasible aggregate plan is disaggregated into all the various models and options necessary to meet specific customer demand. The time horizon (Figure 6–3) for disaggregate planning often ranges from hours to months. The first step in disaggregation is the creation of a *master production schedule* (MPS) from the aggregate production plan. An example in Figure 6–5 from a manufacturer of fans illustrates the disaggregation process. The units listed under the aggregate production plan are similar models of exhaust fans with similar components, motors, and production processes; therefore, aggregation of these units is possible. The master production schedule in Figure 6–5, the first step in disaggregation, is required to plan the production of the individual units at the model and option level. Note that the sum of all the disaggregate quantities in a given period should be equal to the original aggregate plan quantity.

Disaggregate Planning Options

The *material requirements planning* (MRP) strategy included in the manufacturing planning and control (MPC) model in Figure 6–4 is one of several options available for component planning at the disaggregate level. The MPC strategy used is a function of the characteristics of the manufacturing system. Review the manufacturing

Figure 6–5   Disaggregation of Production Plan for Fans.

(a) Aggregate level, production plan

	Month				
	Jan.	Feb.	Mar.	Apr.	May
Planned output[a]	1500	2000	1800	2500	3000

[a]*Aggregate units.*

(b) Disaggregate level, master production schedule

Planned output[a]	Month				
	Jan.	Feb.	Mar.	Apr.	May
24″ 110 V	500	550	525	650	675
24″ 220 V	150	450	450	500	700
36″ 220 V	475	550	500	650	725
48″ 220 V	175	200	150	325	475
54″ 220 V	200	250	175	300	350

[a]*Fans by size and voltage.*

characteristics for the five manufacturing systems in Figure 2–3 and note how *process speed* and *product complexity* vary under different manufacturing systems. Now compare those manufacturing characteristics in Figure 2–3 with the MPC disaggregate planning options included in Figure 6–6. The MPC method used to establish the master production schedule and to perform component planning at the disaggregate level is a function of manufacturing characteristics, such as process speed and product complexity.

Continuous process systems use a flow-type MPC management system to disaggregate the production plan since the product has few component parts and tracking is straightforward. The management of raw materials and distribution of finished goods often is the most complex issue in continuous process systems.

An assembly-type management system is used to disaggregate the production plan in the line and repetitive manufacturing systems, where typical products include watches, automobiles, microcomputers, and pharmaceuticals. Component complexity is higher than flow, and management of parts is required. However, the management issues in developing the master production schedule are less severe because components are coordinated through the assembly process.

The material requirements planning (MRP) system addresses the need for parts management of complex products and product mixes with relatively high rates of production. Figure 6–6 indicates that the MRP management system is used in some line, repetitive, and job shop settings. Products with hundreds of parts and productions systems with thousands of purchased and manufactured items use MRP to disaggregate the production plan. The MPC model in Figure 6–4 supports the line, repetitive, and job shop production systems using MRP to disaggregate the production plan. A different model would be required to describe the operation of flow and project manufacturing systems.

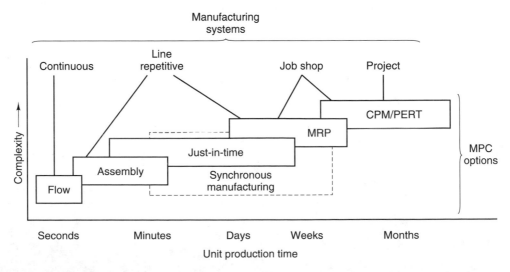

Figure 6–6   Comparison of MPC Disaggregate Options and Manufacturing Systems.

The just-in-time (JIT) and synchronous manufacturing management systems overlap the disaggregation options that support the line, repetitive, and job shop manufacturing systems. JIT and SM systems are described in detail in Chapter 9; however, for the present overview, *just-in-time* is defined as a component planning process that encourages shorter cycle times, lower inventory, and reduced lead and setup times. *Synchronous manufacturing* or *drum–buffer–rope* (DBR) is a planning technique for components or products moving through multiple production stations. The DBR system, a combination of JIT and MRP, uses a pull-type production system at most work cells, but a push scheduling system to build buffers where bottlenecks occur at critical points in the production flow. In general, moving the disaggregation strategy from right to left in Figure 6–6 promotes *order-winning criteria* because unit production time decreases. The practice repeated in many companies starts with implementation of MRP to gain control of the component planning activity. With MRP in place, specific parts or products are shifted to JIT and synchronous manufacturing to increase competitiveness.

The final disaggregation strategy shown in Figure 6–6, CPM/PERT, is used primarily for project work. While the project system has a large design component and often a large number of parts, the major problem is to schedule the delivery of parts at the required time over the extended assembly time. The commercial construction industry and manufacturers of ships and petrochemical plants would use this type of disaggregation from an aggregate plan.

6–3 MANUFACTURING PLANNING AND CONTROL MODEL— MANUFACTURING RESOURCE PLANNING (MRP2)

The MPC model in Figure 6–4 indicates that all production planning is either global at the aggregate level or detailed in the disaggregate group. The *operational level column* emphasizes the change in planning scope, with intermediate-range business planning occurring at the top and detailed component planning, scheduling, and tracking located in the disaggregate group. The aggregate and disaggregate processes are divided into two sections: *operations planning* and *resources planning*.

A model for a formal planning system that links high-level business decisions, the planning of material and physical resources, and the detailed schedule for the shop floor was developed in the early 1980s. APICS (formerly known as the American Production and Inventory Control Society, now shortened to the acronym APICS) has worked to develop and document this model. The model continues to be a solid depiction of the integrated central planning process. Several variations have emerged, and new methods and techniques such as just-in-time (JIT) manufacturing and enterprise resource planning (ERP) have been developed since the original model shown in Figure 6–7 was conceived. The model has been able to stand the test of time. It depicts clearly many of the functional relationships among plans, actions, databases, and priority schedules. High-level planning starts at the top. Plans for material and physical resources are in

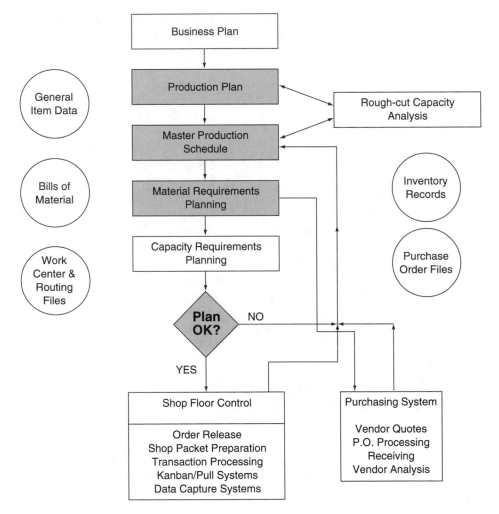

Figure 6–7   A Model for a Manufacturing Resource Planning (MRP2) System.

the middle of the diagram. Detailed output of the planning processes, such as the shop schedule, purchase orders, and transaction reporting, is found at the bottom of the model. The model does not describe the mechanics of how detailed plans are to be made or how the shop floor is to be managed and controlled. The focus is on the relationship of the critical planning and scheduling functions and the data that support them.

Note that the model does not try to show all the connecting relationships between the planning functions (the boxes in the model) and the primary databases (shown as circles in the model). Modern computer-based systems use database structures that provide the data elements to all the planning functions. An attempt to show all the links would make Figure 6–7 unnecessarily complicated.

Aggregate Planning

The aggregate operations strategy, *business planning,* is performed at the top of the process. The business planning process uses *market planning* information as a basis for the aggregate business plan. The business plan drives the aggregate *fixed resource planning* strategy, where intermediate-term requirements for facilities and production machines are determined. In addition, the business plan, along with a forecast of customer demand from the information resources column, provides projected output rates, inventory levels, and back orders per period necessary for *production planning.* The data in Figure 6–5 are examples of aggregate production planning information.

The production planning strategy provides the data to plan the *variable resources,* which include full and temporary employment levels, total labor hours per period, and number of subcontractors. In addition, the production plan, along with forecasted customer demand, provides the aggregate information from which the disaggregate *master production schedule* (MPS) is produced.

Disaggregate Planning

The development of master production schedule data is the start of disaggregate planning; however, the level of detail remains at the product group level, as the operational level column in Figure 6–4 indicates. The MPC model in this figure indicates that the master production schedule is the bridge between the aggregate plan and the detailed disaggregate plan. The MPS also drives the *rough-cut capacity planning,* where the available fixed and adjustable resources are compared to the scheduled production rates for each model and product option. The rough-cut capacity requirements of any master production schedule are usually estimated by one of the following techniques: *capacity planning using overall planning factors* (CPOF), *capacity bills,* or *resource profiles.* The differences between the MPS and available gross capacity are reconciled at this level before further disaggregation occurs. The heart of the disaggregation process for the MPC model in Figure 6–4 is a material planning and production scheduling process that includes *material requirements planning, detailed capacity planning, production activity control,* and *input/output control.* Each of these disaggregation areas is described in the following sections.

6–4 MATERIAL REQUIREMENTS PLANNING

In addition to driving the rough-cut capacity planning area, the master production schedule also provides the input data for *material requirements planning* (MRP). The MRP operation model shown in Figure 6–8 illustrates the interaction between the MRP function and other units in the enterprise. The MRP process starts with the master production schedule providing the quantity of each model or part required per period (Figure 6–5). This required production quantity of parts and products constitutes the gross requirements in the MRP system.

Figure 6–8   MRP Operational
Model.

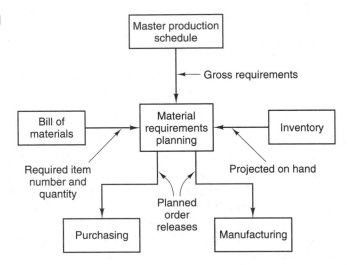

MRP System Input Data

Two additional inputs, *bill of materials* (BOM) and *current inventory*, provide critical information for an effective MRP system. The inputs from these two sources must be *timely* and *accurate* for the formal MRP system to work. Updates of the inventory control system for changes in inventory due to part movement in manufacturing or purchasing must be continuous. For example, in some manufacturing operations, parts from vendors arrive daily, so the inventory control system is updated when parts arrive to provide timely information for planners.

A *product structure diagram* graphically represents the bill of material for the end product in terms of all required component parts. For example, in Figure 6–9, product A is produced by assembling one of part B, one of part C, and one of part E. Part C is made by assembling one of part D and one of part E. The items in the product structure diagram are identified by levels, with the 0 level representing the finished product. The product structure diagram illustrates clearly the sequence required to build the product. For example, the first step in manufacturing product A is the production and assembly of parts D and E. The uncomplicated product structure shown in Figure 6–9 has three levels and five parts; however, a product structure for a riding lawnmower could have many levels and hundreds of parts.

In most MRP installations, a BOM accuracy of 95 percent or better is required, and a location and count accuracy of 98 percent is necessary for specific parts in the inventory system. The bill of materials (Figure 3–9) provides the MRP system with the *part number* and *quantity* of all parts required to build and assemble the product. The inventory control systems supplies the MRP system with the *projected on-hand balance* of all parts and materials listed on the BOM.

The planning system, whether manual or computer-aided, is driven by the data supplied to it. The critical data needed by the MRP planning system includes

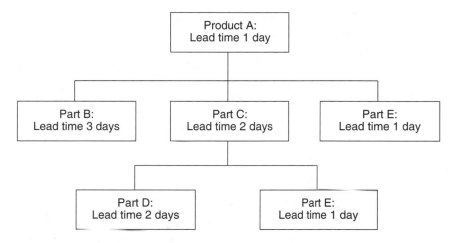

Figure 6–9 Product Structure Diagram.

manufacturing order information that incorporates the critical information in the master schedule: the specific quantities of products to be made and when they are due, and common item-related data such as the item number, description, unit of measure, group technology code, lead times, costs, and who is responsible for the item. Product structure information describes the relationships of items and the products that the company makes. Inventory records tell how much material is on hand, where it is stored, and the status of the material. Purchase order data provides information on what material is currently on order and the status of open purchase orders.

The primary lead times are of particular importance to the MRP planning process. Manufacturing time is made up of several elements that are drivers of the planning system. Setup and run times, the expected move and queue times, and the time required to get material out of inventory and to the appropriate manufacturing area are typically defined in the planning database. Purchasing lead times for buyer analysis and order placing, time for vendor processing and delivery, and time for receiving and any needed inspection are elements that are entered and maintained in the database.

MRP System Outputs

The time-phased material plan generated by the MRP planning process is a driver of the capacity plan. The formal planning system can provide a fairly detailed look at the requirements for machines and people that will be needed to produce the company's plans. Capacity requirements planning (CRP) works with the system data to complete the capacity calculations and provide a view of the capacity requirements to the system users. CRP systems often generate load profile information that shows the detailed scheduling of the work at the defined work centers. CRP does its work following the MRP run.

The MRP run produces the requirements for purchasing and production that are needed to complete the master schedule. Planned purchase orders are evaluated and finalized by the purchasing function and are then released to material suppliers. Planned manufacturing orders are evaluated by the production control function and released to specific work centers. Release of an order to a supplier or to a work center provides the authorization to do the specific task by a scheduled completion or due date. A full description of the MRP process, including the terminology and a review of the planning logic and calculations, is provided in Chapter 8.

Detailed Capacity Planning

Capacity planning and material planning are equally important in the manufacturing planning and control model (Figure 6–4). A good material plan has no value if the capacity is planned with either insufficient or excess manufacturing capability. Insufficient capacity leads to deteriorating delivery performance, escalating work-in-process inventories, and frustrated manufacturing personnel who quickly abandon the formal MRP system. On the other hand, excess capacity is inefficient and works against a cost order-winning criterion.

The *detailed capacity planning* activity in the resources planning column in Figure 6–4 uses planned order releases and scheduled receipts data from MRP to allocate the production resources in the most efficient manner. In addition, detailed capacity planning uses work-in-process and routing information as a basis for calculating time-phased capacity requirements. The most frequently used detailed capacity planning technique is *capacity requirements planning* (CRP).

Part Routing and Lead Times

Routing and lead-time data for every part in the product are essential inputs to the CRP system. The routing sheet, prepared by the industrial engineering department or generated by computer software, specifies each production operation and the work center location. A sample routing sheet (Figure 6–10) for parts D and E in the product structure diagram (Figure 6–9) lists the time required for each operation and the total hours in each work center. Study the routing data until you understand how the total lead time was determined. The lead time for each part, included in the product structure diagram in Figure 6–9, is the same value as that used in the MRP records to calculate the planned order release dates.

As the routing sheet in Figure 6–10 indicates, lead time includes four elements:

- *Run time.* The product of lot size and the operation or machine run time per piece.
- *Setup time.* The time to set up or prepare the work center independent of the lot size scheduled.
- *Move time.* The time required to move the batch of parts or raw material to the present work center from the former work center.
- *Queue time.* The time spent waiting for processing at the work center.

Part D routing (hours)

Operation	Work center	Run time	Setup time	Move time	Queue time	Total time	Rounded time
1	201	1.6	0.5	0.4	2.6	5.1	5.0
2	208	1.5	0.3	0.2	2.8	4.8	5.0
3	204	0.1	0.1	0.3	0.6	1.1	1.0
4	209	1.2	0.8	0.3	2.3	4.6	5.0

Total lead time 16 hr (2 days)

Part E routing (hours)

Operation	Work center	Run time	Setup time	Move time	Queue time	Total time	Rounded time
1	201	1.1	0.4	0.3	1.8	3.6	4.0
2	204	0.2	0.3	0.2	0.5	1.2	1.0
3	205	1.2	0.1	0.4	1.5	3.2	3.0

Total lead time 8 hr (1 days)

Figure 6–10   Part Routing.

The last three elements add cost to the product but no value; as a result, the reduction or elimination of setup, move, and queue times is part of the continuous improvement process in world-class manufacturing. The queue time is frequently the major contributor to lead time and offers the greatest opportunity for lead-time improvement. Small lot sizes are not economical when setup and move times are large compared to the run time because they drive up the cost of each part.

The results of the CRP process are used to determine short-term capacity needs for key work centers and labor skills. Capacity requirements planning works with the system data to calculate the labor and machine time requirements needed to complete the master production schedule and support the overall business objectives. It generates load profile information, which shows the scheduling of work at defined work centers. It is produced following the completion of the MRP processing. A full description of the CRP process is included in Chapter 8.

6–5 INTRODUCTION TO PRODUCTION ACTIVITY CONTROL

Production activity control (PAC), formerly called shop-floor control, manages the detailed flow of materials inside the production facility. *PAC* is defined as follows:

Production activity control (PAC) is the function of routing and dispatching the work to be accomplished through the production facility and of performing supplier control. PAC encompasses the principles, approaches, and techniques needed to schedule, control, measure, and evaluate the effectiveness of production operations. (*APICS Dictionary*, 9th ed., 1998.)

PAC supports *order-winning criteria* by seeking a balance among the following three goals: (1) to minimize inventory investment, (2) to maximize customer service, and (3) to maximize manufacturing efficiency. The production planning and control model (Figure 6–4) illustrates the links between the material plans generated by the MRP system and PAC. Three different processes, *Gantt charts*, *priority rules for sequencing jobs at a work center*, and *finite loading*, are used for scheduling production in manufacturing.

6–6 GANTT CHARTS AND SCHEDULE BOARDS

Gantt or *bar charts* are basic shop floor control tools used by small and medium-size manufacturers. The process starts with the preparation of a *setback chart* (Figure 6–11), which shows the manufacturing start and finish dates based on routing sheet (Figure 6–10) or MRP lead times for all parts in the product (Figure 6–9). Based on the lead times, production must start on May 3, which is 5 days before the product ship date. Study the setback chart and the routing sheet (Figure 6–10) for parts D and E and note anything common between the two parts.

Two events critical to shop floor control are apparent: common work centers and overlap in the production schedule. Both parts use work center 201 as the first operation and work center 204 for a later operation. In addition, the production of parts D and E overlap on May 4; also part E is scheduled for production on two different days. Using the setback or back scheduling chart (Figure 6–11) and data from the routing sheet for each part, the Gantt chart in Figure 6–12 for work center 201 was developed. The chart includes total lead time (run, setup, move, and queue) and indicates that work center 201 can handle both parts. No conflict occurs between the schedule for parts D and E. The cross-hatched area in the bar indicates the queue-time component in the total lead time. If a conflict develops in the work center as other parts are scheduled, the work center schedulers can use

Figure 6–11 Operation Setback, or Back Scheduling, Chart.

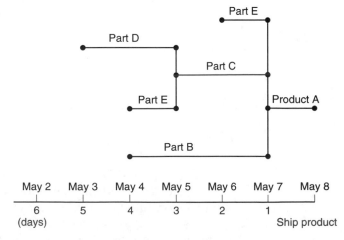

Figure 6–12 Gantt Chart for Work
Center 201: Parts D and E.

Work center 201

May 3											May 4										
8	9	10	11	12	1	2	3	4	5		8	9	10	11	12	1	2	3	4	5	

Part D

Part E

queue time to adjust part schedules by shortening the waiting time and starting
work on the parts ahead of the original schedule. Frequently, the move and queue
times are not included in the work center Gantt charts, so that the chart reflects
work center productivity more accurately.

In some applications, companies put the Gantt chart on a large scheduling
board where the planned activity for every work center is easy to check. Scheduling work with Gantt charts is well understood and effective for a limited number of work centers and part variations. However, keeping the chart current with
the latest schedule change is difficult when the number of parts and work centers increases. In addition, communicating the changes to the shop floor in real
time is a demanding requirement. Some companies have addressed this problem
by using personal computers on the shop floor and software that creates a scheduling board.

6–7 PRIORITY CONTROL AND DISPATCHING TECHNIQUES

The second technique used to control activity on the shop floor includes a set of
priority sequencing rules. Effective application of the rules requires an MPC system
with an *integrated database, dispatching mechanism,* and *priority rules.*

System Requirements

The system requires an integrated database for all manufacturing scheduled receipts with open shop orders that includes part identification data, routing information for all operations, scheduled operation dates, and scheduled completion
date. In addition, the database must accumulate the following data on work-in-
process: actual completion date, part and material move dates, labor utilization,
and scrap.

A second requirement for effective shop floor control is a reliable *dispatching
mechanism,* defined as follows:

> *A dispatching mechanism is a process for the selection and sequencing of manufac-
> turing jobs at individual work centers and a process to assign jobs to workers.*

The dispatching system performs four functions: (1) determines the *relative* priority
ranking through priority rules for all jobs released to manufacturing, (2) communicates job priority rankings to the shop floor through the use of a *dispatch list,*

Figure 6–13 Dispatch List with Look-Ahead.

Plant: 10   Department: 25   Work center: 15
Capacity: 100 hr
Date: 9-5

Plant number	Order number	Quantity	Hrs/unit	Total hours	Due date
12-9201	SO 73421	300	0.2	60	9-20
12-4510	SO 73107	100	0.3	30	9-22
18-2009	SO 73560	150	0.2	30	9-28
				120	

Arrival on 9-6

12-7210	SO 73416	100	0.5	50	10-4
15-0379	SO 73601	100	0.2	20	10-7

(3) tracks the movement of jobs on a real-time basis across the shop floor, and (4) monitors and projects contents of work center queues. The dispatch list, a primary component of the dispatching system, provides a daily listing of manufacturing orders in priority sequence, oriented by work center. Study the dispatch list example illustrated in Figure 6–13 and note the type of data present.

A set of rules for establishing the priority of jobs at work centers is the last requirement for a *priority rule* system. The primary requirement for an effective rule-based system is consistency in the priority rankings. The rule-based processes use several different routines to rank manufacturing jobs in priority order. Some examples of routines include: *shortest operation next, first come/first served, earliest due date, critical ratio, order slack, order slack time per remaining operations, queue ratio rule,* and *operation start and operation due date.* A description and example of each of these rules is provided in the following section.

Priority Rule Systems

The information in Figure 6–14 provides the data used for the following priority scheduling examples. The future due dates and current date are expressed as the number of days from the first day of the year. For example, a due date of February 7 would be listed as a due date of 38 (31 plus 7) for these calculations.

Exercise 6–1

Calculate the priority schedule for the shop orders in Figure 6–14 using the next operation time rule. The order with the shortest operation time has the highest priority, and the one with the longest operation time has the lowest priority.

Solution

The rule ignores the due-date and processing time remaining information because it processes orders based on speed of completion. This rule

Figure 6–14 Shop Order Data for Priority Calculations.

Shop order number	1A	2A	3A
Date arrived	117	115	116
Date due	125	120	130
Lead time remaining (LTR)	8	5	10
Processing time remaining (PTR)	6	4	5
Operations remaining (OR)	1	2	8
Standard scheduled queue (SSQ)	2	1	5
Next operation time	6	3	1

Note: All Times in Days; Today's Date, M-Day 117.

Priority	Order number	Next operation time
1	3A	1
2	2A	3
3	1A	6

maximizes the number of shop orders processed and minimizes the number waiting in queue.

Exercise 6–2
Calculate the priority schedule for the shop orders in Figure 6–14 using first come/first served. The priority is set by the arrival date of the shop order in the work center.

Solution

Priority	Order number	Arrival date
1	2A	115
2	3A	116
3	1A	117

The rule schedules on the basis of the arrival time of the part in the work center and will result in shortest average queue times for parts.

Exercise 6–3
Calculate the priority schedule for the shop orders in Figure 6–14 using the critical ratio rule. The rule is based on the ratio of *time remaining* to *work remaining*. If the ratio is 1, the job is on schedule; if the result is greater than 1,

the job is ahead of schedule; and if the result is less than 1, the job is behind schedule. The priority is set based on the ratio values. The *critical ratio* (CR) *formula* is

$$CR = \frac{\text{due date} - \text{today's date}}{\text{lead time remaining}}$$

Order 1A: $CR = \frac{125 - 117}{8} = \frac{8}{8} = 1$ (on schedule)

Order 2A: $CR = \frac{120 - 117}{5} = \frac{3}{5} = 0.6$ (behind schedule)

Order 3A: $CR = \frac{130 - 117}{10} = \frac{13}{10} = 1.3$ (ahead of schedule)

Solution

Priority	Order number	Critical ratio
1	2A	0.6
2	1A	1.0
3	3A	1.3

The rule identifies the job that is most behind schedule (the one with the lowest ratio) and the job most ahead of schedule (the one with the highest ratio). The rank order of the jobs by ratio makes sure that those behind schedule are performed first.

Exercise 6–4
Calculate the priority schedule for the shop orders in Figure 6–14 using order slack. The rule is based on *slack time,* which is the difference between the remaining production time (due date minus current date) and the sum of setup plus run time. The highest priority is assigned to the part with the lowest slack time. A positive slack value indicates a part ahead of schedule; negative slack, a part behind schedule; and zero slack, a part on schedule. The highest priority is assigned to the part with the lowest slack time. The *slack formula* is

slack = due date − today's date − processing time remaining
Order 1A:
          slack = 125 − 117 − 6 = 2   (ahead of schedule)
Order 2A:
          slack = 120 − 117 − 4 = −1   (behind schedule)
Order 3A:
          slack = 130 − 117 − 5 = 8   (ahead of schedule)

Solution

Priority	Order number	Slack
1	2A	−1
2	1A	2
3	3A	8

The rule identifies the job that is most behind schedule (the smallest or most negative value) and the job most ahead of schedule (the largest or most positive value). The rank order of the jobs by slack value (smallest to largest) makes sure that the jobs with too little slack time are performed first. This priority rule addresses the sequencing of jobs based on the value of work remaining.

Exercise 6–5

Calculate the priority schedule for the shop orders in Figure 6–14 using slack time per remaining operations. The rule is based on the ratio of *slack time* to *total number of operations remaining*. The highest priority is assigned to the part with the lowest slack time for each operation remaining. A positive value indicates a part with slack in the operations, a negative value indicates a part with insufficient operations time to meet the schedule, and a value equal to 1 indicates that the operation times are on schedule. The highest priority is assigned to the part with the lowest value. The formula is

$$\text{slack per operation (S/O)} = \frac{\text{slack time}}{\text{operations remaining}}$$

Note: Values for slack time are obtained from Exercise 6–4.

Order 1A:

$$S/O = \frac{2}{1} = 2.0 \qquad \text{(ahead of schedule)}$$

Order 2A:

$$S/O = \frac{-1}{2} = -0.5 \qquad \text{(behind schedule)}$$

Order 3A:

$$S/O = \frac{8}{8} = 1.0 \qquad \text{(on schedule)}$$

Solution

Priority	Order number	Slack per operation
1	2A	−0.5
2	3A	1.0
3	1A	2.0

The rule identifies the job that is most behind schedule (the one with the lowest ratio) and the job most ahead of schedule (the one with the highest ratio). The rank order of the jobs by ratio makes sure that the jobs with insufficient operations slack are performed first.

Exercise 6–6

Calculate the priority schedule for the shop orders in Figure 6–14 using the *queue ratio* (QR) *rule*. The rule is based on the ratio of *slack time* to *standard scheduled queue time*. The highest priority is assigned to the part with the lowest ratio value. A positive value indicates slack in the queue time, a negative value indicates a part that must shorten the standard queue time for an on-time delivery, and a value equal to 1 indicates that the part is on schedule. The highest priority is assigned to the part with the lowest value. The *queue ratio formula* is

$$QR = \frac{\text{slack time}}{\text{standard scheduled queue}}$$

Note: Values for slack time are obtained from Exercise 6–4.

Order 1A:

$$QR = \frac{2}{2} = 1.0 \qquad \text{(on schedule)}$$

Order 2A:

$$QR = \frac{-1}{1} = -1.0 \qquad \text{(behind schedule)}$$

Order 3A:

$$QR = \frac{8}{5} = 1.6 \qquad \text{(ahead of schedule)}$$

Solution

The rule identifies the job with the least sufficient queue time (the one with the lowest ratio) and the job with the most queue time to use (the one with the highest ratio). The rank order of the jobs by ratio makes sure that those with insufficient queue slack time are performed first.

Priority	Order number	Slack per operation
1	2A	−1.0
2	1A	1.0
3	3A	1.6

To simplify priority sequencing rules, many companies have adopted a process based on *operation start* and *operation due dates*. The due date is generated from the MRP output, and the due dates for all shop orders are updated daily through a *dis-*

patch list (Figure 6–13) or work center schedule. In this type of system, the due dates would not appear on any shop paperwork that travels with the work-in-process inventory. The shop paperwork would show just the static information, such as work standards and routings. This process permits the due dates to be revised frequently because the only reference to the due date is on the daily dispatch list. Confusion in the work center is avoided because the job priority is set from the due date on the dispatch list.

Exercise 6–7

Determine the priority schedule for the shop orders on the dispatch list in Figure 6–13 using the earliest due date. The highest priority is assigned to the part with the earliest due date, and the lowest is assigned to the part with the latest due date.

Solution

Priority	Order number
1	SO 73421
2	SO 73107
3	SO 73500

The rule schedules order priority using the order due date; as a result, the process works to maximize the number of on-time deliveries.

The specific priority rule selected by the production automation control group depends on several business and production factors.

6–8 SHOP LOADING

Loading is defined as the process of committing capacity and implies a scheduling process for work centers and machines. Shop loading is either *infinite* or *finite*.

Infinite Loading

Infinite loading results generated by material requirements planning software establish a work center schedule that does not balance the planned work order resource needs with the capacity of the work center. Study the dispatch list in Figure 6–13 and compare the capacity of work center 15 with the total hours scheduled. The work center capacity of 100 hours is insufficient to handle the 120 hours scheduled. The concepts of loading are illustrated in the load profile diagrams of Figure 6–15a shows the concept of infinite loading, with a work center with a capacity of 80 hours per period. The infinite loading illustrated in this figure was generated by capacity requirement planning (CRP) software and MRP data. Note that the work in process (WIP), called *open shop orders*, and the new jobs scheduled for the work center, call *planned orders*, exceed the 80-hour capacity during the first three periods; in addition, there is a large number of past due open shop orders,

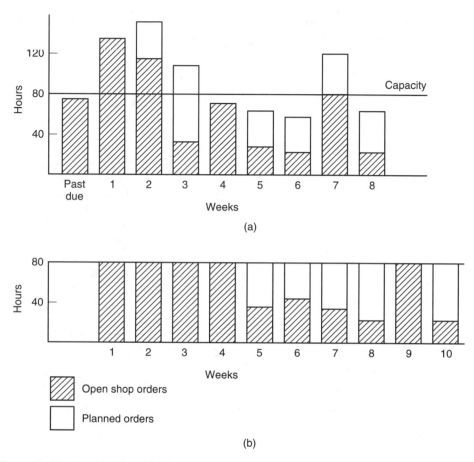

Figure 6–15   Load Profiles: (a) Infinite Loading; (b) Finite Loading.

and the open shop orders exceed the capacity in the first two. All the open shop orders scheduled for period I will not be completed; therefore, deliveries may be late and shop work will be pushed to the following period in a past-due category. The priority for work at the infinitely loaded center is established by production control personnel or supervisors who take control of the work center schedule and set the job schedule using one of the priority rules described previously. Without human intervention, the multiple jobs scheduled into the work center at the same time would cause chaos.

Finite Loading

Finite loading systems produce a detailed schedule for each part and work center through simulated shop order start and stop dates. The result of finite loading (Figure 6–15b) is a work center scheduled at capacity for the first four periods, with open orders and no past-due work-in-process (WIP). One output of the finite loading system is a simulation of how each machine and work center will operate over the planned time horizon. The schedule is built from the following information: jobs presently waiting in the queue, the completion time for operations scheduled at the work center ahead of the current work center, and the priority on current WIP. Based on that information, a decision could be made to let the work center remain idle until the next job scheduled for the center is finished on the previous machine. Two approaches are used to fill a work center: *vertical loading* and *horizontal loading.*

Vertical and Horizontal Loading

In the vertical loading approach, the jobs currently in the work center plus those that will be finished at a previous operation are evaluated from a priority standpoint and the highest priority is loaded. In the horizontal method, an entire shop order or job with the highest priority is completed at each work center in the sequence, with the shop order taking priority over other parts. For example, in Figure 6–10, part E starts in work center 201 and the center would schedule the run time for all the parts plus the setup time as the highest priority for the work center. If the previous job in work center 204 finished before part E was ready to move to the work center, 204 would sit idle waiting for the high-priority job to arrive. The horizontal process often produces "holes" on idle times in the capacity profile for the work center.

In addition, a choice must be made between a forward or a backward scheduling process. The *forward scheduling* process (Figure 6–16a) starts with the current date and builds through the total manufacturing lead time to the completion date. The alternative approach, *backward scheduling* (Figure 6–16b), starts with the order due date and subtracts the manufacturing lead time to arrive at the shop order release date. In both cases, the manufacturing lead time may not be consistent with the time remaining to complete the job; as a result, corrective action is necessary.

6–9 INPUT/OUTPUT CONTROL

The function of *input/output* control in the MPC model (Figure 6–4) is to evaluate the number of production orders that were released to manufacturing versus the work completed. Input/output control provides a check on shop floor efficiency and is a valuable feedback tool for production planning. For example, if the input/output control reports shows that back orders of some products develop whenever orders are released to the shop, that is a reliable indication of some type of production problem.

6–10 AUTOMATING THE MPC FUNCTION

The MPC model in Figure 6–4 was described with an emphasis on the manual implementation of the various aggregate and disaggregate functions. With small numbers of noncomplex products, a manual implementation could be supported;

Figure 6–16
(a) Forward Scheduling; (b) Backward Scheduling.

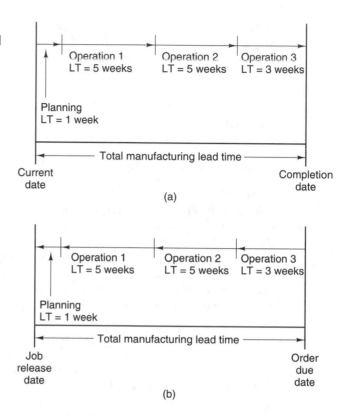

however, the concept of a single image of the product in an enterprise data base would be impossible to achieve. To reap the full benefits of CIM and to support multiple or complex product structures, a computer-based MRP2 system is necessary. Companies today realize that they must be responsive to the marketplace because of increased competition at home and from foreign companies on price and customer service. Manufacturing must be responsive to the needs of the company and the marketplace. Communication and the sharing of information by all departments is critical. Think of the links required between the functional areas shown in Figure 6–17. The names in the boxes may change from one company to another, but the required functions and the need for communication and shared data remain.

Data and information are vital parts of an effective operation. Most companies that start out small and grow will soon outgrow their initial manual systems for managing and controlling their operation. A computer system can help with the communication and exchange of information among the functional areas.

Numerous vendors provide MRP2 software support on all three types of computer platforms. These systems now often run on a minicomputer and can support small to very large enterprise production control databases. A formal system provides management with a tool that can help link the corporate strategy to the operational objectives of the firm. Management's plans and systems to achieve desired objectives address the following:

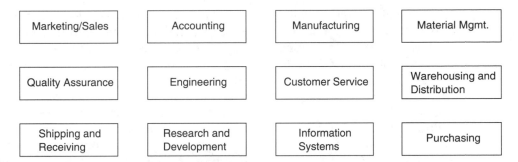

Figure 6–17   Functional Areas in the Enterprise.

Setting goals that describe the management vision for the company.

Developing plans and objectives to achieve the goals.

Executing and implementing the approved plans.

Providing feedback and communication of the results and outcomes.

The planning system must be understood and supported by the functional area managers. The plan must be future oriented, with checks of internal and external resources. It must integrate the expertise of the functional areas. The higher level formal plans become the basis for preparing detailed functional operating plans. A formal planning approach is an ongoing effort. It cannot be done correctly as a one-shot effort that is reviewed only annually.

System packages of software for MRP2 computer applications for controlling material and inventory began to emerge in the 1960s with the release of communications-oriented production information and control system (COPICS) by the IBM Corporation. COPICS initially was not a complete MRP2 system, but over the years the package was modified and expanded beyond the initial material management focus. The advances in technology have led to the design of systems that go well beyond the initial material planning applications to integrate all the functional areas of the company. The new systems help link the strategic plans of the firm to specific operational objectives.

Recent publications by APICS now provide information on nearly 200 software packages for MRP/MRP2 and ERP. Packages are available for companies of nearly any size and market type. The evaluation and selection of a package that best fits a company is a major part of the system implementation process. The selection process requires a thorough understanding of the business and the operations required to support it.

6–11 SUMMARY

In the past, the term *production management* was associated only with the production of hard goods. The growth of the service industries requires a new definition of the management function that governs the generation of goods and services.

The new definition, *production operations,* takes a much broader look at the enterprise, with the recognition that every organization has some production-oriented activities and some service functions. The four areas where the differences occur are consistency of product, point of contract, labor intensity, and level of productivity.

The manufacturing planning and control (MPC) function is defined on the SME CIM wheel. The primary activity is planning for production at both the aggregate and disaggregate levels for the product and the enterprise resources. In addition, MPC is responsible for the generation of production schedules and the tracking of production efficiency. The disaggregate planning options include material requirements planning, just-in-time production, synchronous manufacturing, and CPM/PERT. The MPC model described a process and flow for the generation of production planning and scheduling information. In addition, the model indicated the areas of responsibility and decision points for every step in the planning and scheduling of production for a product.

Production activity control (PAC) is responsible for the control and management of detailed material flow on the shop floor. Three different techniques are used: Gantt charts, priority rules, and finite loading. Product structure diagrams and routing sheets support the shop material flow under PAC. The priority sequencing rules technique for PAC uses a CIM integrated database, a dispatching mechanism, and priority rules to achieve effective material flow on the shop floor. Eight different priority rule calculations are used, depending on the item manufactured and the production preference for setting production priorities. The rules include shortest operation next, first come/first served, earliest due date, critical ratio, order slack, order slack time per remaining operations, queue ratio rule, and operation start and operation due date.

Infinite shop loading, an output of the MRP process, is usually produced in the capacity requirements planning activity. Under infinite loading conditions, the capacity of the individual work centers is often scheduled at a level that exceeds the capacity of the machines. Finite scheduling, a more numerically intensive process, considers machine capacity in every work center in the generation of work center production schedules. Input/output control monitors the effectiveness of both production scheduling and production processing. To reap the full benefits of CIM, the MPC functions must be automated.

REFERENCES

CLARK, P. A., *Technology Application Guide: MRP II Manufacturing Resource Planning.* Ann Arbor, MI: Industrial Technology Institute, 1989.

COX, J. F., J. H. BLACKSTONE, Eds. *APICS Dictionary,* 9th ed. Falls Church, VA: APICS—The Educational Society for Resource Management, 1998.

GROOVER, M. P., *Automation, Production Systems, and Computer-Integrated Manufacturing,* 2nd ed. Upper Saddle River, NJ: Prentice Hall, 1987.

SHRENSKER, W. L., *Computer-Integrated Manufacturing: A Working Definition.* Dearborn, MI: CASA of SME, 1990.

SOBCZAK, T. V., *A Glossary of Terms for Computer-Integrated Manufacturing.* Dearborn, MI: CASA of SME, 1984.

STEVENSON, W. J., *Production/Operations Management,* 3rd ed. Homewood, IL: Richard D. Irwin, 1990.

VOLLMANN, T. E., W. L. BERRY, and D. C. WHYBARK, *Manufacturing Planning and Control Systems,* 4th ed. Homewood, IL: Richard D. Irwin, 1998.

QUESTIONS

1. Compare production management with operations management.
2. How are service operations and manufacturing operations different, and how are they similar?
3. What is the goal of manufacturing planning and control?
4. What is aggregate planning, and how does it fit into MPC/MRP2?
5. Describe disaggregate planning and its function in MPC/MRP2.
6. Compare the three disaggregate planning options.
7. Describe the MPC/MRP2 model presented in the chapter.
8. Describe the MRP system input and output data.
9. What is the relationship between MRP and CRP?
10. Define *production activity control.*
11. Describe the three processes used in PAC.
12. What is the function of a product structure diagram?
13. What is the function of a routing sheet, and what manufacturing is included on the sheet?
14. Describe how Gantt charts and scheduling boards are used for PAC.
15. What are priority sequencing rules?
16. Define *dispatching mechanism.*
17. Describe the four functions performed by a dispatching system.
18. List the priority rule routines described in the chapter.
19. Discuss the concept of infinite versus finite shop loading.
20. Compare vertical and horizontal shop loading.
21. Discuss forward and backward scheduling for shop orders.

PROBLEMS

The shop order information that follows provides the data necessary for the following priority scheduling problems. The future due dates and current date are expressed as the number of days from the first day of the year, and all times are in days. The current M-day is 110.

Shop order number	1B	2B	3B
Date arrived	107	110	109
Date due	118	116	119
Lead time remaining (LTR)	11	5	8
Processing time remaining (PTR)	9	2	6
Operations remaining (OR)	4	1	3
Standard scheduled queue (SSQ)	4	0.5	2
Next operation time	1.5	2	1

1. Calculate the priority schedule for the shop orders from the data above using the following methods:
   a. Shortest operation next
   b. First come/first served
   c. Critical ratio rule
   d. Order slack time rule
   e. Slack time per remaining operations rule
   f. Queue ratio rule
   g. Earliest due date rule
2. Determine the impact on shop order priorities if the due date on the 3B shop order is changed to 115.
3. An engineering change causes the OR and PTR for the 2B shop order to change as follows: OR = 3 and PTR = 5. How do these changes affect the priority schedule for the shop orders?
4. Compare the first come/first served and earliest due date rule with the remaining methods. How well does each method account for process time, number of operations, and lead time?

PROJECTS

1. Using the list of companies developed in Project 1 in Chapter 1, determine the type of MPC model used by the companies.
2. Compare the MPC model presented in the chapter with the model used by one company from the small, medium-size, and large groups, and describe how their operations differ from the model.
3. Determine the type of shop order priority scheduling method used most often by the companies in Project 1.
4. Visit the APICS site on the World Wide Web and research the current information on three commercial systems for MRP2/ERP. Choose your systems based first on the price of the software and include systems that fall in the high-, mid-, and low-price ranges. Remember that MRP/ERP systems are not inexpensive. Even a low-price system may have a software cost of $2000 or more per seat.

5. Identifying the problem of manufacturing: In Part 1 of the problem, you are the manager of a manufacturing company. Brainstorm a list of the critical elements of your operation (the essential actions or activities that must happen or are required for you to be able to serve your customers effectively and stay in business). In Part 2, prioritize the list by identifying your top 5 elements, then draw a circle around the element you feel is most critical. (This activity is best done in small groups but may be done individually.) Be prepared to share your ideas with others.

6. Visit the APICS Education and Research site on the World Wide Web and find the requirements for the Donald Fogarty Student Paper Competition. Select a topic of interest and write a research paper following the contest guidelines. Enter your paper in the contest through your local APICS chapter. Categories are open for part-time and full-time students at the undergraduate and graduate levels.

Introduction to Manufacturing Planning and Control

Manufacturing planning and control (MPC) is part of the production/operations management function described in Chapter 6. MPC systems focus on all enterprise production activities, from acquisition of raw material to delivery of finished products. In addition, the MPC system is closely linked to key management functions and activities, such as strategic planning, fixed and variable resource allocation, all major accounting functions, purchasing, customer service, and marketing. The common links between the MPC function and the manufacturing organization are illustrated in Figure 6–2. MPC is positioned in the heart of the manufacturing organization and is the principal conduit through which product design and marketing information must flow to reach the plant floor. The importance of MPC in the computer-integrated enterprise is evident from the position occupied on the Society of Manufacturing Engineers' CIM model shown in Figure 6–1. The models imply that MPC has responsibility for the *planning and control* of the *shop floor*, production *materials*, production *scheduling*, *quality process*, and *facilities planning*. The title *MPC* implies that two distinct functions are performed: *manufacturing planning* and *manufacturing control*. In this chapter, we focus on the planning and control aspects of the MPC function.

7–1 PLANNING IN THE MPC SYSTEM

The manufacturing planning and control (MPC) model shown in Figure 6–4 was described in Chapter 6. Review the model until you are familiar with the information flow throughout the model. The APICS model was also introduced in Chapter 6, Figure 6–7. The previous discussion of these two models focused on the general development of aggregate and disaggregate plans for the production of finished goods. In addition, the MPC model includes several blocks that address the planning process for the production of products. Forecasting future demand is an important part of planning and is discussed in the next section.

High-Level Planning for the Business

Business planning starts with top management's definition of the business of the firm. The definition includes the products to be sold (in general terms), the markets to be served, plans for market share, net sales, return on investment, profits, and the return on equity. The plans involve deciding how the firm will position itself in the marketplace and the strategies that will be used to compete (high volume, low cost, high service, specialty, etc.). Next come the documentation of measurable objectives, the development of plans to achieve the objectives, and the determination of realistic target dates for the implementation of the plans.

Marketing and financial plans are critical elements of the business planning process. This activity requires the participation of the true leaders of the enterprise. The true leaders control the available resources and determine the direction of the enterprise. The business plan tends to look several years into the future and is typically stated in dollars. A plan for producing and selling dollars is of little direct use to manufacturing, however. The general business plan must be translated into a language and form that the people in the operating areas can understand and use. Figure 7–1 shows an example of the planning progression as the reader moves through the diagram from the top to the bottom.

Manufacturing's operating plan, the production plan, expands on the strategic decisions and marketing plans. It leads to more detailed plans and decisions concerning the production technology to be used, the type and quantity of facilities, the operating organization, quality requirements, operating policies, expected through-put times, and more.

Forecasting Future Demand

The MPC steps in planning the production of a product are detailed in Figure 7–1. Compare the blocks in the diagram in Figure 7–1 with the MPC model illustrated in Figure 6–4. At the highest level in the corporation, the marketing and financial areas pool customer and market data to produce the business plan. The business plan is a *forecast* of expected future demand for all the products produced by the enterprise. The data collected to produce this forecast come from several sources; for example, the historical sources provide sales information, quantity, and percentage change for product families from previous years. In addition to historical information, forecasting techniques use (1) the current economic condition of regional, national, and global markets; (2) the trends present in the current economic data and product markets; and (3) anticipated shifts in consumer demand. The forecast present in the business plan is management's "best guess" for the future demand for a product.

Planning for Production

Production planning is the function of setting the overall level of manufacturing output (production plan) and other activities to satisfy best the

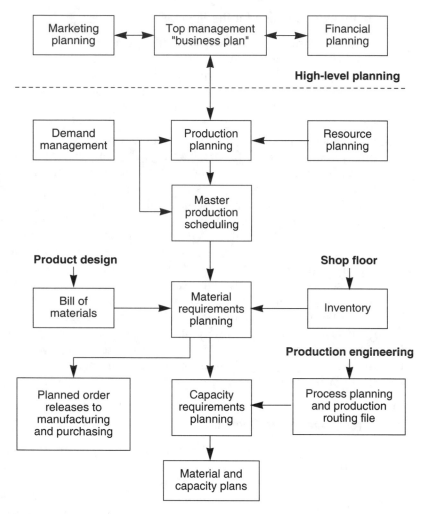

Figure 7–1 Production Planning Function.

current planned levels of sales (sales plan or forecasts), while meeting general business objectives of profitability, productivity, competitive customer lead times, etc., as expressed in the overall business plan. The sales and production capabilities are compared, and a business strategy that includes a sales plan, a production plan, budgets, pro forma financial statements, and supporting plans for materials and workforce requirements, etc., is developed. One of its primary purposes is to establish production rates that will achieve management's objective of satisfying customer demand by maintaining, raising, or lowering inventories or backlogs, while usually attempting to keep the workforce relatively stable. Because this plan affects many company functions, it is normally prepared with information from marketing and coordinated with the functions of manufacturing, engineering, finance, materials, etc. (*APICS Dictionary,* 9th ed., 1998.)

Production planning is not a new concept; the significance of this planning function was documented in the late 1970s. Berry, Vollman, and Whybark described the production plan at a conference of production leaders in 1979: It is a top management responsibility. It is an agreement among marketing, finance, and manufacturing about what is to be produced and made available for sale. It is a clear link between the corporate plans and the master production schedule (MPS). This planning concept is still important today.

The aggregate forecast from the business plan is applied to the production plan (Figure 7–1) together with adjustable resource planning data and demand management information from marketing. The typical *production plan* shown in Figure 6–5 indicates that planned production is expressed in the total number of units, without reference to the mix of products. The aggregate data in the production plan are then broken down into individual products in the *master production schedule* (MPS) example shown in Figure 6–5.

The production planning block diagram in Figure 7–1 indicates that the MPS provides the data needed in *material requirements planning* (MRP) to produce the *time-phased* requirements for all manufactured and purchased parts. The term *time phased* implies that a purchasing and manufacturing schedule is generated that considers the lead time required to buy or manufacture each part needed for the finished products in the MPS. The time-phased schedule is generated using an MRP process that is illustrated in Figure 6–8. The detail in the planning process has increased exponentially at this stage because every part in the finished product could require the generation of a separate MRP record. The MRP function in Figure 7–1 is supported with bill of material (BOM) and inventory data. Note that critical links to other enterprise areas are present. BOM data is supplied by the design department, and inventory levels come from shop floor data sources.

The diagram of the production planning function shown in Figure 7–1 indicates that the MRP record also provides the data required to perform *capacity scheduling.* Note that capacity planning uses process planning and routing data that come from another external source, production engineering. With the capacity and material plans defined, the planning process for production is complete. The next step is the implementation of the plans on the shop floor and the evaluation of production effectiveness.

The production planning diagram in Figure 7–1 shows five sources of information outside the MPC function: *marketing, finance, product design, shop floor,* and *production engineering.* These links to other areas in the enterprise for production data underscore the need for a common database for production information and emphasize that integration of systems across the enterprise is required for an effective MPC system.

The overview of the production planning function provided in this section is a preparation for the detailed analysis of the production planning process in the following sections. Reread this section until the overall process is clear.

7–2 PRODUCTION PLANNING

The production planning function, illustrated in Figure 7–1, provides the key communications link between top management and manufacturing. The production planning function provides a framework for resolving conflict due to changes in product marketing and production resources. Suppose that marketing sees an opportunity to expand into a new market and requests production resources for the new product. With a specific production plan in place, resources for the new product cannot be allocated without reducing production of some other product. The production planning process forces the business plan developed by top management to be consistent with the production capability required for the production plan. After marketing and financial issues are resolved and a production plan for the intermediate term is set, the manufacturing mission for the enterprise is clearly defined. The production plan provides manufacturing planning and the shop floor with the marching orders necessary to meet the objectives of the firm.

The production plan is usually stated in dollars or in aggregate units of output per month. The production plan is *not* a forecast of demand; rather, it is the planned production stated in aggregate terms for which everyone in the enterprise is responsible. For example, the forecast demand may exceed the aggregate units in the production plan. A decision by top management to produce less than the forecasted demand may be made for several reasons: desire to have lower quantity and higher quality, desire to defer investment in fixed or adjustable resources, or a desire to put available resources into other product areas. The production plan provides a vehicle for tough trade-off decisions in setting production goals at the aggregate level.

The Production Planning Process

The production planning process starts with a good sales forecast for the next year that discounts as many of the variables in the marketplace as possible. The demand management (Figure 7–1) issues, such as interplant transfers, distribution requirements, and service parts must also be factored into the production plan. Changes in inventory or backlog levels that affect the overall production rate must also be considered.

Effective production planning processes have reviews at regular intervals with a *time fence* for changes requested in the aggregate production levels. For example, successful firms often review the production plan monthly and make changes quarterly. The time fence frequently sets limits on how late in the planning cycle changes in the aggregate levels can be made. For example, the time fence may dictate that no changes can be made in the current or closest period and that no more than a 10 percent change can be made in the nearest future period. Routine reviews of the production plan keep the communication alive between top management and manufacturing.

Production Planning and Variable Resource Management

The production plan states the aggregate production goals for all products manufactured by the enterprise. For example, consider the production plan stated in aggregate dollars per month for ABC Manufacturing, illustrated in Figure 7–2. Converting this aggregate forecast into a production plan requires a decision on resources. A study of the charts reveals the following:

- Sales peak at $16 million in November.
- June has the minimum sales, $6.6 million.
- Two sales peaks occur, in spring and fall.
- Total sales for the year are $132 million.

Before variable resources can be allocated to meet this production plan, the sales in dollars per period must be converted to labor hours per period. The conversion is performed using an estimate, obtained from company accounting records, that relates the dollar value of sales to hours of direct labor. In a low-technology production situation that relies on manual labor, for example, each hour of direct labor might equate to $30 of sales. In a production situation where high-technology production equipment is used, the figure might be $100 of sales resulting from every hour of direct labor. Using the $100 conversion value, the table in Figure 7–3 was developed for the production plan in Figure 7–2. Study the headings across the top of the table. The Sales column comes directly from the forecast (Figure 7–2). The Labor Hours column is calculated by dividing the sales dollars by the labor conversion rate ($8,600,000 divided by $100 per labor hour equals 86,000 direct labor hours). The Days Worked column indicates the number of working days in each period or month. Note that the plant is closed in June for 10 working days

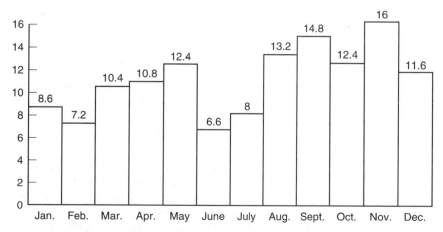

Figure 7–2 ABC Manufacturing Monthly Sales Forecast (Millions of Dollars).

	Sales (millions of dollars)	Labor hours (dollars)	Days worked	Variable workforce	Variable work week
Jan.	8.6	86,000	21	512	30.02
Feb.	7.2	72,000	20	450	26.39
Mar.	10.4	104,000	23	565	33.15
Apr.	10.8	108,000	20	675	39.59
May	12.4	124,000	22	705	41.32
June	6.6	66,000	10	825	48.39
July	8	80,000	21	476	27.93
Aug.	13.2	132,000	22	750	43.99
Sept.	14.8	148,000	21	881	51.67
Oct.	12.4	124,000	22	705	41.32
Nov.	16	160,000	20	1000	58.65
Dec.	11.6	116,000	20	725	42.52
	132	1,320,000	242	682	39.99

Figure 7–3   Chase Production Strategy Data.

for employee vacations. A study of Figure 7–3 indicates that 160,000 direct labor hours are required in November to satisfy the maximum sales value of $16 million, and only 66,000 labor hours are required in June to meet the demand of the lowest sales period. This wide variation in human resource requirements is the basis for planning in the variable resource area. Three different production planning strategies, called *chase*, *level*, and *mixed*, are used to address this variation in direct labor hours required by the production plan. A description of each of these strategies is provided in the following sections.

Chase Production Strategy

The *chase* production strategy requires that the production in each period equals the planned production for that period, which implies that the product inventory level at the beginning of each period would be zero because all the production planned for the period would be produced during the period. A pure chase strategy requires that either the number of employees or the hours worked per week by each employee must change to meet the production planned for the period.

Chase data for the ABC Company, for example, is provided in the last two columns in Figure 7–3. Study the data in Figure 7–3 and the graph of the monthly employment in Figure 7–4 until the chase production strategy is clear. The column in Figure 7–3 labeled Variable Workforce indicates the number of full-time employees required each month to meet the planned production exactly. The number of employees is found by first calculating the total hours worked by each employee in the period (21 days worked times 8 hours per day equals 168 hours

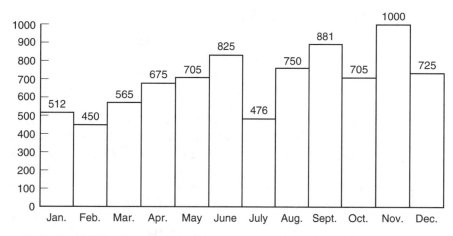

Figure 7–4 Monthly Number of Full-time Employees in the Chase Strategy.

per period per employee). The number of employees required per period is then calculated by dividing the total labor hours for the period (86,000) by the employee hours (168) per period (86,000 labor hours divided by 168 hours per employee equals 512 employees). Stop and calculate the total employees required for the months of February and March to verify that you understand the calculations.

Analysis of this strategy indicates that employee levels vary from a low of 450 in February to a maximum of 1000 in November. A change of this magnitude in the level of full-time employees over twelve months would be difficult to support. The only industries that can use this type of variable resource strategy successfully usually require only a low skill level in the workforce.

Another approach used in the chase strategy keeps the number of employees in the workforce constant and varies the hours worked per week. The change in the hours worked per week for the ABC Company example is listed in Figure 7–3 under Variable Work Week. The average number of employees required for a year's production (682) is indicated at the bottom of the Variable Workforce column in Figure 7–3. This strategy keeps the workforce at 682 and varies the hours worked per week to meet exactly the production planned. Note that the weekly workload varies from a low of 26.39 in February to a high of 58.65 in November. The hours required per work week for the chase strategy are calculated by first finding the hours per month per employee (86,000 labor hours per month divided by 682 employees equals 126.1 labor hours per employee per month). Dividing the previous value by the working days in the month yields the hours worked per day per employee (126.1 divided by 21 days equals 6 hours worked per day per employee). Finally, multiplying the daily hours by 5 days provides the total weekly hours for each employee. Stop and perform the calculation for February and March to verify that you understand how to calculate the total weekly hours per employee in this chase strategy. The swings in hours worked per week is wide, but this strategy is frequently used to chase a production schedule with wide variations.

Level Production Strategy

The *level* production strategy requires that the production in each period equals the monthly average production calculated from the total production value for the year. With this strategy, the workforce and weekly work hours are constant, so that production is roughly the same each month. As a result, in some months products produced are not sold, so that an inventory of parts develops to handle the months where market demand is greater than the production. The monthly production totals are listed in Figure 7–5 in the columns labeled Monthly Production and Inventory Balance. Note that the workforce is held at 682 employees and the work week is 40 hours. The monthly production is calculated by multiplying the number of employees (682) times 8 hours per day times days worked (21) times the sales dollar conversion factor of $100 per employee hour worked established in the variable resource section. The dollar inventory balance is the difference between sales dollars and production dollars. The level strategy is illustrated graphically in Figure 7–6 with the production level and inventory balance plotted. The production is not exactly level because the number of days worked per month varies slightly. The inventory balance increased early and reached a peak in July in preparation for the high product demand in the fall.

Each method, chase and level, has advantages and disadvantages. The advantage of the chase strategy is that no excess inventory is carried; however, the disadvantage of the variable workforce implementation of chase is the high cost of hiring and firing employees. In the variable work week implementation of chase, the disadvantage is the cost of overtime and in implementing shortened work

	Sales (millions of dollars)	Labor hours (dollars)	Days worked	Level workforce	Monthly production (millions of dollars)	Inventory balance (millions of dollars)
Jan.	8.6	86,000	21	682	11.46	2.86
Feb.	7.2	72,000	20	682	10.91	6.57
Mar.	10.4	104,000	23	682	12.55	8.72
Apr.	10.8	108,000	20	682	10.91	8.83
May	12.4	124,000	22	682	12.00	8.43
June	6.6	66,000	10	682	5.46	7.29
July	8	80,000	21	682	11.46	10.75
Aug.	13.2	132,000	22	682	12.00	9.55
Sept.	14.8	148,000	21	682	11.46	6.21
Oct.	12.4	124,000	22	682	12.00	5.81
Nov.	16	160,000	20	682	10.91	0.72
Dec.	11.6	116,000	20	682	10.91	0.04
	132	1,320,000	242	682	11.00	6.31

Work week = 40 hrs.

Figure 7–5 Level Production Strategy.

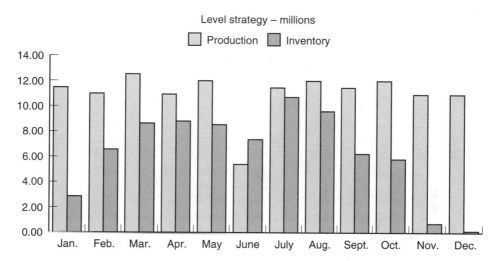

Figure 7–6   Level Strategy Graph.

weeks. The advantage of the level strategy is the constant workforce and work week, but that constancy creates a disadvantage in the cost of carrying the excess inventory created. As a result, many companies use a combination of the chase and level strategies in implementing the production planning requirements.

Mixed Production Strategy

The *mixed production strategy* uses the best parts of the chase and level strategies. Study Figure 7–7, which illustrates the process. Note that the workforce is held level at 610 employees for the first seven months and then increases to 809 employees for the last five months. The numbers in the chart assume a constant 40-hour work week over all periods. Using this method, the average inventory is reduced from $6.31 million for the level strategy (Figure 7–5) to $3 million for the mixed approach. The savings in carrying cost for the inventory must offset the cost of changing the employment levels twice a year. The mixed strategy often includes a varying work week, which might reduce the number of employees changed. In most manufacturing situations, the mixed strategy is used to adjust the variable resources to meet the production planning goals.

The aggregate values in the production plan drive the *master production schedule* (Figure 7–1), where the first level of disaggregation of the product plan occurs. The development and operation of the master production schedule function is described in the following section.

Considering Resource Capacity

The production plan must be a realistic statement of what is expected to be available to supply the demands of customers. The plan must consider the capacity of identified critical resources. A technique known as rough-cut capacity planning is

	Sales (millions of dollars)	Days worked	Level workforce	Monthly production (millions of dollars)	Inventory balance (millions of dollars)
Jan.	8.6	21	610	10.25	1.65
Feb.	7.2	20	610	9.76	4.21
Mar.	10.4	23	610	11.22	5.03
Apr.	10.8	20	610	9.76	3.99
May	12.4	22	610	10.74	2.33
June	6.6	10	610	4.88	0.61
July	8	21	610	10.25	2.86
Aug.	13.2	22	809	14.24	3.89
Sept.	14.8	21	809	13.59	2.69
Oct.	12.4	22	809	14.24	4.52
Nov.	16	20	809	12.94	1.47
Dec.	11.6	20	809	12.94	2.81
	132	242		11.23	3.00

Work week = 40 hrs.; Jan. starting inventory balance is zero.

Figure 7–7 Mixed Production Strategy.

often used at the production plan level to evaluate expected critical resources. These key resources may include machines, labor, floor space, vendor, or even the dollars required. The purpose of the rough-cut analysis is to evaluate the feasibility of the plan before attempting additional detailed planning and implementation activities. The idea is to identify problem areas early in the planning process and resolve conflicts and issues early by taking a planning approach.

7–3 MASTER PRODUCTION SCHEDULE

The *master production schedule* (MPS) is the basic communications link between the business "game plan" and manufacturing.

Master production schedule (MPS) can be defined in one of two ways: (1) the anticipated build schedule for those items assigned to the master scheduler. The master scheduler maintains this schedule and, in turn, it becomes a set of planning numbers that drives material requirements planning. It represents what the company plans to produce expressed in specific configurations, quantities, and dates. The master production schedule is not a sales forecast that represents a statement of demand. The master production schedule must take into account the forecast, the production plan, and other important considerations such as backlog, availability of material, availability of capacity, and management policies and goals. It is also called the *master schedule*. (2) The result of the master scheduling process. The master schedule is a presentation of demand, forecast, backlog, the MPS, the projected on-hand inventory, and the available-to-promise quantity. (*APICS Dictionary,* 9th ed., 1998.)

The MPS must be realistic, attainable, and reasonable. An overstated master schedule becomes a wish list, not a true production schedule. Once the operating departments determine that the master schedule is unrealistic, they are forced to develop alternative informal systems for planning what is really produced. The informal system plans will lead to confusion and chaos.

Note in Figure 7–1 how the MPS is positioned strategically between manufacturing and business/production planning. The MPS is an anticipated build schedule for end products; as a result, the MPS states production plans, not market demand. The MPS is not a forecast. In most cases, the MPS and the forecast have different values because the MPS represents what can be built given the enterprise resources, whereas the forecast portrays anticipated market demand.

The MPS and Production Strategies

The MPS must support the production strategies engineer to order, make to order, assembly to order, and make to stock, described in Chapter 2 and illustrated in Figures 2–4 and 2–5. Study these figures and review the characteristics of each strategy. The engineer-to-order (ETO) and make-to-order (MTO) strategies use a similar process in the MPS implementation. The assembly-to-order (ATO) and make-to-stock (MTS) processes are quite different. The primary impact on the MPS for all of these production approaches is in the choice of units for the MPS.

In the ETO and MTO company, where no finished-goods inventory exists, all products are built to customer orders. This type of production is used because of a large number of possible product configurations and knowledge that the customer will wait for complete manufacture and assembly of the specific end item. In the case of the engineer to order, the customer will also allow the time required for basic product engineering to be included in the delivery time. The units used for these two strategies are specific end items or sets of products that comprise a customer order. Since production often starts before the end item is completely designed or specified, development of the MPS is difficult. Frequent adjustments to the MPS in this type of production setting are necessary and expected.

The assembly-to-order production strategy is used when there is a wide variety of end-item configurations built from a broad range of basic components and subassemblies. In this case the required customer delivery time is shorter than the production lead times for components, so production is started in anticipation of customer orders. The large number of end-item configurations makes forecasting of specific end-item configurations difficult and stocking of finished goods risky. As a result, assembly of common base components and building of subassemblies are started into production before the customer order is received. The final assembly to a specific end configuration is started only when the customer order is recorded. Therefore, an ATO company does not include specific end items in the MPS. The MPS units reflect an average product requirement based on percentage estimates of typical market demand. For example, in the automotive industry, the MPS schedules into production the average number of four-door versus two-door vehicles with the average number of four-cylinder versus six-cylinder engines. The percentage of different options are also averaged, so that subassemblies of the options are available to build into the car during final assembly. The averages and percentages, used to determine the number of units in the MPS, come from a *planning bill of materials*. The planning bill of materials has common parts and options as its components because the development of a specific bill of materials for all possible combinations of finished products is impractical.

The units for the MPS in the make-to-stock operation are all end-item catalog numbers, which is reasonable because MTS companies usually use batch production techniques and carry finished-good inventories for most, if not all, of the end items. In this case the MPS is a statement of how much of each of the standard products to make and when to produce them. In some companies, the specific products are aggregated into model groupings so that planning of common components achieves a better economy of scale. The items are then broken into specific products at the latest possible time. For example, in furniture manufacturing a *consolidated item number* is used in the MPS for furniture that is identical except for finished color. Then a separate system allocates lot size in the MPS for each possible color at the last step in the production process.

MPS Techniques

The principal method used to represent MPS data is the *time-phased record*, illustrated in Figure 7–8. The record is used to show the relationship between the *rate of output, sales forecast*, and the expected *inventory balance*. The number of periods indicated on the record is a function of the individual industry and product. The *Forecast* row represents the number of end-item units that the business anticipates will be sold in each period. The end items or numerical values put in each period in the Forecast row represent a disaggregation of the product data in the production plan. As the information flow diagram in Figure 7–1 indicates, the MPS time-phased record is developed directly from the production plan. Depending on the production strategy, the units listed in the Forecast row will be either actual catalog product numbers for MTS or information from planning bills of materials for ATO.

The Available row represents the inventory balance at the *end* of the period or the number of units available for sale in the next period. The MPS row indicates

	Period number									
	1	*2*	*3*	*4*	*5*	*6*	*7*	*8*	*9*	*10*
Forecast	5	5	5	5	5	5	20	20	20	20
Available	26	32	38	44	50	56	47	38	29	20
MPS	11	11	11	11	11	11	11	11	11	11
On hand	20									

Figure 7–8   MPS Time-Phased Record.

the number of units scheduled for production during the period and available to satisfy the forecast for the period. The On Hand value is the number of units present in inventory at the start of the first period. To avoid confusion between the On-Hand and Available values for period 1, remember that On Hand is the inventory coming into the first period, and Available is the inventory leaving the first period. The time-phased record is a method for visualizing how the MPS relates to forecast and inventory.

The time-phased record in Figure 7–8 shows all the data for ten periods of production. Study the data in the figure and try to determine if the production process is level or chase. Note that the forecast calls for the sale of five units per period through period 6 and twenty units per period through period 10. The production process is determined from the entries in the MPS row. Since the plan calls for the production of eleven units in every period, a level production process is present.

The master production scheduler starts with a time-phased record that has the Forecast row and On-Hand value present. Based on company policy and available resources, a decision to use a level, chase, or mixed production strategy is reached. For the record in Figure 7–8, a level production rate was chosen. The number of MPS units per period was determined by adding the forecast for 10 periods (110 units) and dividing by the number of periods (10), arriving at a period production rate of 11. Note that inventory is built up in the first six periods to cover the higher sales rate in the last four periods. With the MPS row now inserted into the record, the scheduler calculates the Available row, starting with period 1. The Available value at the end of period 1 equals the On-Hand balance plus the units produced in period 1 minus the forecast sales for the period (20 + 11 − 5 = 26). The same process is used for period 2, except that the Available value in period 1 becomes the On-Hand value for period 2 (26 + 11 − 5 = 32). Stop and make the calculations for periods 3 through 10 to verify that you know how to complete an MPS time-phased record.

In many production situations, the products and parts are produced in *lot sizes.* In the MPS record in Figure 7–8, the assumption was made that eleven units was an economical lot size for the product. If a different lot size is required to manufacture the product economically, that requirement must be reflected in the time-phased record. To demonstrate this concept, assume that a lot size of thirty units is the smallest production level used. The new time-phased record is illustrated in Figure 7–9. Note that the MPS would call for production of thirty units

	Period number									
	1	2	3	4	5	6	7	8	9	10
Forecast	5	5	5	5	5	5	20	20	20	20
Available	45	40	35	60	55	50	60	40	20	0
MPS	30			30			30			
On hand	20									

Figure 7–9 MPS Time-Phased Record with Lot Sizing.

in periods 1, 4, 7, and 10. Compare the Available row of the records in Figures 7–8 and 7–9. Note that inventory levels are higher when larger lot sizes are necessary. Also, no MPS value is present in period 10 because the twenty available in period 9 satisfies the forecast for twenty in period 10. However, that means the inventory for period 11 is zero. If there is a minimum level of inventory or *safety stock* required by manufacturing, an MPS order for thirty units in period 10 would need to be established.

If no lot size is required, the production system produces *lot for lot* or just what is required; however, because of production machine setup time and several of the other cost-added operations present in manufacturing, many manufacturing operations set a minimum lot size for production.

Computer-based production scheduling systems used in CIM implementations capture the same data presented in the time-phased record illustrated in Figure 7–8 and make all the calculations for future periods. However, the computer database displays the information in report formats different from those used for hand calculations.

Order Promising

The time-phased records illustrated in Figures 7–8 and 7–9 do not allow for production situations where customers place orders for future delivery. The earlier record format assumed that sales planned for each period would occur only in that period, when in reality sales planned for future periods may occur as an order from a customer in an earlier period. To handle this additional variation in planning the master schedule, the time-phased record is expanded to include two more rows: *ATP* (available to promise) and *Orders*. The record in Figure 7–10 illustrates this concept. The Orders row indicates all orders received for future products that will be consuming the forecast. The quantity listed in the Forecast row is expected to be sold and the values in the Order row represent firm sales of the product. The calculation for the Available row is unchanged.

The ATP value for period 1 is calculated by adding the On-Hand and MPS values and subtracting the total orders received up to period 3 $(10 + 30 - 6 - 5 - 3 = 26)$.

	Period number									
	1	2	3	4	5	6	7	8	9	10
Forecast	10	10	10	10	10	15	15	15	15	15
Orders	6	5	3							
Available	30	20	10	30	20	5	20	5	20	5
ATP	26			30			30		30	
MPS	30			30			30		30	
On hand	10									

Figure 7–10   MPS Time-Phased Record with Order Promising.

Space does not permit coverage of the many variations on the ATP concept practiced by companies using a time-phased MPS system, but it is important to know that order promising is an integrated part of the master production scheduling process.

Automating the MPS Function

The master scheduling function and the master production schedule have appeared in several figures in Chapters 6 and 7. Refer to Figures 6–7 and 7–1 and note the location of master production scheduling. In automated systems, there is often a program module or set of screens that helps support the entry and maintenance of the master production schedule. Functions supporting the input and maintenance of the production plan, its family level schedules, and rough-cut capacity analysis are often included with the master scheduling software tools. Modern system tools often provide the capability for what-if analysis that lets the master scheduler test different versions of the production plan and master schedule. These variations can then be evaluated to determine the best plan to send as an input to the detailed material and capacity planning functions.

7–4 INVENTORY MANAGEMENT

The American Production and Inventory Control Society (APICS) defines *inventory* as follows:

> Inventory consists of those stocks or items used to support production (raw materials and work-in-process items), supporting activities (maintenance, repair, and operating supplies), and customer service (finished goods and spare parts). Demand for inventory may be dependent or independent. Inventory functions are anticipation, hedge, cycle (lot size), fluctuation (safety, buffer, or reserve), transportation (pipeline), and service parts.

Inventories consist of *raw materials, component parts, work-in-process,* or *finished products and goods.* The singular reason for carrying a finished-goods inventory is to satisfy a customer lead time that is less than the production lead time. In many situations, it is not possible to eliminate the finished-goods inventory without losing market share. As long as the need for the inventory is an order-qualifying criterion for every competitive manufacturer, the cost of the inventory is included in each vendor's product prices. In that case, the battle to gain new customers and increase market share is fought in the production area.

Two fundamental reasons exist for carrying the raw material, component parts, and work-in-process inventory items listed earlier. The primary justification is the elimination or reduction of disturbances in the production cycle. These disturbances have many causes, including unreliable part supplier deliveries, poor quality, poor scheduling, and undependable production operations. The second reason to carry production inventory is to take advantage of the economies of larger order sizes from suppliers. Data from world-class companies indicate that *order-winning* advantages are possible for the industry willing to change how manufacturing functions. Therefore, the focus for the rest of this section is on better management of the production inventory.

The level of production inventory is a function of the complexity of the product and the layout of the manufacturing process. Flow-oriented production systems that use effective group technology require less inventory in both raw materials and work-in-process. Regardless of the system used, however, many manufacturing operations require some level of production inventory. Therefore, the critical issue in minimizing the cost-added effects of inventory is to know how much inventory is present and where it is located.

Benefits of Inventory Accuracy

The benefits of accurate inventory include fewer missed manufacturing schedules and delays in shipping because incorrect inventory counts created shortages of critical parts. In addition, safety stock can be eliminated or reduced significantly because inventory records indicate exactly the parts on hand. From a financial viewpoint, inventory accuracy means correctly stated inventory cost reports, less costly material expediting, and reduced losses due to obsolete and excessive inventory in stockrooms. Although the benefits of accurate inventory are numerous, the most important reason to solve the inventory accuracy problem is that material requirements planning and the other manufacturing planning and control software modules fail to function if the inventory values are less than 98 percent accurate.

Inventory accuracy is most often measured by one of two methods: inventory dollar basis or a physical count of the parts. The dollar method compares the dollar value of the parts in inventory with the dollar value stated in the financial records. The degree to which the two totals agree is a measure of the accuracy of the inventory part count. This method has two serious problems. First, the ability to identify poor inventory practices is reduced because the records often have offsetting errors. For example, consider an inventory error on two parts, one that has a $100 value and a second that has a $10 value. If the count were 10 high on the $10 part and 1 low on the $100 part, the dollar value of the inventory would be unchanged. The second serious problem is related to the first. Due to offsetting errors, manufacturing believes that the inventory stated in dollars means that the parts are available. In fact, when manufacturing obtains the last of the $100 parts from the stock area to finish an assembly, no part is found because the records were wrong. As a result, a shipment date is missed and a customer is not satisfied.

The only method that guarantees parts will be there for manufacturing is the *physical count method*. Inventory management is successful only if the inventory records and the physical count of parts in the stockroom have more than 98 percent agreement. For all parts tracked, the records and count must agree on both total parts present and location in the warehouse or plant. Therefore, the important questions for inventory management are: What needs to be counted? And how often does it need to be done?

Cycle Counting the ABC Part Classification

An accurate inventory record of every part in a product is not necessary as long as a sufficient quantity is available for final assembly. For example, it is not necessary to keep accurate on-hand balances of nuts, bolts, and washers that cost less than a few cents each. On the other hand, the part count and location of high-cost components (e.g., a $150 part) must have an accurate inventory record. The 80–20 rule applies in this situation. Experience shows that 20 percent of the parts account for 80 percent of the inventory value. Figure 7–11 illustrates this concept and indicates that accurate records for all of the A parts and probably some of the B parts is a good inventory management process.

For the balance of the parts in classifications B and C, a *two-bin* inventory system satisfies the production requirements. In such a system, parts are divided into two bins, the size of each being determined by the production rate and re-order lead time of the part. Parts are drawn from the production inventory bin to satisfy manufacturing; when this bin is empty, the purchasing department orders more parts. Production uses the second bin, which holds a sufficient number of parts to last through the lead time on the replacement parts order. The cost of parts in category C is low, so the burden of carrying the excess inventory is negligible.

To manage the more costly parts effectively, a physical count and verification of warehouse location are required. The method preferred is called *cycle counting*.

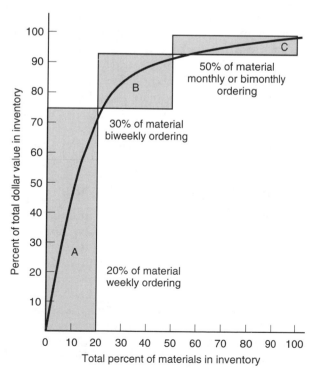

Figure 7–11 Inventory Classification.

In cycle counting, every item in classification A is checked regularly to verify that the quantity and location on the inventory record agree perfectly with the number of parts at a specific location in the warehouse. The cycle counting activity is assigned to specific persons with training in accurate cycle counting techniques. For example, some companies have cycle counting teams check a different group of part numbers each day. This practice ensures that all parts in classification A get counted every two or three production periods. In this type of system, accuracy is determined as follows:

$$\text{Inventory accuracy} = \frac{\text{total hits}}{\text{total counts}}$$

where the term *total hits* refers to every part number checked that had the inventory present within the required quantity range and all the parts in the correct location. For some parts, there is an allowed range in the count number or error where the count still falls within the definition of a *hit* or an accurate count. These parts usually are lower-cost items or parts where the count is determined through a weight scale reading. The *total counts* is the total number of parts checked through cycle counting. All discrepancies between the inventory record and the count are investigated thoroughly to improve continuously the reliability of the inventory management system. It is not uncommon for companies that implement cycle counting to achieve an inventory accuracy of 0.99. Under these ideal production conditions, 99 percent of the time when a part is checked, the count value falls within the range on the inventory record and the location is correct. In these situations, manufacturing knows that parts are available if the inventory record indicates parts are on hand.

The *annual physical inventory* is intended to prove that the dollar value of the inventory is correct as stated in the financial records. However, material requirements planning (MRP) and the manufacturing planning and control (MPC) software system require accurate inventory status on a day-to-day basis, not just once per year. The goal of cycle counting is to demonstrate a level of inventory accuracy that makes MRP effective and eliminates the need for the costly annual physical inventory.

Automating the Inventory Management Function

Refer again to Figures 6–7 and 7–1 and note the location of the inventory data files. In automated systems, there is often a program module or set of screens that helps support the entry and maintenance of the inventory data. Many functional areas use the inventory data for various calculations and reports. Software packages for inventory management often provide information on quantity of material on hand, the location and status of the material, and age/lot control information. Data in the inventory database are often used for cost analysis and inventory management decisions. Modern systems provide for analysis of inventory value by ABC classification, age of inventory (receipt date), and producer lot ID. Support for inventory cycle counting is often a feature of modern systems.

7–5 PRODUCT DATA MANAGEMENT

Management of the product data is critical for the MRP system to be effective because the accuracy and reliability of the output information is related directly to the quality of the data going into the system. The bill of materials and inventory levels are critical items in the management of product data. For example, bill of materials (BOM) accuracy of more than 97 percent and inventory information with more than 98 percent accuracy is critical for dependable MRP system generation of planned orders for manufacturing and purchasing. A technique called *cycle counting* used to achieve highly reliable and accurate inventory information was described in Section 7–4. A description of the bill of materials process follows.

Bills of Materials

World-class manufacturing organizations maintain just one *bill of materials* (BOM) for a product; however, the BOM is stated in different terms for different departments in the organization and has several representations. The BOM originates in product design when the parts to produce the product are either designed or purchased from vendors. The representation of the BOM in design engineering is often just a list of parts and subassemblies necessary to build the product. The information associated with each part includes quantity, part number, and specifications necessary for manufacturing or purchase. Figures 3–9 and 3–10 illustrate a part design and the BOM.

Production engineering, responsible for planning the total manufacturing and shipment of the product, frequently adds to the design bill boxing and packaging items along with raw material requirements. In manufacturing planning and control (MPC), the bill is represented as either a *product structure diagram* or an *indented bill of materials*. The product structure diagram and indented BOM for a simple product, a table, are provided in Figure 7–12. These two representations of the BOM are important in MPC because they describe how the product must be manufactured and assembled. For example, the legs, long rail, and short rail must be manufactured before a leg assembly can be produced. In addition, the product structure diagram also indicates that every leg assembly requires *two* short rails, *two* long rails, and *four* legs. While both the product structure and indented bill contain the same information, the representation as an indented bill is much easier to capture in MPC computer software. However, the product structure diagram helps to explain how the time-phased material requirements planning (MRP) records are used to plan the production of the table and its components.

In other types of production operations, *planning* bills of material are generated to represent product families with large numbers of end-item configurations. The planning bill is not a different BOM, just a different representation of the design bill. Bill of materials accuracy in the 97-plus percent range is possible when three conditions are satisfied: (1) responsibility for maintenance of the bill rests in one department, (2) a formal process for engineering change approval is established and religiously followed, and (3) a single image of all product data and BOM information is maintained in a central CIM database.

Figure 7–12 Product Structure and Indented Bill of Material.

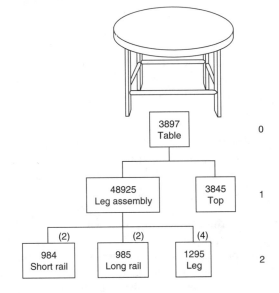

| | 3897 Table | 0 |

Indented bill of materials (BOM)

3987 Table (1 required)

    48925 Leg assembly (1 required)
      984 Short rail (2 required)
      985 Long rail (2 required)
      1295 Leg (4 required)

    3845 Top (1 required)

Item	Low-level code
Table	0
Leg assembly	1
Top	1
Short rail	2
Long rail	2
Leg	2

Transient Items in the Product Structure

Transient items are often used to represent items in the product structure and help keep engineering and manufacturing synchronized. Traditionally, the bill of material structure shows how various parts come together to make subassemblies, and how the subassemblies come together with other parts to make a final product. Manufacturing may determine that it is more efficient to combine steps and operations in a way that appears to change the product structure. The coding of levels of the structure that are to be combined at the request of manufacturing as "transient items" allows the structure to meet the needs of engineering and manufacturing. Transient items exist on the shop floor for a short time before they are consumed or converted into a higher level product. This type of item is also known as a "phantom item" or a "blow-through item." Transient items exist for such a short time that there is essentially no need to track them closely and report on them. They allow the structure diagram to show the various levels of assemblies and subassemblies, but they do not create additional work for manufacturing. Transient items have two important characteristics: they have no manufacturing lead time, and manufacturing orders are not created by the system for these items. The use of transients leads to the reduction of the effective levels in the bill of material. When transients are used in a material requirements planning system, the results are more efficient planning, less processing time, fewer planned orders for subassemblies, and much less reporting and transaction processing. In addition, the lower number of transactions leads to less possible reporting errors and improved system accuracy. A great source of additional information about the use of transients and how to improve bills of material and product structures can be found in the book *Bills of Material, Structured for Excellence*, by R. D. Garwood.

Automating the Product Data Management Function

Bills of material are often entered and maintained in the formal system using software for product data management (PDM). The PDM software often includes the functional ability to create items and describe the key item planning data that become part of the item master file. Once an item is defined or created in the planning system, it may become part of a product structure. PDM software typically allows the linking of items to a higher level item that they make up. The product structure information shown graphically at the top of Figure 7–12 is typically entered into the formal system through the product data management module or function. This input to the system specifies the quantity of each lower level item that goes into the higher level or "parent" item in a single level relationship. Complex structures are built up in the PDM software from these single-level relationships. PDM software typically allows the generation of several views of the product structure data, including single level and indented bills of material, as well as "where used" and "costed" bill of material reports. This software module is typically a requirement of the formal system and is addressed and installed first during the system implementation project. The common interfaces to the PDM system are shown in Figure 7–13.

7–6 SUMMARY

Manufacturing planning and control (MPC) is at the heart of the manufacturing organization and is the channel through which product design and marketing data must flow to reach the production floor. In the SME CIM model, MPC has responsibility for the planning and control of the shop floor, production materials, production scheduling, the quality process, and facility planning. Manufacturing planning and manufacturing control are the functions that MPC provides.

The planning element starts with the business plan, a forecast of expected future demand. The data to build this plan come from historical sources and forecasting techniques. The business plan is the basis for the development of the production plan, with output expressed in total number of units. The aggregate production plan data are broken into individual products, and the first disaggregate plan, called the master production schedule (MPS), is developed. A more detailed production schedule is produced using MPS data and a material requirements planning (MRP) process. Time-phased requirements for all manufactured and purchased parts and

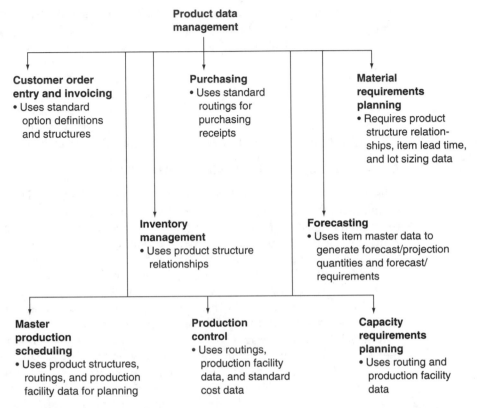

Figure 7–13 Common PDM Interfaces.

assemblies flow from the MRP planning system. In addition to the time-phased requirements, the MRP also provides data for the development of a capacity schedule, called capacity requirements planning (CRP).

Implementation of the production plan requires efficient application of variable resources. Three different production planning strategies, called chase, level, and mixed, are used to resolve the variation in direct labor hours required by the production plan. The chase strategy attempts to maintain zero inventory by varying the labor hours directly with sales volume. The level strategy maintains a constant workforce by using inventory built up in low-sales-volume periods to carry the higher product demand in high-volume periods. The mixed strategy is a combination of the chase and level processes.

The method used to develop the master production schedule (MPS) is called a time-phased record. The record shows the relationship among the rate of output, sales forecast, and expected inventory balance. Frequently, two other rows, called *orders* and *available-to-promise*, are added to the MPS record to allow order promising. Software to automate the MPS function is available from a wide variety of MPC software systems.

Inventory management works to minimize costs associated with inventories in the following areas: raw materials, component parts, work-in-process, and finished goods. One of the principal inventory management goals is inventory accuracy, which means the correct quantity in the correct location. Accuracies of more than 99 percent are recorded by some world-class manufacturers. The primary process used to achieve an accurate inventory is cycle counting. The cycle counting process is often used with an ABC parts classification technique. Software to automate the inventory management function is also available from a wide variety of MPC software systems.

Product data management (PDM) and the control of the bill of materials (BOM) are at the heart of an effective material requirements planning (MRP) system. An accuracy of more than 97 percent for the BOM is necessary for accurate time-phased production schedules. Software is available to automate the PDM function by transferring BOM data from the design drawings directly into the MPC software system.

REFERENCES

BERRY, W. L., T. E. VOLLMANN, and D. C. WHYBARK, *Master Production Scheduling Principles and Practice.* Falls Church, VA: APICS, 1979.

CLARK, P. A., *Technology Application Guide: MRP II Manufacturing Resource Planning.* Ann Arbor, MI: Industrial Technology Institute, 1989.

COX, J. F., and J. H. BLACKSTONE, Eds., *APICS Dictionary,* 9th Ed. Falls Church, VA: APICS—The Educational Society for Resource Management, 1998.

COX, J. F., J. H. BLACKSTONE, and M. S. SPENCER, *APICS Dictionary,* 7th ed. Falls Church, VA: American Production and Inventory Control Society, 1992.

GARWOOD, R. D., *Bills of Material, Structured for Excellence.* Dogwood Publishing Co: Marietta, GA, 1988.

SOBCZAK, T. V., *A Glossary of Terms for Computer-Integrated Manufacturing.* Dearborn, MI: CASA of SME, 1984.

STEVENSON, W. J., *Production/Operations Management,* 3rd ed. Homewood, IL: Richard D. Irwin, 1990.

VOLLMANN, T. E., W. L. BERRY, and D. C. WHYBARK, *Manufacturing Planning and Control Systems,* 4th ed. Homewood, IL: Richard D. Irwin, 1998.

QUESTIONS

1. Where does the MPC function reside in a manufacturing operation?
2. Describe the production planning process using the model in Figure 7–1.
3. How does MPC link to the CIM common database and other departments in the enterprise?

4. Describe the resources used to forecast production requirements.

5. Describe the production planning process.

6. Describe the three different production planning strategies: chase, level, and mixed.

7. Describe the two options available when the chase strategy is used.

8. Compare the advantages and disadvantages present when the chase and level production strategies are adopted.

9. What is the master production schedule (MPS)?

10. Compare and contrast the MPS with the production plan and the forecast and describe similarities and differences.

11. What changes occur in the MPS for the various production strategies listed in Figure 2–5?

12. Describe the function of the MPS time-phased record and include a description of the terms.

13. What impact does the introduction of lot sizes have on an MPS record?

14. Compare lot-for-lot and lot-size production and describe the differences.

15. What is order promising, and how does it affect the MPS?

16. How is the MPS function automated?

17. Define *inventory* and list the sources of inventory.

18. Describe the causes of inventory and discuss the positive and negative aspects of inventory balances in each of the sources.

19. Describe the two methods used to measure inventory accuracy, and discuss the advantages and disadvantages of each.

20. What is cycle counting?

21. Describe the ABC parts classification method of inventory control.

22. How do cycle counting and the ABC method provide accurate inventory control?

23. How is the inventory function automated?

24. What is *product data management* (PDM), and how do the accuracies of the bill of materials (BOM) and inventory affect PDM?

25. How do BOMs differ in different departments in the enterprise?

26. Describe the relationship among BOMs, product structure diagrams, indented BOMs, and planning bills.

27. What conditions must be met to achieve BOM accuracy in the 97-plus percent range?

28. How is the PDM function automated?

PROBLEMS

1. Develop an electronic spreadsheet template to generate the chase production strategy data in Figure 7–3. Input the sales and days-worked values from the

figure and use the $100 of sales per direct labor hour to verify the values in the columns for Labor Hours, Variable Workforce, and Variable Work Week.

2. Develop an electronic spreadsheet template to generate the level production strategy data in Figure 7–4. Input the sales and days-worked values from the figure and use the $100 of sales per direct labor hour to verify the values in the columns for Labor Hours, Workforce Level, Monthly Production, and Inventory Balance.

3. The demand and production rates for a company that permits back orders is illustrated below. With a beginning inventory at the start of period 1 of 150 units, calculate the back orders per period, average inventory, and ending inventory.

Period	Demand	Production
1	5500	5350
2	8200	5800
3	3100	6000
4	4500	4000

4. A company using a level production policy has the following demand for six periods:

Period	Demand
1	8
2	6
3	14
4	6
5	9
6	5

With a zero starting inventory balance, determine the following:

a. The level production rate that will result in a zero inventory balance after period 6.

b. The inventory and back orders per period with the production rate in 4.

c. The level production rate that would eliminate back orders.

5. Atlas Furniture Company has the following annual sales (times 100,000) forecast: January, $6; February, $7.5; March, $8.2; April, $9.5; May, $13.8; June, $10.5; July, $7; August, $5.8; September, $6.7; October, $7.6; November, $9.6; December, $12.3. Analyze the chase, level, and mixed production strategies using the spreadsheets developed in Problems 1 and 2. Use the same values for days worked and $50 of sales per direct labor hour in the calculations. Discuss the strengths and weaknesses of each strategy.

6. Use the graphing function of the spreadsheet to generate graphs to support the conclusions from the analysis in Problem 5.

7. Complete the following MPS record, assuming a level production strategy and a safety stock of 10.

					Period					
	1	2	3	4	5	6	7	8	9	10
Forecast	5	5	5	5	10	10	20	20	20	20
Available										
MPS										
On hand	15									

8. Complete the MPS record in Problem 7 while assuming a minimum lot size of 25. Would the resulting MPS require a level, chase, or mix production strategy?

9. A company inventory accuracy standard has the following tolerance on cycle counts (percentage variation from recorded inventory value): class A items, ±0.5 percent; class B items, ±1 percent; class C items, ±3 percent. All class A items are hand counted, and the class B and C items, which are weight counted, have an additional 1 percent of error permitted. Determine which of the following items have cycle count data within specifications.

Item	Class	Count basis	Recorded inventory	Counted inventory
1	A	Hand	56	55
2	C	Scale	10,425	10,950
3	B	Hand	2,850	2,825
4	A	Hand	430	431
5	B	Scale	8,530	8,745

10. The following inventory data apply to the product structure for the table assembly shown in Figure 7–12. There are no scheduled receipts for any parts, and the lead time for each is one week.

Item number	Inventory
3897	15
48925	20
984	40
3845	25
985	45
1295	72

For each of the following situations, determine how many tables could be delivered to the customer at the end of the period.

   a. No problems; the inventory data are accurate and all parts meet quality specifications.

   b. Ten of the 985 part numbers are defective.

   c. A cycle count determines that the inventory for part 984 is 35 parts and not 40.

PROJECTS

1. Using the list of companies developed in Project 1 in Chapter 1, determine the type of production method (chase, level, or mixed) used for each of the products they produce.

2. Compare the production planning flowchart in Figure 7–1 with the process used by one company from the small, medium-size, and large groups, and describe how their operations differ from that of the model.

3. Compare the bill of materials and inventory accuracy of the companies selected in Project 2 with the standards established in the chapter. Describe the inventory counting method used by each company and, discuss how that method contributes to the accuracy of inventory data.

4. Consider the problem of manufacturing from the materials management perspective: getting the right materials to the right place, at the right time, to satisfy the customer's requirements.

   a. Who are the people (functional job titles) involved?

   b. What data related to a specific item or part need to be documented in the manufacturing database?

   c. What information needs to be included in the product structure?

   d. What information related to inventory is needed, and what level of accuracy is needed?

   e. What additional information is needed for effective material planning and the issuing of purchase orders?

5. Consider the problem of manufacturing from the manufacturing management perspective: getting the right products delivered to the right place, at the right time, to satisfy the customer's requirements.

   a. Who are the people (functional job titles) involved?

   b. What information is required to define clearly the manufacturing resources (capacity) needed for production?

   c. What is the critical information that should be included on the manufacturing order to be released to the shop floor?

CASE STUDY: PRODUCTION SYSTEM AT NEW UNITED MOTOR MANUFACTURING, PART 1

On February 17, 1982, General Motors Corporation and Toyota Motor Corporation reached agreement on the formation of a joint company, later named New United Motor Manufacturing, Inc., to produce a subcompact car. The location chosen for the operation was Fremont, California, at the site of a GM plant that was closed because of poor production and labor troubles.

Background

Each company wanted to achieve specific objectives from the joint venture. General Motors wanted firsthand experience with the cost-effective and efficient *Toyota production system* and wanted a high-quality vehicle for the Chevrolet Division. Toyota needed experience working with American unionized labor and wanted immediate access and information to U.S. suppliers. The venture was an experiment in the production of an automobile that blended the Toyota production philosophy with U.S.-supplied parts and a unionized workforce.

After two years and an estimated $350 million for renovations and improvements to the existing facility, the first Nova cars were produced in December 1984. The Nova was produced from 1984 to 1988, then production was shifted to the Toyota Corolla, and eventually to the Chevrolet Geo Prizm. In 1989, the workforce included 2300 hourly and 400 salaried team members organized into approximately 340 teams with 100 group leaders. Production totaled 485 Geo Prizms and 335 Corolla sedans per day, with approximately 1100 vehicles in the system at one time. At the present time, the total time to produce a vehicle is about 19 hours, so the facility has the capacity to produce 220,000 cars annually.

The Production System

The production system at New United Motor covers 60 acres with over 3 million square feet of covered production space. The final assembly line is 1.3 miles long and includes the installation of over 3000 automotive parts. Parts are supplied from U.S. and Japanese vendors, with the engine and transmission coming from Toyota's Japanese production facility. The joint venture uses 74 U.S. parts vendors, with 17 located in California. Just-in-time production requires forty-five truck shipments daily and four ships from Japan weekly.

Most of the body parts are produced in the stamping plant at the site, where twenty-six stamping presses produce eighty-one different body parts from approximately 300,000 pounds of steel per day The body shop assembles the body-in-white with 95 percent of the welds on each vehicle performed using automation. Flexible automation utilizing 210 robots produces 70 percent of the 3800 welds on each vehicle.

The function of the system is to manufacture cars with quality as high as that anywhere in the world while ensuring that product costs are the most competitive of any manufacturer. The philosophy of the production operation is that quality should be ensured in the production process itself. The system is built on the concept of *work teams*, with an average team composed of six members. Team meetings, led by the *team leader,* are held in forty team rooms placed across the production facility. The team leader is directly responsible for the performance of the individual team. The team leader is a member of the bargaining unit and an integral part of the team-building process. Three to five teams are organized into a group with a *group leader*. This person is the first line of salaried supervision at New United Motor. The key responsibility of the group leader is to ensure open two-way communications between team members and managers.

In addition to production responsibilities, team members also follow the four *S's: seri* (clearing), *seiton* (arrangement), *seiketsu* (cleanliness), and *seiso* (sweeping and washing) These four Japanese words help team members focus on the practice of work-cell order and control. If the workplace is clean, well organized, and neat, production will be more efficient, easier, and safer. To achieve this efficiency in production, three concepts are emphasized: *just-in-time* production; *jidoka*, a Japanese term meaning *the quality principle;* and full utilization of the worker's abilities. A description of these principles and other aspects of the Toyota Production System are included in the continuation of this case study at the end of Chapter 8.

Material Planning, Production Scheduling, and Operating Systems

The manufacturing planning and control process in the CIM enterprise is responsible for the aggregate and disaggregate planning of production and scheduling of manufacturing resources. The aggregate planning process starts with a production plan stated in broad product specifications. The first disaggregate plan, broken into specific product models, is called the master production schedule (MPS). The MPS states the production plan for each model over several production periods using an MPS record. The output of the MPS record provides the data for the material requirements planning (MRP) scheduling system.

8–1 MATERIAL REQUIREMENTS PLANNING

The American Production and Inventory Control Society (APICS) defines *material requirements planning* (MRP) as follows:

> Material requirements planning (MRP) is a set of techniques that uses bill of material data, inventory data, and the master production schedule to calculate requirements for materials. It makes recommendations to release replenishment orders for material. Because it is time-phased, it also makes recommendations to reschedule open orders when due dates and need dates are not in phase. Time-phased MRP begins with the items listed on the MPS and determines (1) the quantity of all components and materials required to fabricate those items and (2) the date that the components and material are required. Time-phased MRP is accomplished by exploding the bill of material, adjusting for inventory quantities on hand or on order, and offsetting the net requirements by the appropriate lead times. (*APICS Dictionary,* 9th ed., 1998.)

The description of the MRP operational model was given in Chapter 6 and is supported by Figure 6–8. Study this figure, and verify that the model is compatible with the MRP definition.

The relationship of MRP to the bill of materials and use of the MRP record to calculate the time-phased release of orders for manufacturing are fundamental to the operation of the manufacturing planning and control (MPC) system.

Therefore, the function and creation of the MRP record is an appropriate place to start a study of material requirements planning.

The MRP Record

The record in Figure 8–1 represents the production plans for the part number under study. The accepted convention considers period 1 the current period, and periods 2 through 10 to be in the future. The period is a specified production time used for planning. For example, some companies equate each period to five working days, with the period starting on Monday and ending on Friday. When this period convention is used, all the entries in the record have the value listed at either the start of the period, the beginning of the workday on Monday, or at the end of the period, the end of the workday on Friday. An understanding of the MRP process starts with a working knowledge of all the words used in the record. Study the following definitions until you understand the terminology for MRP record data.

- *Period number.* The period number is the time duration used in the MRP planning process. Usually, one period represents a day, week, or month; however, it could be any number of days or hours.
- *Part number.* The part number identifies the specific part being planned with the MRP record.
- *Gross requirements.* Gross requirements equal the anticipated future demand, both independent and dependent, for an item, stated period by period, not aggregated or averaged. The gross requirements consist of a statement of the exact number of parts needed in each of the periods covered by the MRP record.
- *Scheduled receipts.* The scheduled receipts represent all orders released to manufacturing (production, manufacturing, or shop orders) or to suppliers through purchase orders. Another term used for scheduled receipts is *open orders.* The quantity listed as a scheduled receipt arrives at the *start* of the period in which the item quantity number appears. Remember, sched-

		Period number							
Item: tube steel #246784		1	2	3	4	5	6	7	8
Gross requirements					100				
Scheduled receipts									
Projected on hand	140	140	140	140	40	40	40	40	40
Planned order receipts									
Planned order releases									

Order Policy = Fixed lot   Order Quantity = 200   Lead time = 5

Figure 8–1 MRP Record.

uled receipts represent orders that have been placed with either manufacturing or a vendor and *do not* include orders for parts or raw material that will be placed in the future.

■ *Projected on hand.* The projected available balance represents the calculated inventory for the item projected through all the periods on the record. The projected available balance is the running sum of on-hand inventory minus gross requirements plus scheduled receipts and future planned order releases. The value for the projected available balance represents the inventory at the *end* of the period in which it appears. Therefore, the projected available balance is the inventory available at the start of the next period. A special value of the inventory is provided in the box immediately after the term "Projected on hand." This value is the inventory at the end of the previous period or the on-hand inventory balance ready for use in period 1.

■ *Planned order receipts.* The planned order receipts entry in the record indicates when a planned order would be received if the planned order release date is exercised. As soon as the order is placed, the value in planned order receipts moves up to a scheduled receipt at the beginning of the period.

■ *Planned order releases.* The planned order releases are the suggested order quantity, release date, and due date generated using MRP software. Orders are released at the beginning of a period. These suggested planned orders are calculated by the MRP software based on the gross requirements for the period and the inventory balance available to satisfy the gross requirements. When a planned order is finally released, it converts into a scheduled receipt.

■ *Lead time.* The lead time is the time between the release of an order and the completion or delivery of the order. For a manufactured item, lead time represents the time required to produce the quantity of parts in the order. In a subassembly, lead time represents the time required to complete the assembly.

■ *Lot size.* The lot size, when a specific value is given, is the required minimum order quantity determined by the economics of the production process. Usually, if the net part requirements exceed the lot size, the planned order is specified in multiples of the minimum lot size. If the lot size is listed as *lot for lot*, the planned order quantity is equal to the *net requirements* for the period. Lot-for-lot capability indicates that production efficiencies permit any quantity to be manufactured economically.

■ *Safety stock.* Safety stock is the lowest level of inventory allowed in the projected on-hand line. Safety stock protects against variations in delivery from manufacturing or from vendors due to production or quality problems.

Under normal operation, seven of the entries—part number, gross requirements, scheduled receipts, projected on-hand inventory available for period 1, lead time, lot size, and safety stock—are known quantities in the MRP record. Calculation with computer software of the remaining data in an MRP record is performed in an MRP run. After the MRP run, the projected on-hand balance and planned order releases are calculated for all periods covered by the record. The manufacturing planning and control (MPC) system uses the MRP records to generate a time-phased production plan for each part in the product structure.

MRP Calculations

The MRP record in Figure 8–2 illustrates the values present in the record before the start of MRP calculations. The *gross requirements* (16 units) come from either the next-highest level in the product structure diagram or from the master production schedule. The *scheduled receipts* value (three times the lot size of 5, or 15 units) represents planned order releases that became firm orders when the orders were placed with either manufacturing or a vendor in the last period. The *on-hand* inventory for the start of the first period (4 units listed in the box) is determined from the MRP record from the last period and often verified by a cycle count of part inventory. The *lead time, lot size,* and *safety stock* are values set by purchasing and manufacturing.

The calculations for the record start with the first period and proceed to the last. The calculations determine the *projected on-hand balance* and the need for a *planned order release.* If the projected inventory balance is positive and above the *safety stock* level, no action is required for that period. However, if the projected inventory balance is negative or less than the required safety stock, a planned order release is required and must be included in the inventory balance calculations. The calculations are easy to understand when illustrated by an example using the starting data in Figure 8–2. Check the fully completed MRP record in Figure 8–3 as you read through the calculations that follow.

	Period number									
Part number	*1*	*2*	*3*	*4*	*5*	*6*	*7*	*8*	*9*	*10*
Gross requirements	16		8	15	21		12	15		28
Scheduled receipts	15									
Projected on hand 4										
Planned order receipts										

Planned order releases
  Lead time = 1
  Lot size = 5
  Safety stock = 0										

Figure 8–2 Basic MRP Record with Starting Values.

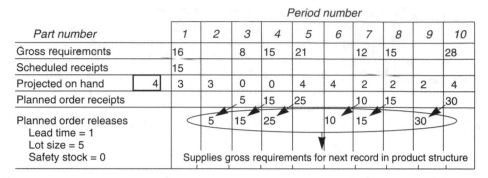

Figure 8–3 Completed MRP Record.

Period 1 Calculations:

$$\underset{\text{inventory}}{\text{Starting}} + \underset{\text{receipts}}{\text{scheduled}} - \underset{\text{requirements}}{\text{gross}} = \underset{\text{balance}}{\text{projected available}}$$

$$4 + 15 - 16 = 3 \text{ units}$$

The projected available balance of 3 units would be available at the start of the second period.

Period 2 Calculations:

$$\underset{\text{inventory}}{\text{Starting}} + \underset{\substack{\text{receipts}\\ \text{(or planned}\\ \text{order}\\ \text{receipts)}}}{\text{scheduled}} - \underset{\text{requirements}}{\text{gross}} = \underset{\text{balance}}{\text{projected available}}$$

$$3 + 0 - 0 = 3 \text{ units}$$

The starting inventory for period 2 is the ending inventory for period 1. Period 2 could have either a scheduled receipt or a planned order receipt, depending on the production needs and the lead time. In this situation, neither is present. The equations for all subsequent periods will be the same as the equation given for period 2.

Period 3 Calculations:

$$3 + 5 - 8 = 0 \text{ units}$$

The inventory balance at the end of the period is zero units because the gross requirement for 8 units consumes the inventory balance of 3 plus the planned delivery of 5 additional units.

Period 4 Calculations:

$$0 + 15 - 15 = 0 \text{ units}$$

Period 5 Calculations:

$$0 + 25 - 21 = 4 \text{ units}$$

Continue to use the formula for the last five periods and verify that you can understand the calculations.

The Product Structure and the MRP Record

Study the product structure diagram and the indented bill of materials illustrated in Figure 7–12 and determine how many different assemblies, subassemblies, and parts are represented. The table assembly in the product structure diagram, part number 3897, is constructed from five components and subassemblies. Every box in the product structure diagram is covered by an MRP record; as a result, the MRP records are linked, as Figure 8–4 indicates. A study of the links between the MRP records in the figure indicates that the gross requirements for records at level 2 are generated from the planned order releases from level 1. For example, the gross requirements for the short rail, number 984, came from the planned order releases from the 48925 leg assembly. Figure 8–5 shows the planned order release for the 48925 leg assembly flowing into the MRP record for the 984 short rail. The planned order releases for the leg assembly in period 1 is 30 units; therefore,

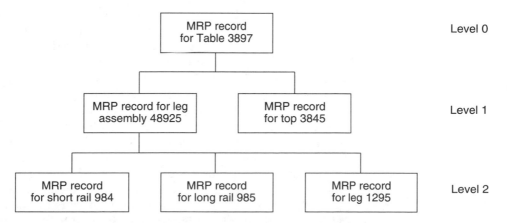

Figure 8–4   MRP Records for Table Product Structure.

Product 48925 Leg assembly	Period number									
	1	**2**	**3**	**4**	**5**	**6**	**7**	**8**	**9**	**10**
Forecast	10	10	10	20	10	10	20	20	20	20
Available	25	15	5	30	20	10	65	45	25	5
MPS	(30)			(45)			(75)			
On hand	5									
			2 x			2 x			2 x	
Part number 984 Short rail	**1**	**2**	**3**	**4**	**5**	**6**	**7**	**8**	**9**	**10**
Gross requirements	(60)			(90)			(150)			
Scheduled receipts	40									
Projected on hand 23	3	3	3	3	3	3	3	3	3	3
Planned order receipts				90			150			
Planned order releases		90			150					

Lead time = 2
Lot size = 5
Safety stock = 0

Note: 2 of the 984 short rails are required for each 48925 leg assembly

Figure 8–5   MRP Record Interface.

the gross requirements for the short rails in the same period is 60 units. The gross requirements for the short rails is twice the number of required leg assemblies because the product structure (Figure 7–12) indicates that it takes two short rails for every leg assembly. Check periods 4 and 7 and verify that you understand how the planned order releases from one record flow into the gross requirements of the record at the next lower level.

The final assembly MRP record, part 3897, does not have a higher level record to dictate the gross requirements for finished parts. Therefore, the part number at the zero level in the product structure diagram (Figures 7–12 and 8-4) uses the *master production schedule* (MPS) to determine the quantities for the gross requirements entry. The MPS and MRP records in Figure 8–6 show how the MPS quantities flow down to the final assembly MRP record. The MPS is used to determine the MRP gross requirement quantities in each period for the table, part number 3897. With production information for the table available, the MRP record at the highest level (Figure 8–4) in the product structure is completed. With the highest level planned, the MRP records at level 1 and then at level 2 are completed, and planning is complete for all subassemblies and components of the table. These calculations for all items in the product structure are performed by MRP software when an MRP run is executed. Study the product structure and MPS/MRP records in Figures 8–1 through 8-6 until this concept is understood.

Product Table		Period number									
		1	2	3	4	5	6	7	8	9	10
Forecast Available		10	10	10	20	10	10	20	20	20	20
MPS		25	15	5	30	20	10	65	45	25	5
On hand		(30)			(45)			(75)			
		5		1 x			1 x			1 x	
Part number 3897 Table		1	2	3	4	5	6	7	8	9	10
Gross requirements		(30)			(45)			(75)			
Scheduled receipts		40									
Projected on hand	5	15	15	15	10	10	10	15	15	15	15
Planned order receipts					40			80			
Planned order releases			40			80					
Lead time = 3											
Lot size = 20											
Safety stock = 0											

Figure 8–6   MPS and MRP Record Interface.

Computer-Assisted MRP

The previous example uses a simple product with few subassemblies and components; as a result, the calculations could be performed manually. However, imagine the product structure diagram for a riding lawn mower, automobile, videocassette recorder, or computer. There are often over 30 levels in the bill; hundreds of different parts and subassemblies, each requiring an MRP record; and thousands of individual items. In addition, there are usually multiple models of the final assembly, which share common parts and subassemblies. As a result, the MRP records for these common items have gross requirements coming from different sources that must be combined before the final production plan is completed. As an added burden, the MRP plan is never static; gross requirements, lead time, and on-hand balances change frequently. It is obvious that manual MRP calculations on final assemblies of this size are impossible; as a result, computer software is used.

Even with computer assistance, it is not uncommon for the MRP run to take hours for the computer to complete. As a result, MRP runs on products with large parts counts are often performed monthly or at best once a week. Changes in the gross requirements, lead times, and on-hand inventory, which can occur daily, make the monthly or weekly MRP recommendations for planned order releases inaccurate. To overcome this problem, MRP software has a feature called *net change*. When executed, the net change option recalculates the MRP records affected by the changes and produces a revised set of recommendations for the planned order releases. For example, if an inventory cycle count changed the starting on-hand balance for the top (part number 3845 in Figure 7–12), the net change calculations would require recalculation of the MRP record for the top. However, if the starting inventory balance changed for the leg assemblies (part number 48925 in the figure), the net change option would need to recalculate the MRP record for the leg assemblies and possibly for all three components at level 2. An example of the change in an MRP record due to a change in information is illustrated in Figure 8–7. The MRP calculations from the previous period indicated that an on-hand balance of 4 units was expected for period 1. However, an inventory cycle count turned up 9 units instead of 4. The impact on the MRP record is illustrated in the figure. Notice that the net result is the cancellation of a planned order release for 5 units in period 2.

The Benefits of MRP

The last example clearly illustrates one of the benefits provided by MRP. The correction of the on-hand inventory from cycle count data is common in manufacturing. The use of the MRP system allows production control personnel to view the effect of the change in future periods. They can see that a planned order recommended by MRP for period 2 could now be canceled. The effect of that cancellation on parts lower in the product structure is also easily analyzed. The primary benefit is that problems in manufacturing due to disturbances in the production system are solved early when a greater number of alternatives are available to the planner.

The second substantial benefit from implementing MRP results from the preparation for the installation. As stated earlier, an accurate bill of materials (BOM) and a cycle-count process to guarantee reliable inventory records are minimum

Part number		Period number									
		1	2	3	4	5	6	7	8	9	10
Gross requirements		15		8	15	21		12	15		28
Scheduled receipts	9	15									
Projected on hand	4	4 9	4 9	1	1	0	0	2	2	2	4
Planned order receipts				5	15	20		15	15		30
Planned order releases			5	15	20		15	15		30	
Lead time = 1					↓						
Lot size = 5	Supplies gross requirements for next record in product structure										

Figure 8–7   Impact of Change on an MRP Record.

conditions for successful MRP operation. The self-study used to improve the BOM and inventory tracking uncovers other operations that do not add value to the product. The correction of these problems and the improvements in inventory and BOM add substantially to the profitability and quality of the products. The following list of improvements in the operation of the enterprise are frequently attributed to implementing MRP.

- Improved customer service.
- Reduction in past-due orders.
- Better understanding of capacity constraints.
- Significant increases in productivity.
- Reduction in lead time.
- Reduction in the inventory for finished goods, raw material, component parts, and safety stock.
- Reduction in work-in-process (WIP).
- Elimination of annual inventory.
- Significant drops in annual accounting adjustment for inventory problems.
- Usually, a doubling of inventory turns.

Automating the MRP Function

Material requirements planning (MRP) is often a program module or a defined function in a formal system software package. MRP software is somewhat mysterious. The software typically follows the model of the system shown in Figure 6–8. The MRP function pulls schedule data from the master schedule, production forecast, or customer orders. The structure or recipe is pulled in from the bills of material. The on-hand balances are found in the inventory data files, and general item data are found in the part master files. The module then works to complete the following:

1. The calculation of the gross material requirements for each production order.
2. The consideration of on-hand inventory and open purchase orders for the items needed to complete the production orders.
3. The calculation of the remaining material that must be obtained to complete the production orders.
4. Time phasing or back scheduling from the due date of the scheduled production orders to determine the critical dates for the planning, release, and scheduled receipt of new materials and intermediate subassemblies.
5. Identification of problems and conflicts and the generation of exception messages and reports to the responsible people working with the system.
6. Generation of dependent item subassembly schedule dates that become input to the capacity requirements planning function.

Figure 8–8 describes the general data flows and system interfaces related to the MRP system.

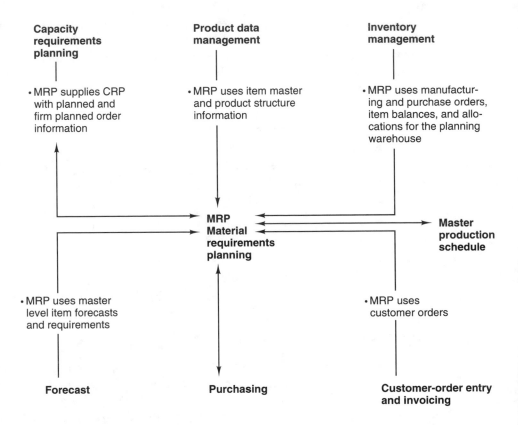

Note: MRP's net change planning reflects activity by any of the interfacing applications.

Figure 8–8   MRP System Interface.

8–2 CAPACITY REQUIREMENTS PLANNING

A review of capacity management and control in the manufacturing planning and control model in Figure 6–4 reveals the fact that capacity planning occurs at three different points in the production control process: (1) variable resource planning supports the production planning process, (2) rough-cut capacity planning aids in the development of a valid master production schedule (MPS), and (3) *capacity requirements planning* (CRP) supports the material requirements plan (MRP). Priority schedules for work cells are set through the production activity control process discussed in Chapter 7. The effectiveness of work-cell scheduling in utilizing capacity is monitored by the capacity control function using input/output control reports.

The CRP process produces the infinite shop scheduling and loading discussed in Chapter 6. The shop load, usually expressed in hours of work per work center per period, is produced by the CRP software using open shop orders and planned order releases generated in MRP. Infinite loading occurs because CRP schedules capacity for each part independently and then sums the planned capacity for each work center for every part processed. If a work center is used by many of the parts, there is a good likelihood that the planned capacity exceeds the hours available in the period. Study the CRP process described by the flow-chart in Figure 8–9. The process starts with a tentative MPS that is converted into planned order releases through the MRP process. Using the routings to identify

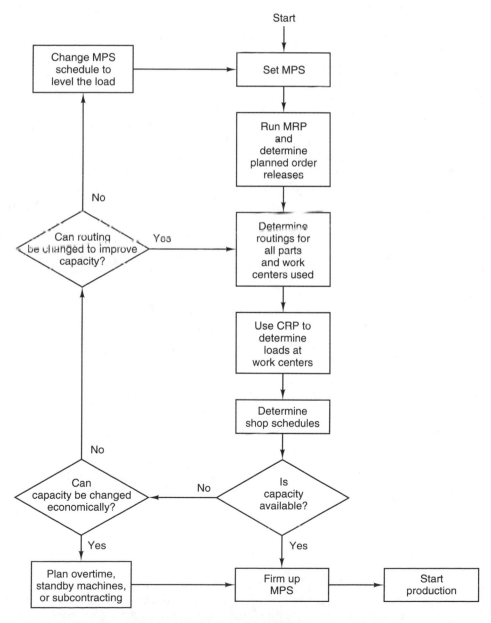

Figure 8–9 Process to Establish MPS for Available Capacity.

the work centers required, the material requirements are converted into labor and/or machine loads at the work centers. The machine loads are scheduled through CRP and a check on planned capacity versus available capacity is made. Work centers with sufficient hours to produce the parts have the orders released. However, work centers with loads that exceed capacity require a change in either the routings or MPS to meet the existing capacity, or an increase in capacity through overtime, additional shifts, use of additional machines, or subcontracting of the work. Study the flowchart until you understand the interaction among the MPS, MRP, and CRP functions in achieving a workable schedule for the shop floor.

Automating the CRP Process

Capacity requirements planning (CRP) is often a program module or a defined function in a formal system software package. The CRP function can be found in Figures 6–7 and 7–1. CRP is a logical process for calculating the requirements for specific resources. It works to plan physical resource requirements in much the same way that MRP works to plan material resource requirements. The CRP function receives schedule data from the master schedule and the subassembly schedules generated by MRP. Data files are established to provide work centers with specific information on labor requirements, scheduling, and the expected capacity available for use. The resource requirements for the items to be produced are determined from the routing documentation of the specific work operations to be completed and the work center where that work is to be done. General item data are pulled from the part master files. The CRP module then works to complete the following:

1. The calculation of the workload for each operation required to complete the production orders using the time standards stored in the item master files.
2. The generation of load profile information showing the calculated amount of work to be done at each workstation. This generation is done most often without restraints on the scheduled load in any time period, creating a so-called infinite load situation.
3. Identification of problems, schedule conflicts, and resource shortages, and the generation of exception messages and reports to the responsible people working with the system.
4. Generation of dependent item subassembly schedule dates that become input to the capacity requirements planning function.

Figure 8–10 describes the general data flows and system interfaces related to the CRP system.

8–3 FROM REORDER-POINT SYSTEMS TO MANUFACTURING RESOURCE PLANNING (MRPII)

A study of the scheduling and planning systems used in manufacturing production and control (MPC) over the last forty years will illustrate the impact of computer technology on manufacturing in the global market. The evolutionary

Figure 8–10   CRP System Interfaces.

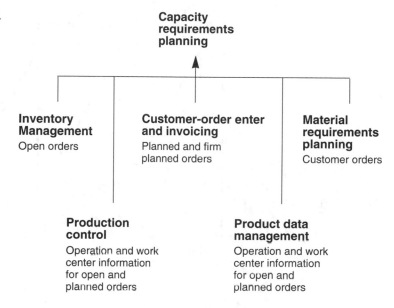

change in the production and planning process was fueled by advances in technology that permitted manual planning processes to use the computational power of the computer for increased control. Initially, order timing decisions were based on a *reorder-point system*.

Reorder-Point Systems

The timing of replenishment orders under the *order-point rule* is determined by the use of trigger levels. The inventory level was monitored continuously and a replenishment order for a fixed quantity was issued when the part count dropped to a specified level. The graph of inventory levels in Figure 8–11 illustrates the order-point concept. The sawtoothed solid lines in the figure represent the inventory level at any time during production. The inventory starts at the highest level, M, on the graph and drops at a uniform rate as parts are used in production. When the inventory level reaches the reorder level marked by the dash line labeled R, an order is placed for the quantity of parts indicated in the figure. The lead time required to produce the parts is highlighted in the figure between points 1 and 2. Note that the trigger point is selected so that a new order arrives as the current inventory reached the safety stock level, S. Usually some safety stock is carried on critical parts to prevent stock depletion.

The selection of the reorder point is influenced by four factors: the production demand rate for parts, the lead time required for replenishment inventory, the degree of uncertainty in the demand rate and lead time, and the management policy concerning inventory shortages. If the demand rate and lead time have a high degree of certainty, there is no need for safety stock and the reorder point is easily set. In most manufacturing operations, however, demand is rarely constant, and

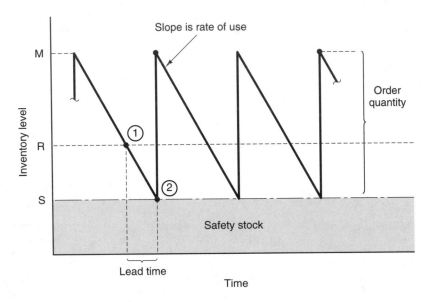

Figure 8–11   Order-point System.

the lead time, especially from external vendors, is affected by market conditions. Thus, the need arose for a system more responsive to actual and forecasted demand—not one based solely on historical data. The result was the development of the material requirement planning (MRP) process.

Open-Loop MRP

The open-loop MRP process described earlier in the chapter provided an alternative planning mechanism for parts and subassemblies in the product structure. These *dependent demand* items are identified by exploding the schedule for the finished product through the bill of materials and accounting for orders due and inventory on hand. The demand for these items is related to the product structure; for example, the quantity of parts in level 2 of the product structure shown in Figure 7–12 is set by the production quantities identified in level 1. Therefore, the MRP process was used to schedule items that had a dependent demand, and the reorder-point system was used to manage finished-goods inventory where the demand was *independent* and set by historical data and forecasting techniques.

Closed-Loop MRP

The open-loop MRP system answered the *when* and *what* aspects of the make-or-buy questions. In addition, the open-loop system had well-defined links between master production scheduling (MPS), material requirements planning (MRP), and capacity requiremts planning (CRP). However, the system was static and not tuned to the dynamic nature of most manufacturing operations. Problems with production machines, labor, quality, and late vendor deliveries made the MRP-generated schedules irrelevant, and manual updating of order status was difficult and error-prone.

The problem was solved by interfacing the purchasing and production activity control modules with the MRP module. The purchasing interface provided MRP with a dynamically updated order status from which new planned order releases and suggested new schedules could be generated. The links to the shop floor were critical for collecting daily production and operator time sheet data so that production status could be determined. The use of automatic tracking devices such as bar codes permitted a near real-time production activity control interface to MRP and a responsive closed-loop planning system. Study Figure 8–12, which illustrates the feedback process in the closed-loop MRP system. The loop includes shop production data and the planning data present in MPS, MRP, and CRP.

> Closed-loop MRP is a system built around material requirements planning that includes the additional planning functions of sales and operations planning (production planning), master production scheduling, and capacity requirements planning. Once this planning phase is complete and the plans have been accepted as realistic and attainable, the execution functions come into play. These functions include the manufacturing control functions of input-output (capacity) measurement, detailed scheduling and dispatching, and anticipated delay reports from both the plant and suppliers, supplier scheduling, etc. The term *closed loop* implies not only that each of these elements is included in the overall system but also that feedback is provided by the execution functions so that the planning can be kept valid at all times. (*APICS Dictionary,* 9th ed., 1998.)

Companies must be able to take prompt corrective action when conditions require a change. There must be feedback throughout the organization to communicate the exceptions and plan changes to everyone concerned. The discipline to keep the information in the system correct and up to date must be instilled in the company's employees at all levels. Confidence that the information in the formal system is correct must be developed and maintained. A formal system that lacks employee confidence is of little value to any company.

Figure 8–13 compares the reorder-point system and the MRP system for several of the processes present in manufacturing. Study Figures 8–12 and 8–13 until the differences among reorder point, open-loop MRP, and closed-loop MRP are clear.

Manufacturing Resource Planning

The evolution in the planning and scheduling processes used for dependent demand items has been continuous since the early 1960s. However, a major change occurred in the 1970s that expanded the interface between enterprise operations significantly. The revised process, called *manufacturing resource planning* or *MRP II,* was described and promoted by Oliver Wight, a manufacturing consultant. MRP II links a broader range of enterprise departments into the production planning process. The closed-loop MRP system was expanded to include interfaces to all the financial

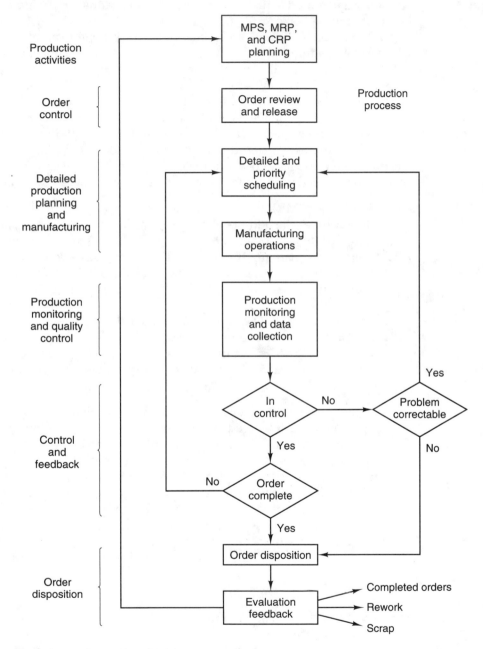

Figure 8–12 Closed-Loop MRP System.

	MRP	*Order point*
Demand	Dependent	Independent
Order philosophy	Requirements	Replenishment
Forecast	Based on master schedule	Based on past demand
Control concept	Control all items	ABC
Objectives	Meet manufacturing needs	Meet customer needs
Lot sizing	Discrete	EOQ
Demand pattern	Lumpy but predictable	Random
Types of inventory	Work-in-process and raw materials	Finished goods and spare parts

Figure 8–13   Comparison of MRP and Order Point.

areas including purchasing with accounts payable, sales order processing with accounts receivable, and inventory and work-in-process with general ledger. The MRP II system is illustrated in Figure 8–14. Note that MRP, material requirements planning, is just one part of MRP II. As a result, material requirements planning is often call "little mrp" and manufacturing resource planning is called "big MRP."

Together with the development of MRP II, Oliver Wight produced a set of evaluation questions to measure the degree of compliance between an enterprise's operation and the ideals defined in MRP II. The first set of evaluation questions was published in 1977 and revisions to the questions have occurred regularly. The current evaluation process, published in 1993, divides the questions into the following five basic business functions: strategic planning processes, people/team processes, total quality and continuous improvement processes, new product development processes, and planning and control processes. The overview questions for each of the five business functions are listed in Appendix 8–2. Read through these questions so you can understand the level of operation required for world-class manufacturing operation. The complete checklist has more detailed questions for some of the overview questions to help determine the best answer for the overview questions. The ABCD checklist process requires a company to answer each of the overview questions based on a rating scale. If a company is perfectly executing the activity described by the overview question, then an "excellent" rating is chosen and a 4 point value is earned for that question. On the other hand, if the company has not yet started to work in the area addressed by the overview question, a score of "not doing" is assigned and the point value is 0. After all sixty-eight questions are answered, the average for the numeric score is found and a letter grade is assigned. The operational descriptions for companies working at the four grade levels are also listed in Appendix 8-2.

Implementing MRP II

The implementation process for MRP II requires a highly structured approach that involves every employee, with a minimum of 90 percent of the workforce trained, including management. A team made up of representatives from all the functional

Figure 8–14 MRP II System.

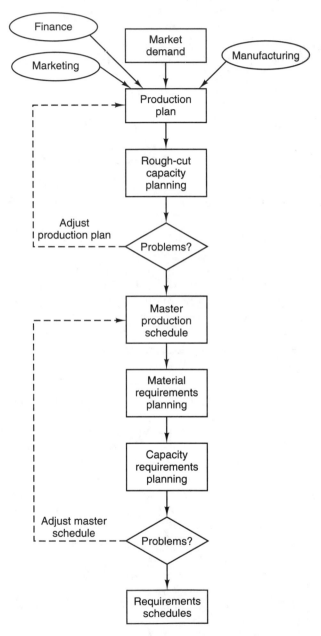

areas often leads a successful implementation of formal systems. Some team members may be full-time; others may participate only as needed. The project team typically reports to a high-level steering committee made up of the executive managers who control the business functions. Figure 8–15 identifies many of the planning considerations that must be addressed by the project team and the steering committee. Decisions that cannot be made at the project team level will be taken to the steering committee. The steering committee ultimately resolves any conflicts.

Essential prerequisites
- Knowing what systems are currently in place and how they work.
- Identification of features required of future systems.

Organizational issues
- Who are the key players, and what are their roles?
    —The members of the steering committee?
    —The members of the implementation project team?

Project management plans
- Preparing a realistic plan of action.
- Plotting milestones and tracking progress toward them.
- Initial measurements (who, what, when, and how).
- Auditing progress along the way and following completion of the project.

Developing the project justification and implementation plans
- What are the expected results?
    —Tangible benefits.
    —Intangible benefits.
- What are the expected system costs?
    —Software and hardware.
    —Implementation costs.
    —Education costs.
    —Any additional personnel costs related to the implementation.

Education plan details for all employees
- Who, what, when and why it is needed.
- How the education will be delivered.
- Identification of the education resources needed and currently available.

Plans for a pilot test of the new system and software
- System requirements and timing of availability.
- Setup and organization of test data and exercises.
- Location and accessibility.
- Assessment and follow-up.

Figure 8–15   System Implementation Planning Considerations.

The process used in many successful implementations, illustrated in Figure 8–16, is called the *proven path*. Note that education is the first step in the process and is continuous throughout the implementation. Top management commitment and involvement in the MRP II program is critical. Management must understand the MRP II process and comprehend the costs and effort required to install and operate the process fully. In addition, management must know how MRP II will affect every department and the benefits that will result from successful implementation.

The installation process is divided into four phases: *initial, preparation, implementation,* and *operation*. After the initial phase, the project team is in place with a

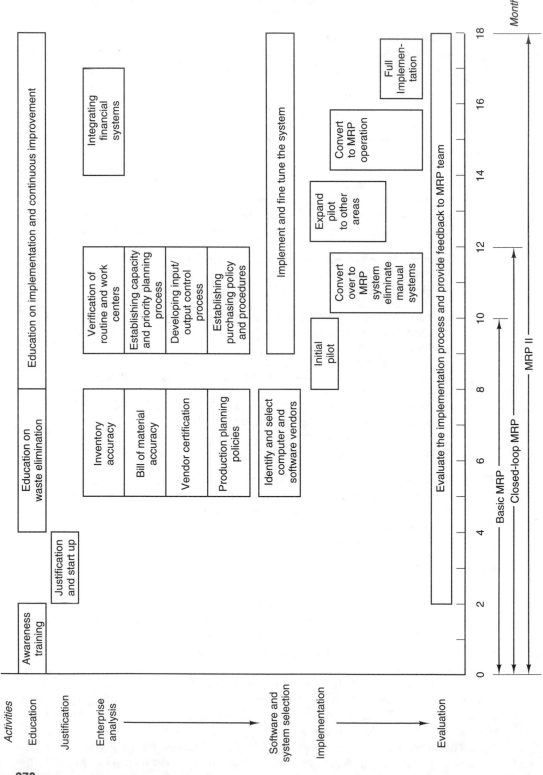

Figure 8–16 MRP II–Proven Path Method.
Source: Courtesy of Industrial Technology Institute.

full-time project director, and work on problem analysis begins. It is not uncommon for the project team to identify between 50 and 500 problems that must be addressed before the system is installed. Problems are divided into functional areas and prioritized, then teams of employees from the areas start working on solutions. For example, the inventory group could be assigned to work on inventory count accuracy, location accuracy, and damaged goods problems. As Figure 8–16 indicates, MRP II implementation can take eighteen months in a medium-size company, with the first eight to twelve months used to get the current manufacturing system in proper order for the implementation of the hardware and software. Study the proven path model until you are familiar with all the activities.

The cost of the implementation is directly proportional to the size and type of company. The costs are usually divided into four categories: (1) consulting (10 percent), (2) education and problem analysis (40 percent), (3) hardware (20 percent), and (4) software (30 percent). The cost of the software is a function of the computer hardware. For example, costing base and scheduling software for a job shop operation that runs on a microcomputer is in the range of $20,000 to $30,000. MRP II software for a microcomputer-based system would be less than $50,000, while software for a mini- or mainframe computer is usually over $100,000. While roughly half of the cost for MRP II is in hardware and software, the resources spent on education and problem analysis are the most critical for successful system implementation.

Implementation costs for a formal system are significant. The costs extend well beyond the cost of software and hardware that will run it. An MRP implementation is not a small task, even for companies that already have an MRP system installed. A mainframe-based system implementation, including software, hardware, and the human resources needed to implement them in a company, can easily cost more than $1 million. Modern systems that use the currently available technologies for minicomputers and personal computer networks may reduce the cost of hardware and software by 60 to 70 percent, but the cost of a new formal system remains high.

Benefits from MRP II

A formal manufacturing planning system can be a powerful tool that helps all the functional areas of the operation work more effectively. MRP II systems have proven to be valuable to companies and have produced real, "tangible" cost savings in several areas. Reductions in inventory have reduced the amount of money spent on investments for material and the cost of keeping that material on the shelf until it is needed. Better planning in communications often leads to reductions in the amount of inventory declared obsolete and written off as a loss. Manufacturing effectiveness is improved and savings result from having better schedules with fewer delays, resulting in reduced direct labor costs and reductions in unplanned overtime. Better scheduling provides the purchasing function with better visibility of future requirements and allows better planning with suppliers and lower purchasing costs. Component waste is reduced through better scheduling of similar items, planning of changeovers, and improved control of material issues to the shop floor. Customer service levels are improved by shipping more complete orders on time and with improved responsiveness to customer requirements.

In addition, companies have found other intangible savings that cannot easily be assigned a dollar value. Although difficult to quantify, these intangible savings are still considered significant. Changing requirements and business conditions are realities in manufacturing. The formal plan becomes a basic reference point supporting the planning for change. Companies skilled in planning can often respond better to changing conditions. It is difficult to put a value on the benefit of an increased ability to develop and execute plans for the operation and to deal with change. Additional intangible benefits often include improved customer relations, reductions in expediting costs, improved financial planning, and improved employee morale.

Achieving class A status through the implementation of MRP II requires a commitment by management to an 18- to 24-month plan, a great deal of hard work, and the investment in computer hardware and software. As with any investment in business, the payoff must justify this level of effort and cost. A study performed by the Oliver Wight companies asked participating companies to indicate their MRP II rating level and the percentage improvement for operations in four areas: customer service, productivity, purchase cost, and inventory turnover. The results are listed in Figure 8–17. Note that class A companies are three times better than class D, twice as good as class C, and almost 50 percent better than class B. As a result, companies that implement MRP II successfully achieve the world-class metrics described in Figure 1–6, have the order-winning criteria necessary to expand market share, and receive tangible dollar benefits ranging from 10 to 25 percent of gross sales. Review the case study in Chapter 2, which describes a company moving from class D to class A through a structured improvement program. Note the benefits obtained by reaching the class A status. Success in using MRP II concepts is frequently accompanied by work in another production process, *just-in-time* manufacturing, which is covered in Section 8–4.

Cost Justification

Companies and their planning teams must take time to consider carefully the expected benefits and the costs associated with a formal system implementation. There are no simple solutions and turning-point implementations. Planning and implementation decisions will have an ongoing impact on the cost effectiveness of

Figure 8–17 Percentage Improvement by MRP II Class Ratings.

	Percentage improvement for class:			
	A	B	C	D
Customer service	26	18	13	8
Productivity	20	13	9	5
Purchase cost	13	9	6	4
Inventory turnover	30	21	13	8

the new system. A meaningful justification requires careful analysis and a conservative look at the benefit and cost figures. The savings have proven to justify the high cost of implementation for many companies. Many companies have approached the implementation of a formal planning system with a "quick fix" approach that has produced few tangible benefits. Management commitment must be real and demonstrated for a successful implementation.

Manufacturing System Education

Employee education is the "glue" that holds the implementation project together. It is more than the training of people in the operation of the new system. Education prepares people to deal with the changes that come with the implementation of new tools and new ways to perform their jobs. The need for employee education does not end with the implementation of a new system. The project team should develop a detailed education plan that meets the needs of system users at all levels of the organization.

An effective way to develop and display the education plan is to build a spreadsheet. The spreadsheet lists all employees by functional group and all the topics to be covered in training sessions. An example is shown in Figure 8–18. The spreadsheet shows the plan for education by the topic area that the project team recommends for each employee. It becomes the basis for planning detailed and overview education, time requirements, class size, the number of sessions, and more. The spreadsheet is a powerful tool for education planning.

8–4 JUST-IN-TIME MANUFACTURING

The term *just-in-time* (JIT) is defined in the APICS dictionary as follows:

> Just-in-time (JIT) is a philosophy of manufacturing based on planned elimination of all waste and on continuous improvement of productivity. It encompasses the successful execution of all manufacturing activities required to produce a final product, from design engineering to delivery, and includes all stages of conversion from raw material onward. The primary elements of just-in-time are to have only the required inventory when needed; to improve quality to zero defects; to reduce lead times by reducing setup times, queue lengths, and lot sizes; to revise incrementally the operations themselves; and to accomplish these activities at minimum cost. In the broad sense, it applies to all forms of manufacturing—job shop, process, and repetitive—and to many service industries as well.

JIT encompasses every aspect of manufacturing, from design engineering to delivery of the finished goods, and includes all stages in the processing of raw material. Other names used to describe the JIT process are *short-cycle manufacturing, stockless production,* and *zero-inventory manufacturing.*

Education Plan Matrix	Overview of Mfg. Systems	Bills of Material	MRP logic	Planning using MRP	Product Data Management	Master Scheduling	Rough Cut Capacity Planning	Other topics to be determined . . .
Planners:								
Bill Jones	X	X	X	X	X	X	X	
Mary Davis	X	X	X	X	X	X	X	
Karen Hopping	X	X	X	X	X	X	X	
Accounting:								
Mary Johns	X	X						
Larry Captain	X							
Engineering:								
Bob Cramer	X	X			X		X	
Jane Williams	X	X			X		X	
Sales/Marketing:								
Robert Keyes	X				X	X	X	
Joe Smith	X	X			X	X	X	
Manufacturing:								
Johnny King	X	X		X	X	X	X	
Mike Mitchell	X	X		X	X	X	X	
Sue Brock	X	X		X				

Figure 8–18   Education Plan Spreadsheet.

Just-in-time is much more than a material ordering plan that schedules deliveries at the time of need. JIT supports the new approach to value-added manufacturing. The concept, developed in Japan following World War II, focuses on the elimination of all waste. Robert Hall's 1987 classic *Attaining Manufacturing Excellence* presents a summary of the "seven wastes" that become the target of elimination in a

JIT process (see Figure 8–19). A just-in-time approach leads to operations having only what is needed, nothing more. JIT takes the "problem of manufacturing" to the extreme level of having *only* the right materials, parts, and products in the right place at the right time. It is a relentless approach where waste at any point in the operation (even at the management level) is not tolerated.

The definition of JIT outlines the two fundamental objectives of waste elimination and *kaizen,* an attitude of continuous improvement. Supporting these objectives are three JIT elements that help management keep the JIT focus and foster an environment conducive to successful implementation: technology management, people management, and system management (Figure 8–20). The graphic in the figure was developed by David W. Buker, a management consultant on MRP II and JIT implementation. The following description of the JIT elements was adapted from a manual published by David W. Buker, Inc., *7 Steps to JIT.*

JIT Elements

The first JIT element, *technology management,* calls attention to the production environment and emphasizes the need for a responsive manufacturing system. A responsive production system results when the following four areas are addressed.

Waste of over production: Make only what is needed now—reduce set-up time, synchronizing quantities and timing between steps, compacting layout.

Waste of waiting: Synchronizing work flow as much as possible, balance uneven loads by flexible workers and equipment.

Waste of transportation: Establish layout and locations to make transport and handling unnecessary. If possible reduce what cannot be eliminated.

Waste of processing itself: Question why this part should be made at all — why is this process necessary?

Waste of stocks: Reduce stocks by reducing set-up times and lead times, reducing other wastes reduces stocks.

Waste of motion: Study motion for economy and consistency. Economy improves productivity. Consistency improves quality. Be careful not to just automate a wasteful operation.

Waste of making defective products: Develop process to prevent defects from being made. Accept no defects and make no defects. Make the process "fail safe."

Figure 8–19   The "Seven Wastes."
Source: Robert W. Hall, Attaining Manufacturing Excellence, *The McGraw Hill Companies, 1987.*

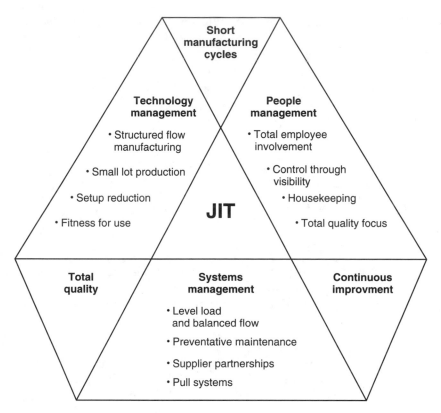

Figure 8–20 Elements of a JIT Implementation.
Source: Courtesy of David W. Buker, Inc.

Structured Flow Manufacturing. In flow manufacturing, the machines and work cells are organized and grouped to maximize the velocity of parts through production and to minimize the transportation and queue time for parts. The three types of production layouts shown in Figure 8–21 illustrate this concept. In the initial layout (a) the ratio of value-added work to part movement is very low because machines are grouped by function. The structured flow layout (b) provides increased throughput because the machines are organized by product process and assembly requirements. In the last layout (c) the structured flow had additional refinements with consideration of group technology principles.

Small-Lot Production. Reducing lot sizes to the smallest quantity possible is supported by structured production and short setup times. The goal is a lot size of 1, or the smallest customer order.

Setup Reduction. Setup time is the total time from the completion of the last piece of the previous production run to the first good part on the new production job. Reduction in setup time increases capacity and production capability while reducing inventory.

Figure 8–21   Structured Production Flow.

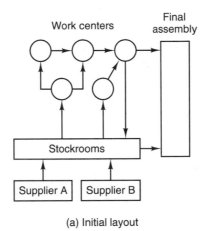

(a) Initial layout

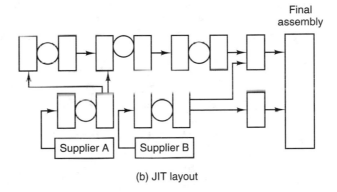

(b) JIT layout

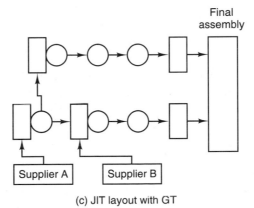

(c) JIT layout with GT

Fitness for Use.   Fitness for use means that the product satisfied the customer's requirements perfectly. Customers are the external users of the finished products or the next workstation in the structured flow production line. Every operation in the enterprise is a customer for someone and a vendor for someone else. In each case, the needs of the customer must be met precisely.

The second area, *people management,* is critical for the continuous improvement objective in JIT. This element creates an environment in which all employees, from the president to the hourly workers, have the responsibility and authority to suggest and implement improvements to the production system. Creation of this type of environment requires the following items.

Total Employee Involvement.   The company that has every employee working on solutions to performance problems will outperform the competition. The Japanese term for continuous improvement, *kaizen,* recognizes that all employees are valuable resources for the solutions to problems. JIT is built on the premise that everyone works on continuous improvement of the process through functional and cross-functional corrective action teams.

Control Through Visibility.   Control through visibility uses simple visible means to communicate goals and identify problems: for example, progress charts for work-center goals, control charts for tracking critical process variables, and flashing lights to indicate a machine problem that needs immediate action.

Housekeeping.   Housekeeping focuses on the work center or workstation with an emphasis on cleanliness, simplification, discipline, and organization to eliminate wasted time, motion, and resources.

Total Quality Focus.   Total quality focus addresses broad quality issues from suppliers to customers at every element of the production chain. The emphasis is on the quality of the process at every function and work center in the enterprise because the quality of the product is determined by the quality of the process.

The third element, *systems management,* addresses the effective distribution and application of the limited enterprise resources. This integration element is supported by the following items.

Level Load and Balanced Flow.   Both of these elements focus on effective utilization of manufacturing resources. The first, level load, deals with scheduling products in roughly equal quantities for a given period, such as a week or month. The second, balanced flow, works toward a continuous flow of products through manufacturing.

Preventive Maintenance.   Preventive maintenance works to eliminate equipment failure as a source of process defects by maintaining machines at the highest level of operational performance. Preventive maintenance, a requisite for structured flow manufacturing and quality products, ensures that machines operate on demand and that operational performance meets specification levels.

Supplier Partnerships. Supplier partnerships are key elements because a healthy supplier and user relationship is critical to a JIT operation. Long-term vendor partnerships cut cost for all members of the partnership through shared quality goals and design cooperation, frequent product deliveries, and a total cost perspective.

Pull Systems. *Pull system* is synonymous with *JIT manufacturing* because it describes in one word, *pull,* how JIT manufacturing works. Parts are produced only when the next workstation in the structured flow production system indicates that parts are required. This requirement implies that parts are produced on demand with very short lead times. As a result, some of the technology management elements, such as rapid setup and small lot production, are important.

These 12 JIT elements in the three management groups represent the environment that must be present for a JIT system to function effectively. A JIT implementation requires that each element be addressed through a systematic plan.

Implementing JIT

JIT concepts are implemented for two reasons: first, to improve current manufacturing efficiencies with no intention of implementing JIT across all operations. For example, a company may have an effective MRP II system but may need to get lot sizes smaller and to increase the flexibility of the production system. Working on the JIT elements would improve an already operational MRP II system. The second reason is to install an operational JIT manufacturing system. The installation may be one or just a few work cells from a large manufacturing operation. These work cells may be judged critical for rapid response to customer needs. Another JIT installation may focus on a product line or cover the entire factory. In each case the work cell, product line, or factory is changed from a *push*-type manufacturing system that produces inventory per some schedule for future use to a *pull*-type system that produces parts only when needed in the next level of the bill of materials. A seven-step implementation process developed by David W. Buker, Inc. illustrates the critical success factors for the installation. The steps include:

1. A four-stage education plan covers every employee. Education ranges from a working overview of JIT for top management to focused group education on how to implement a system.

2. Assessment of the twelve JIT elements in the technology, people, and systems management areas is a major step in the implementation process. The assessment process determines the readiness level for JIT. In addition to the twelve elements, the level of management commitment and effectiveness of education and training must be evaluated and the results of the entire assessment published.

3. When the assessment is complete, an implementation plan is developed that uses the results of the self-study to target areas requiring work. The plan includes: implementation activity, status of activity, person responsible, start and completion dates, and the resources required to complete the activity.

4. A recommended practice in JIT installations is the use of a small pilot project as the initial implementation. The smaller project permits analysis of the im-

pact of JIT at a departmental or unit level. In addition, the pilot permits unexpected problems that arise to be solved before JIT is applied across the enterprise.

5. Continuous improvement activities critical to the long-term health of the JIT implementation are begun. One technique, called small group improvement activities (SGIA), is frequently used to organize and provide structure to the continuous improvement process. In this step, it is critical to provide the leadership and education needed to work in the group setting, to limit the scope of projects so that the group can achieve success, and to develop a method for evaluating the success of the process.

6. The success of the implementation must be measured through a performance evaluation process. Baseline measurements are usually taken in the following areas: customer service, elimination of cost-added operations, product and process cycle time, inventory levels and turns, quality, number of employee suggestions, manufacturing output, and employee productivity.

7. In the final step, the process is moved to other parts of the enterprise. This internalization and companywide transition is a signal to the organization that a journey has been started to bring the organization in line with world-class standards—and that it will never end.

Many U.S. and offshore companies have implemented JIT successfully for the production of various products. One of the most advanced JIT production systems was developed by Toyota. The Toyota JIT model is illustrated in Figure 8–22. The system has two objectives: *cost reduction* and *increase in capital turnover ratio*. Elimination of waste, called *unnecessaries* in the Toyota model, is the mechanism to be used to reach these objectives. The production method utilizes a continuous-flow process with two elements: JIT and *self-stop automation*. If a quality problem is detected, any employee has the power and authority to stop the production line if necessary. The JIT production system has two components: *production methods* and *information systems*. The items listed under production methods are consistent with the definition of JIT provided in this section of the book. The information system used for JIT is called *kanban*.

Kanban

The information system block in the Toyota model (Figure 8–22) represents all of the manufacturing production and control (MPC) functions necessary to run the JIT system. Within the MPC system, *kanban* controls the flow of production material. *Kanban* is a Japanese word that means *card,* with one- and two-card kanban systems in common use. Toyota uses a two-card kanban system to move material through production. The first card is a *transport* or *conveyance* card, and the second is a *production* card. Sample kanban cards are illustrated in Figure 8–23. The *process* refers to the production work centers, and the *issue number* represents the number of containers released. The JIT pull system using kanban is best described by an example.

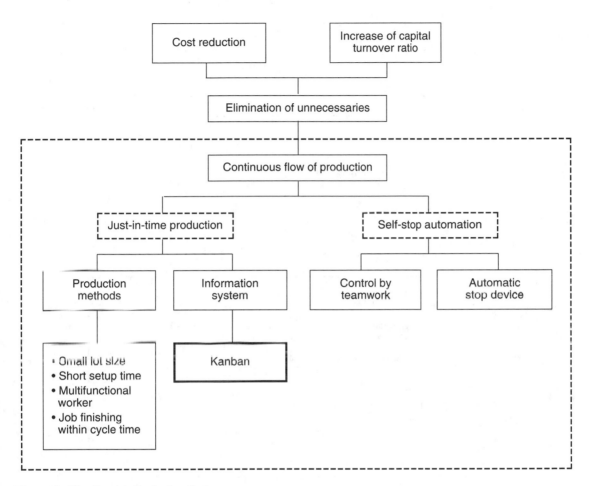

Figure 8–22   Toyota's Production System.

Source: From the European Working Group for Production Planning and Inventory Control, Lausanne, Switzerland, July 1982.

The kanban system used by Toyota is illustrated in Figure 8–24; study the figure until work centers and container labels are familiar. Each work center has some raw material inventory on the left and finished products inventory on the right in the figure. The work cells are labeled A and B, and kanban boxes are located in the raw material inventory, work center, and finished product inventory areas. Each raw material container has a *conveyance* card, and a finished-goods container has a *production* card. The work centers in Figure 8–24 are both idle at the start of this example. The process starts with movement of a finished parts container from work center B to work center C, where the parts are used for an assembly operation. Changes in the work cells as a result of this movement are illustrated in Figure 8–25. Understanding the movement of the cards is critical for insight into the kanban production process. The sequence of card movement is as

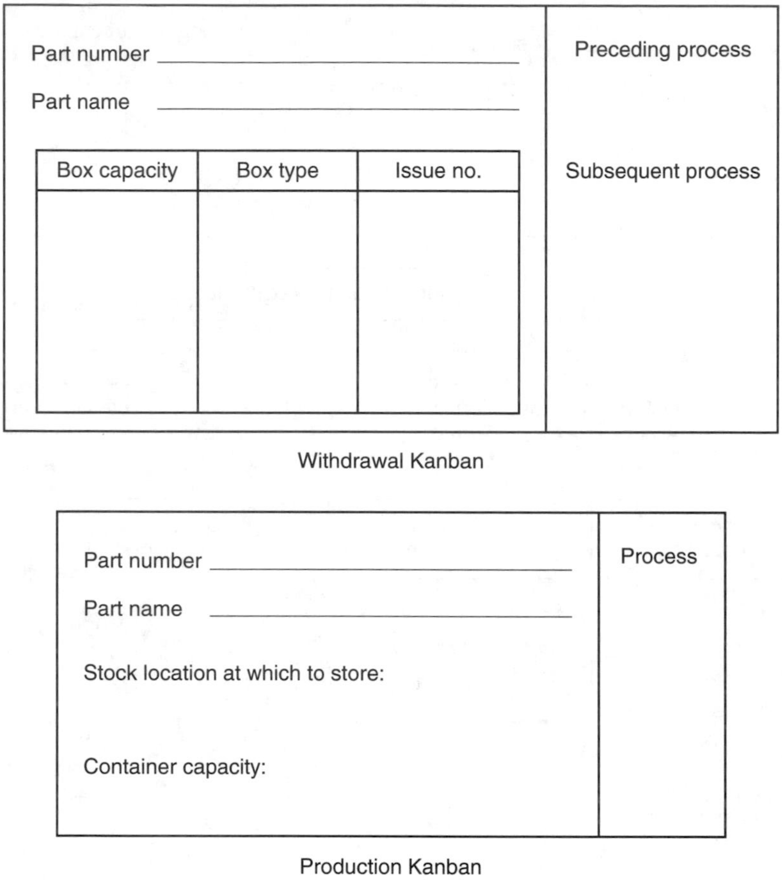

Part number _____	Preceding process
Part name _____	

Box capacity	Box type	Issue no.	Subsequent process

Withdrawal Kanban

Part number _____	Process
Part name _____	
Stock location at which to store:	
Container capacity:	

Production Kanban

Figure 8–23 Production and Conveyance Cards for a Two-Card Kanban System.

follows (the circled numbers in Figure 8–25 correspond to the kanban sequence below):

1. The production card is removed from the production container before the container with finished parts is moved from cell B to cell C. The card is placed in kanban box B1.

2. An inventory specialist moves the production card from kanban box B1 to the work cell kanban box B2. The arrival of a production card authorizes the work-cell operator to produce another container of finished parts to replace the container just removed. The production card is retained by the operator.

3. Production is started by moving a container of raw materials into cell B from the input side. The transport or conveyance card in the raw material container is removed and put into the kanban box B3.

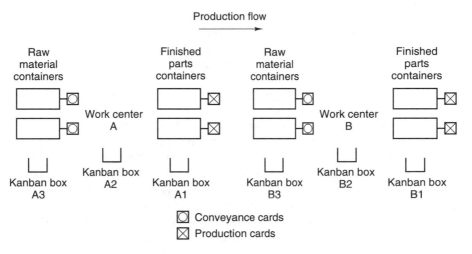

Figure 8–24   Two-card Kanban System.

4. The operator in cell B finishes production on the container of parts and moves the finished-goods container to the output side of the cell. The production card held by the operator is placed on the finished container.
5. The inventory specialist retrieves the conveyance card from kanban box B3. The card authorizes the movement of a finished container from the output

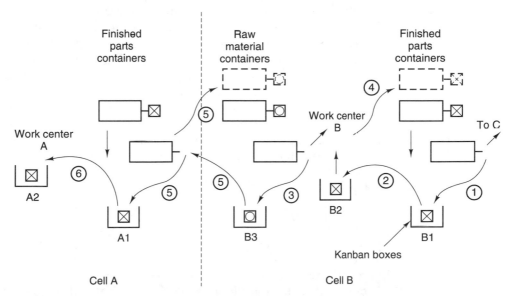

Figure 8–25   Movement of Cards and Inventory in a Kanban System.

side of cell A to the input side of cell B. The production card in the container is placed in the kanban box A1, and the conveyance card is placed in the container on the input side of cell B.

6. A similar process is followed in cell A.

The system is a *pull* type of production system because a work center is authorized to produce parts only when the operator has a production card. A card is present only when the next work center in the production sequence pulls finished parts away. Work is paced by the flow of the kanban system, and no work center is allowed to work just to keep operators busy. In the past example, the input and output queue had two containers; however, depending on the demand rate, multiple containers could be used for raw material and finished parts. In that case, additional cards for each container would be necessary.

The cards replace all work orders and inventory move tickets. Note that production cards just circulate around the output side of the work cell from a finished-goods container to the production cell operator and then back to the finished-goods container. The conveyance cards circulate at the input side of the work cell in a similar fashion. The system is visual, is manual in operation, and operates with less inventory, and the problem of sequencing jobs is greatly reduced. Experience in the Toyota system indicates that movement of cards can also extend to external suppliers.

JIT is used most often in a high-volume *repetitive* manufacturing operation; however, low-volume *nonrepetitive* production systems can apply JIT with some modifications. The most difficult problem in low-volume applications with many product models is leveling the work load at the cells.

Claims for JIT manufacturing include reduced inventory levels, reduced work-in-progress inventory, shorter manufacturing lead times, and increased responsiveness to customers. Shorter lead times help compress schedules and lead to less work-in-progress material. Elimination of any non-value-added activity leads to spending less time and money. The streamlined process works better and faster because the waste has been removed.

JIT will not work if management only announces that the operation is now JIT. Successful JIT manufacturing is the result of dull, boring, incremental, methodical, but highly effective work in the plant and with the material suppliers that support production.

MRP II and JIT

Elements in the operational philosophy for MRP II and JIT are frequently opposed. For example, inventory is minimized in JIT and is planned in MRP II. MRP II is a push system producing parts inventories for future use, and JIT is a pull system. However, the two are not mutually exclusive because the manufacturing system improvements necessary for JIT, such as reduced setup time, would also make MRP II more efficient. In addition, some parts of an MRP II production system could have cells that operate with a JIT philosophy.

8–5 SYNCHRONIZED PRODUCTION

Synchronized production is defined by APICS as follows:

> Synchronized production is a manufacturing management philosophy that includes a consistent set of principles, procedures, and techniques where every action is evaluated in terms of the global goal of the system. Both kanban, which is part of the JIT philosophy, and drum–buffer–rope, which is part of the theory-of-constraints philosophy, represent synchronized production control approaches. Synchronous manufacturing is a synonym.

Kanban, based on the JIT philosophy, and drum–buffer–rope (DBR), based on the theory-of-constraints philosophy, are two synchronized production control approaches. However, many manufacturing systems cannot adopt the JIT philosophy because disturbances in the system do not permit effective implementation. Synchronous manufacturing using DBR techniques can be used in these situations because the production is synchronized by a philosophy that is slightly different from JIT.

History of the Problem

In the past, product cost was the dominant metric used by manufacturing managers to determine the best manufacturing process; as a result, efficiency took precedence over product flow. To keep product cost down, the product cost of individual end items was the target of activity. To achieve this end, large batches with few setups were the rule. However, this approach is in direct conflict with the demands for fast, smooth material flow that requires smaller batches and more setups. In addition, the premise that reducing individual product cost will reduce total enterprise cost is not valid because of the complex interaction present in manufacturing operations. Therefore, the procedures used to evaluate and manage synchronized manufacturing in operations had to be changed. A broader view of manufacturing beyond single products is necessary where the global goals of the system are considered.

One approach to achieving this result is to change the measurements from individual product cost to *throughput, inventory,* and *operational expense.* In most situations, production managers have little control on the throughput requirements received from the business planning area. Therefore, the goal for managing the synchronized production flow system is restated as *meeting the throughput requirements while efficiently managing inventory and production expenses.*

Managing Inventory and Operational Expense

Inventory and operational expenses generally have an inverse relationship. For example, operational expenses in production can be reduced if inventory is allowed to rise. Certainly, larger batch sizes and fewer setups reduce operational expense but result in larger work-in-process inventory. On the other hand, inventory costs are reduced to a minimum level when all parts are produced in a manufacturing system that supports a lot size of one. However, the lot size of one system often has the highest operational expense. The goal is to increase throughput while decreasing inventory and operational expenses. A manufacturing control system supporting synchronous manufacturing is key to reaching this goal.

Drum–Buffer–Rope System

The *drum–buffer–rope* (DBR) system is a synchronous manufacturing process that executes a planned production flow over a given period of time and compensates for the disturbances commonly found in most production flow systems. The disturbances usually result from three *critical constraints: market demand, capacity,* and *material limitations.* The production plan, developed around these critical constraints, must (1) not exceed projected market demand, (2) ensure a sufficient supply of materials, and (3) ensure that the planned production flow does not overload the processing capabilities of the resources.

The DBR process starts with a preliminary production plan and identification of *capacity constraint resources* (CCRs). The production flow is analyzed using input from operators to determine what operations in the flow for a family of products have serious capacity constraints. Operations with constraint problems are designated CCRs, and the remaining operations are labeled non-CCRs. The preliminary production plan is modified based on the schedule for the CCR, and the resulting document, called the *master production schedule* (MPS), is used to schedule actual production on the shop floor and establish customer orders' delivery dates. The MPS sets the beat for production flow in the synchronous manufacturing system; as a result, the process used to establish the MPS is referred to as the *drum.*

Every manufacturing system has disruptions to the process; as a result, the actual flow is often different from the planned flow. To ensure that customer delivery dates are honored, a protective cushion or *buffer* is associated with the operations most likely to cause a disturbance in the flow. The planned lead time is then the sum of the setup and process times plus the time buffers required.

The product structure, process sheets, buffers, and production procedures are used to set the planned production at each resource. Therefore, the planned production schedule at each production resource is tied to the tempo set by the MPS or *drum beat.* All non-CCRs are synchronized to support fully the product rate set by the MPS. Since the MPS was established from the CCRs' schedules, a link between the non-CCRs and the CCRs is built. Linking the resources together in this fashion is analogous to the practice of tying mountain climbers together with ropes so that they climb at the same rate and support each other. This process of linking CCRs and non-CCRs is the *rope* in DRP.

DBR Process

Techniques and processes are well established to identify the CCRs in a flow production system and to determine the production plan, master production schedule, and buffers required for a functioning DBR synchronized manufacturing system. A detailed analysis of the process used to develop a fully functional DBR system is available in several texts covering the topic. In short, the process is as follows:

1. Identify CCRs.
2. Set the *drum beat* by developing an MPS that effectively schedules the CCRs. The MPS plans sequence and lot size that do not overload operations while maximizing throughput and minimizing work-in-process inventory and operational expenses.
3. Set the *buffers* by identifying the time amount of inventory in the front of each CCR to avoid disruptions in the material flow.
4. Identify schedule release points at both the CCRs and non-CCRs.
5. Set the *rope* by developing schedules for release points that are back-scheduled from the MPS using planned lot sizes for the CCRs. Schedule non-CCRs using simple flow control to determine sequence and transfer lot size.

Synchronous Manufacturing: DBR Versus JIT/Kanban

JIT requires a change in the management philosophy for successful implementation because in time, every work center is affected. This type of wholesale change in management is difficult for many top managers to accept. However, DBR's focus on the bottlenecks in the production flow is closer to the emphasis found in many U.S. companies. In addition, U.S. managers take a project management approach to improvements in production and DBR fits this type of thinking better than JIT.

DBR overcomes many of the following limitations present in JIT/kanban:

- JIT/kanban is better suited for a highly repetitious process environment with a limited number of processes.
- Disruptions in the process flow are disastrous for a JIT/kanban implementation.
- JIT/kanban implementation is often lengthy and difficult.
- The JIT continuous improvement process has a systemwide focus, so that critically constrained resources with the greatest possibility for productivity are not singled out.

8–6 THE EMERGENCE OF LEAN PRODUCTION

Competition between manufacturing companies in Japan and the United States reached a startling new level in the late 1980s. U.S. companies found themselves losing market share, even whole product segments, to foreign competition that did not exist just a few years before. Hundreds of executives and managers made the trip to Japan to see firsthand what was taking place. What they saw was the result of a production system that was significantly different from the systems used in the United States. Companies scrambled to figure out what was happening and to try to replicate the Japanese system.

The Japanese system has become known as the "Toyota production system." It was created at Toyota by Taiichi Ohno. When pressed by U.S. executives about the source of this revolutionary system, Ohno is reported to have laughed and said that he learned it all from Henry Ford's book (*Today and Tomorrow*, first published in 1926). Ford's book was republished in 1988. Ford's factories implemented a new system of production and produced quality automobiles at prices that nearly every worker could afford. The Ford system led to fantastic productivity improvements that allowed car prices to be cut in half and worker wages to double. Norman Bodek of Productivity Preps commented in the preface to the new printing of Ford's book:

> [Ford] insisted that work environments be spotlessly clean; that business leaders think in terms of serving their communities and society at large; that production techniques not be taken for granted but continuously change and improve. He said that primary industries should help their suppliers and service industries to produce cheaper and better products in less time; and that managers should not remain in their offices but should walk around, know their workers, and be capable of doing the work themselves. He emphasized that workers should be trained and have the opportunity to better themselves and make product improvements.

Taiichi Ohno explains the motivation for the Toyota production system as an attempt to catch up with the automobile industries of the advanced Western nations following World War II. Remember that in the postwar period, the Japanese economy was trying to recover from extensive bombing. Resources in Japan were limited at best. Ohno knew that to become a true competitor to the West, his operations had to become more productive and produce quality goods at low cost. His system focused on the ferocious elimination of waste in all manufacturing operations and on the effects demonstrated through actual practice on the shop floor. There was no recipe showing how to be an effective producer Toyota learned how to be effective by practicing, making lots of changes, and learning along the way.

The production processes developed and used in the Toyota factories have been documented by Yasuhiro Monden, a professor at the University of Tsukuba, Japan. Monden's book *The Toyota Production System* was first released in 1983, and the second edition was published in 1993. It describes in detail the procedures, techniques, and tools that have been developed at Toyota. The editions of Monden's book are important sources of information on the Toyota system and the tools that the Japanese developed. This system appears to be too simple, just an extension of commonsense. There is much more to the Toyota system, however, than simple tools and techniques. It is important to look deeper at the work of Ohno and Monden.

James Womack and a team of educators from MIT completed a major study of the automobile industry and published their findings in a landmark book, *The Machine That Changed the World*, in 1992. This outstanding book provides a look at the history of the auto industry and the tremendous changes that have taken place since 1975. They pull many concepts that have been mentioned in this book together under the overarching title of "lean manufacturing."

Lean manufacturing is a philosophy of production that emphasizes the minimization of the amount of all the resources (including time) used in the various activities of the enterprise. It involves identifying and eliminating non-value-adding activities in design, production, supply chain management, and customer relations. Lean producers employ teams of multiskilled workers at all levels of the organization and use highly flexible, increasingly automated machines to produce volumes of products in potentially enormous variety. *Lean production is a synonym.* (APICS Dictionary, 9th ed., 1998.)

The heart of the concept is the aggressive elimination of waste throughout all parts of the operation and the organization. It is a great testimonial to the power of the Toyota approach and tools.

James Womack and Daniel Jones teamed up again in 1996 to write a follow-up book titled *Lean Thinking*. Womack and Jones have helped to package the important concepts invented by Ohno and documented by Monden. Their second book provides case study examples and lays out the five principles that summarize the lean production concept:

1. Specify precisely the value of a certain product.
2. Identify how the value is realized by the customer.
3. Make value flow without interruptions.
4. Allow the customer to pull value from the producer.
5. Pursue perfection and improve continuously.

Lean thinking addresses all the areas of waste presented earlier in the discussion of JIT; however, the lean approach is more focused. An expert in waste elimination known a *sensei* is a critical part of the lean approach. The experience of the *sensei* leads to the identification of problems and the design of solutions in a short time. The elimination of waste and the redesign of a work cell, including the relocation of machines, is often done in one day. The benefits of the efficient production cells are immediately visible and often lead to additional changes in other areas. The five steps sound fairly simple and a lot like the commonsense ideas that we should all embrace. Taken together and used as the basis for a plan of action with the help of a skilled *sensei,* the five steps listed above have helped companies remove amazing amounts of waste from their operations in a short time. The cases presented in *Lean Thinking* demonstrate how the five principles have been applied in small, medium, and large companies in the United States, Germany, and Japan. The examples build a credible argument for a move toward lean production, and the concept of lean production should remain popular in manufacturing for years to come.

Jim Womack is the founder and leader of the Lean Enterprise Institute. Through the institute, Womack and his followers lead the crusade for lean production. Visit the Web site at *www.lean.org* for additional news and information.

Americans are notorious for searching for the quick fix and not taking the actions to ensure problem elimination. Too often people skip the critical work of describing or mapping the value stream and really understanding it. The Lean

Enterprise Institute has started to develop training materials to help people do lean the right way. The first how-to book in a series planned by the institute is *Learning to See*, by Mike Rother and John Shook. *Learning to See* provides detailed instructions on how to map the value stream to add value and eliminate waste. Womack has seen too many companies start on a lean production project that never really comes together the way it should. A crisis arises and a champion leads the charge. Too often a company does a lot of work related to implementing the lean techniques and tools but does not realize the benefits they were hoping for. Womack encourages the use of an experienced waste fighter called a *sensei*. The *sensei* provides a jump start to the improvement. The people involved in the improvement process replace a long learning curve with the insight and knowledge of the *sensei*. The improvement team picks something they think is important and starts removing waste immediately.

American companies tend to be impatient with process changes. The critical steps of understanding a product's value and mapping the value stream for all product families is often omitted or done poorly. As Rother and Shook state in their book's title, it all begins with "learning to see." Companies have seen some short-term savings by applying the tools of lean production. An important key to long-term success is truly understanding the product's value and the related value stream from the customer back to the producing company and the company's suppliers. Weaving the beliefs and concepts of lean thinking into the fabric of the organization leads to ongoing improvements and savings that are impossible any other way. You can expect to hear more about the lean concepts and the Lean Enterprise Institute in the future.

8–7 SUMMARY

Material requirements planning (MRP) is a time-phased process that uses the bill of materials, inventory data, and the master production schedule (MPS) to calculate requirements for materials. The MRP uses a record to calculate planned order releases and inventory over some number of future production periods. Capacity planning occurs at three different points in production control: at the aggregate level, in support of the MPS as rough-cut capacity planning, and at the detailed level called capacity requirements planning (CRP).

The reorder-point planning system uses a trigger level to order the replenishment of production materials. The trigger levels are set using parts demand rate, lead time, degree of uncertainty in demand rate and lead time, and management policy. Reorder-point systems are used to plan production on products that are driven by independent demand conditions. Open-loop MRP systems are best suited for dependent demand items where demand is set by exploding the bill of materials for a product to determine the time-phased schedule and production requirements. When the open-loop MRP system adds feedback from the shop floor to correct the MPS based on production problems, the system is called a closed-loop MRP production planning system. Manufacturing resource planning or MRP II is a further enhancement to the closed-loop MRP system and includes links to other critical areas across the enterprise, such as accounting, purchasing, and finance. To ensure a successful implementation of an MRP II system, most companies use a technique called the proven path method.

Just-in-time (JIT) production is a philosophy of manufacturing based on the planned elimination of all waste and the continuous improvement of productivity. JIT requires that three management areas (technology, people, and systems) be addressed for a successful implementation. Within each management area, four subareas require work to prepare and change the organization and production system for the JIT operation. The implementation process for JIT is a seven-step process that starts with education.

Kanban, a Japanese word meaning "card," is a technique used to implement JIT in a pull production system. The two-card system uses a conveyance card and a production card. The cards control all movement of material on the shop floor and the production in the work centers.

Synchronous manufacturing, a management philosophy that looks at production from a global enterprise perspective, has an implementation process called drum–buffer–rope (DBR) based on the theory of constraints. In the DBR process, the critically constrained resources (CCRs) are identified and used to develop a workable master production schedule. The non-CCRs are planned from the revised MPS, and time buffers are placed at the CCRs to ensure a smooth, even flow of work through production.

REFERENCES

BUKER, D. W., *7 Steps to JIT,* 2nd ed. Antioch, IL: David W. Buker, Inc. & Associates, 1991.

CLARK, P. A., *Technology Application Guide: MRP II Manufacturing Resource Planning.* Ann Arbor, MI: Industrial Technology Institute, 1989.

COX, J. F., and J. H. BLACKSTONE, Eds., *APICS Dictionary,* 9th Ed. Falls Church, VA: APICS—The Educational Society for Resource Management, 1998.

COX, J. F., J. H. BLACKSTONE, and M. S. SPENCER, *APICS Dictionary,* 7th ed. Falls Church, VA: American Production and Inventory Control Society, 1992.

EVANS, J. R., and LINDSAY, W. M. *The Management and Control of Quality.* Cincinnati, OH: South-Western College Publishing, 1999.

FORD, H. *Today and Tomorrow.* Portland, OR: Productivity Press, 1993.

GODDARD, W. E., "ABCD Rankings and Your Bottom Line." *Modern Materials Handling,* March 1989, p. 41.

HALL, R. W., *Attaining Manufacturing Excellence.* Chicago, IL: Dow Jones-Irwin, 1987.

LUNN, T. and S. NEFF, *MRP: Integrating Material Requirements Planning and Modern Business.* Chicago, IL: Irwin Professional Publishing, 1992.

MONDEN, Y. *The Toyota Production System.* Institute of Industrial Engineers. Norcross, GA: Industrial Engineering and Management Press, 1993.

ROTHER, M., and J. SHOOK, *Learning to See.* Brookline, MA: The Lean Enterprise Institute, Inc., 1998.

Sobczak, T. V., *A Glossary of Terms for Computer-Integrated Manufacturing*, Dearborn, MI: CASA of SME, 1984.

Srikanth, M. L., and M. M. Umble, *Synchronous Manufacturing Principles for World Class Excellence*. Cincinnati, OH: Southwestern Publishing Co., 1990.

Vollmann, T. E., W. L. Berry, and D. C. Whybark, *Manufacturing Planning and Control Systems*, 4th ed. Homewood, IL: Richard D. Irwin, 1998.

Wight Co., *Survey Results: MRP/MRP II JIT*. Essex Junction, VT: The Oliver Wight Companies, 1990.

Wight, O. *Production and Inventory Management in the Computer Age*. Boston, MA: CBI Publishing, 1974.

Womack, J. P. and D. T. Jones, *Lean Thinking*. New York: Simon and Schuster, 1996.

Womack, J. P., D. T. Jones, and D. Roos. *The Machine That Changed the World*. New York: HarperPerennial, 1990.

QUESTIONS

1. Define *material requirements planning*.
2. Define an MRP record, describe all the terms present, and identify those that are normally given at the start of an MRP run.
3. Clearly define the difference between planned order receipts and scheduled receipts.
4. What is the relationship between the product structure diagram and the MRP records for parts and assemblies?
5. Use the product structure diagram in Figure 8–4 to describe how the gross requirements for the MRP records of each component are generated.
6. List five benefits achieved as a result of an effective MRP system.
7. Compare the level of capacity planning detail associated with the three places in manufacturing planning and control where capacity planning is performed.
8. Describe the interaction between the master production schedule (MPS) and the shop-floor capacity scheduling process that is used to achieve a reliable MPS.
9. Define *reorder-point system*.
10. Describe the four factors that influence the reorder point.
11. Describe why the reorder point is used for independent demand items and why material requirements planning (MRP) is used for dependent demand items.
12. Compare and contrast open-loop MRP, closed-loop MRP, and manufacturing resource planning (MRP II).
13. What are the key elements in the implementation of MRP II using the proven path method?
14. What distinguishes a class A company from the rest of the classes, and what benefits does a class A company have over companies in lower classes?
15. Define *just-in-time* (JIT) manufacturing.

16. Describe the two fundamental objectives associated with JIT.
17. Describe the three JIT elements that must be addressed for successful JIT implementation.
18. Describe the four areas in technology management that must be considered for successful JIT implementation.
19. Describe the four areas in people management that must be considered for successful JIT implementation.
20. Describe the four areas in systems management that must be considered for successful JIT implementation.
21. Describe the seven-step process used to implement a JIT solution.
22. Compare the production systems illustrated in Figure 8–21 and describe the advantages of continuous flow.
23. Describe a two-card kanban system.
24. Compare and contrast a pull production system and a push production system.
25. Define *synchronous manufacturing* in your own words.
26. Compare and contrast the operation and limitations of the kanban and drum–buffer–rope (DBR) production systems.
27. Describe a DBR system in your own words.

PROBLEMS

1. Using the production information for a product in the table below, complete an MRP record to determine planned order releases and inventory balances. On-hand inventory is 15 units, lead time is 2 periods, lot size is 25, and safety stock is 0.

Period	Gross requirements	Scheduled receipts
1	20	25
2	10	0
3	5	0
4	35	0
5	15	0
6	25	0

The following problems use the production data in Problem 1 with a single value changed. Change only the data required and do not carry the changes forward to other problems.

2. A cycle count determines that the on-hand balance in Problem 1 is incorrect and should be 5 units instead of 15. How does that affect the MRP record, and what changes in the planned orders are required?

3. During the first period, marketing adjusts the gross requirements for the data in Problem 1 from 35 units in period 4 to 15. Recalculate the MRP record for data in Problem 1. How were planned orders affected?

4. Recalculate the MRP record for Problem 1 with a lot size of 30. How are planned order releases affected? How does the average inventory for the two lot sizes compare?

5. Recalculate the MRP record for Problem 1 with a lot-for-lot production capability. How are planned order releases affected? How is the average inventory affected?

6. Recalculate the MRP record for Problem 1 with a safety stock of 15 units per period. How are planned order releases affected? How is the average inventory affected?

7. The MPC system at a company performs a weekly update of the MPS and MRP production planning data. At the start of week 1, the MPS for products A and B are as listed below.

Period (weeks)	Product A	Product B
1	5	15
2	15	0
3	0	20
4	25	5
5	0	10
6	15	0

One unit of component C is required to manufacture products A and B. The lead time for C is 1 week, safety stock is 0, lot size is 10 units, the on-hand balance of C at the start of week 1 is 15 units, and a scheduled receipt of 10 units of C is due in week 1. Complete the MRP record for component C for the six periods, and determine planned orders, inventory balance, and average inventory level.

8. Develop a spreadsheet for an MRP record and repeat Problems 1 through 7 using a computer to generate inventory balances.

PROJECTS

1. Using the list of companies developed in Project 1 in Chapter 1, determine which companies use open-loop MRP, closed-loop MRP, MRP II, JIT, or synchronous manufacturing systems.

2. Select one of the companies in Project 1 that uses MRP II or JIT and compare the world-class standards achieved by that company with the standards presented in Chapter 1.

3. Select one of the companies in Project 1 that uses MRP II or JIT and describe the process used to achieve continuous improvement.
4. Develop a spreadsheet to calculate the MRP requirements for all the parts in the product structure of the table assembly in Figure 7–12. Link the MRP records so that MPS requirements for tables over six periods are entered along with beginning on-hand inventories and scheduled receipts. Make assumptions, as necessary, for other production parameters.

CASE STUDY: PRODUCTION SYSTEM AT NEW UNITED MOTOR MANUFACTURING, PART 2

The basics of the Toyota production system used at the joint venture between General Motors Corporation and Toyota Motor Corporation are described in Part 1 of this case study in Chapter 7. The joint venture company, called New United Motor Manufacturing, relies on the following three concepts for effective and efficient operation: *just-in-time* (JIT) production; *jidoka*, a Japanese term referring to the quality principle; and full utilization of *worker's ability*.

Just-in-Time Production

The philosophy of JIT at New United Motor is not to sell products produced but to produce products to replace those that are sold. The first process, selling products produced, implies that cars are made for a finished-goods inventory and represents a *push* type of production system. The latter approach, producing cars to replace those that are sold, suggests that cars are produced only when a demand for the vehicle is present and represents a *pull* type of production system.

In the production environment, JIT is a concept designed to supply the right parts at the right time in exactly the correct amount during each step in the production process. The primary tool used to control the production system at New United Motor is called *kanban*. Kanban is an information system that controls production and manages the JIT pull manufacturing system. As a result of the JIT implementation, *muda*, a Japanese term meaning "waste," is significantly reduced and the product is delivered at the lowest possible cost.

The non-value-added operations in production usually fall into one of seven categories: *corrections, overproduction, processing, conveyance, inventory, motion*, and *waiting*. When *muda* is implemented effectively, all slack is removed from the production system; as a result, problems are exposed that may otherwise be hidden by excess inventories. Production teams use another Toyota method, *kaizen*, that searches on a continuous basis for production improvement. To be successful, *kaizen* must include the workers in the process, using their ideas and suggestions.

Jidoka: The Quality Principle

A basic principle at New United Motor is that quality should be ensured by the production process itself. When equipment and operators function under normal working conditions, they focus on approaching a zero quality problem operation. The principle applied to the production system is called *jidoka* and means "the quality principle." *Jidoka* refers to the ability of production machines or the production line to shut down automatically when abnormal conditions are present. For example, if a machine starts to produce parts that fall outside the allowed tolerances, the system is shut down. In this system, no defective parts move to the next work center because the system stops production. A current phase used to describe this technique is "quality at the source." In addition to halting the production system with automated quality checking sensors, every worker has the responsibility of checking quality problems and stopping the line to prevent poor-quality products from leaving the work center. The objectives of *jidoka* are (1) 100 percent quality at all times, (2) prevention of equipment breakdowns, and (3) efficient use of every worker.

Full Utilization of Worker's Ability

Team members have the authority to make decisions in their work area; are expected to be multifunctional and to solve problems; and are treated with consideration, with respect, and as professionals. Job rotation is a regular part of the learning and training process. In many older production systems, the machines run the workers; however, at New United Motor, the team member operates the machine. Wherever possible, automation has been introduced to eliminate the use of workers in monotonous, difficult, and dangerous operations.

Production Techniques and Methods

The goals of high quality and low cost are addressed with various production techniques and methods. The major techniques and methods include *kanban, production leveling, standardized work, kaizen, baka-yoke, visual control,* and the *team concept.*

Kanban

Kanban is the Japanese word for "card." The card is designed to prevent overproduction and to ensure that finished parts are pulled through the production system as needed. Using the card system, one process produces only enough parts to replace those drawn from the following process. The kanban: (1) gives work instructions, (2) provides for visual control of the production volume, (3) prevents overproduction, and (4) identifies problems for correction.

Production Leveling

Production leveling attempts to average the highest and lowest variations in orders so that the production resources remain relatively constant and at a low cost level. The production volume changes, normally associated with automotive production, cause waste at the work site. At New United Motor, the level process focuses on three causes of volume changes: the total volume, the models of cars produced, and options added. Without this leveling process, the kanban implementation of JIT would not function because of disruptions caused by volume variations. The leveling goes beyond the Fremont plant and includes the many vendors supporting the New United Motor operation.

Standardized Work

Standardized work is defined as follows:

> *Standardized work is work done at highest efficiency, with a minimum of waste, as a result of all tasks at the work site being organized into perfect sequences.*

The three goals for standardized work are (1) high productivity, (2) line balancing for all processes from a production timing standpoint, and (3) elimination of excessive work-in-process inventory. Each team member is trained in standardized work processes and principles; as a result, teams are responsible for the efficient layout of work assignments. Therefore, there is no need for industrial engineers in that function at the facility.

Kaizen

Visitors from the United States who toured manufacturing plants in Japan in the 1980s found companies that were driven to improve quality. This effort was initially thought to be focused on product quality. The philosophy of kaizen is much richer. Kaizen drives improvement in all functional areas of the business. It lifts quality above the manufacturing floor and involves all areas in an effort to enhance the overall quality of the company. The drive to improve all aspects of the business continually has been woven into the fabric of the entire company. Employees no longer have to stop and think about quality decisions; quality improvement is ingrained in their thinking.

The successes driven by kaizen are often the result of the implementation of many small suggestions. The sum of this accumulated effort, over time, has produced outstanding results. Kaizen programs are built on the principles of documented operating practices, the total involvement of every employee, and a commitment from the company to provide training in the philosophy and tools of the kaizen process.

Kaizen continues to gain popularity in the United States and it has proven to be a successful approach for many companies. Visit the following Website for additional information: **http://www.kaizen-institute.com**.

Baka-yoke

Baka-yoke is a Japanese word that means "machine sensors" and is used to identify malfunctions in production machinery. These devices improve in-process quality and serve as a backup in the event of human operator error. When problems are detected, machines are stopped automatically.

Visual Control

Visual control means that the status of production can be determined with only visual inspection of the manufacturing operation. The concept applies to the work of both team members and production machines. The goal of visual control is to spot problems as quickly as possible and correct them immediately. The principal device used in plants for visual control is the *andon signboard,* an electrical board that shows the current state of production operations by means of lighted indicators. The indicators on the board identify the type and location of the problem, and chimes or a musical melody are often played to alert team members to the presence of a problem. The andon signboard is triggered by either sensors on production machines or by team members who spot a problem. Additional visual control occurs in the form of graphs, charts, data, and status displays indicating production effectiveness and efficiency.

The Team Concept

Successful problem-solving in manufacturing today often requires the inputs of people from different functional areas. Individual effort has traditionally been the focus in many organizations. The dynamic environment of manufacturing operations today makes team-based solutions more important. A single person in an organization will seldom have all the answers. Thinking of the best solutions often involves people from many parts of the organization. Members of problem-solving teams must work together effectively, and successful teamwork is more difficult to achieve than one might think. Companies have found that training and development in team dynamics and teamwork skills are often required.

Teams may take on many formats in an organization. James Evans and William Lindsay have identified several of the most common: quality circles, problem-solving teams, management teams, work teams, project teams, and now even "virtual" teams. Teams should be formed to meet or solve a particular task or problem. The goal should be clear, and team members should support each other. The team needs a level of empowerment that allows it to get the job done.

The subject of teams is too large for detailed discussion in this text. However, the significance and power of teams cannot be overlooked. The days of the lone problem solver are gone in most cases. Decisions require the inputs and consideration of others in the organization. Readers with an interest in a career in manufacturing should make a point to study more about teams and teamwork skills. The Internet offers access to various Websites related to teamwork, team-building, and team-skills development. Try researching these key words using a search engine such as Alta Vista, **www.altavista.com.**

Summary

The Toyota production system used at New United Motor effectively integrates the Japanese production system into the U.S. labor and supplier market. The experiment has many more successes than failures and indicates that many of the production strategies used by Asian manufacturing companies can be adopted by Western management.

APPENDIX 8-1: WIGHT'S BICYCLE EXAMPLE

Oliver Wight is recognized as one of the innovators of the tool that has become known as material requirements planning (MRP). He presented an example of the MRP planning logic using a bicycle as the product. Almost everyone can relate to this basic product. This exercise keeps the MRP calculation logic fairly simple. Later examples add to the complexity of the calculation, but the logic behind the planning stays the same.

The bicycle used in the example is a simple, single-rider device. It has two wheels, two tires, a frame, a seat, and one handlebar. The handlebar in this example is manufactured from tube steel that is cut and formed. The bill of material for the handlebar is shown in Figure 8–26. One handlebar is required for each bicycle produced.

Figure 8–27 details the MRP planning records that link the finished bicycles to the handlebars and to the tube steel. Gross requirements for the finished bicycles have been predetermined and are not part of this example. The lead time for assembly of bicycles is one time period, and there is no stock on hand of finished bicycles. Note the one-period offset that links a requirement for forty finished bicycles in period 2 and a planned order release for forty units into assembly at the start of period 1.

Planned order releases into final assembly drive lower level requirements for material. Note that the planned order release for forty in period 1 becomes a gross requirement for forty handlebars in period 1. The MRP planning logic considers the sixty handlebars on hand before period 1, a lead time of four periods to convert raw material into finished handlebars, and a planning requirement for lot quantities of 120 units. The 120 handlebars shown as "scheduled receipts" in period 3 are the result of the release of a past order to the shop, with 120 expected to be available for use at the start of the period. Work through the projected available balance line from left to right. The projected available balance considers the quantity on hand at the start of the period, plus any new receipts expected, minus any

Figure 8–26 Bill of Material for the Handlebar.

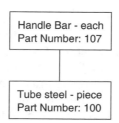

Item: bicycles		Period								
		1	2	3	4	5	6	7	8	9
Gross requirements (from MPS)			40		50			60		60
Scheduled receipts										
Projected available balance	0									
Planned order release		40		50			60		60	

Order policy – lot for lot; order quantity – variable; lead time = 1.

(a)

Item: handlebars		Period							
		1	2	3	4	5	6	7	8
Gross requirements		40		50			60		60
Scheduled receipts				120					
Projected available balance	60	20	20	90	90	90	30	30	90
Planned order release				120					

Order policy = fixed lot; order quantity = 120; lead time = 4.

(b)

Item: tube steel		Period							
		1	2	3	4	5	6	7	8
Gross requirements					120				
Scheduled receipts									
Projected available balance	140	140	140	140	20	20	20	20	20
Planned order release									

Order policy = fixed lot; order quantity = 200; lead time = 5.

(c)

Figure 8–27   MRP Planning Records.

gross requirements. Note that in week 8, not enough stock is on hand and there are no scheduled receipts. This condition forces the planning of a new order release for 120 units at the start of period 4.

Tube steel is a purchased item. The planned order release for handlebars at the start of period 4 drives the gross requirement for 120 pieces of tube steel in period 4. In this example, there is little activity at the tube steel level because there are 140 pieces of steel on hand at the start of period 1 and only one requirement for 120 pieces in period 4.

APPENDIX 8–2: ABCD CHECKLIST*

1. Strategic Planning Processes

Qualitative Characteristics

Class A   Strategic planning is an ongoing process and carries an intense customer focus. The strategic plan drives decisions and actions. Employees at all levels can articulate the company's mission, its vision for the future, and its overall strategic direction.

Class B   Strategic planning is a formal process, performed by line executives and managers at least once per year. Major decisions are tested first against the strategic plan. The mission and/or vision statements are widely shared.

Class C   Strategic planning is done infrequently but provides some direction to how the business is run.

Class D   Strategic planning is nonexistent, or totally removed from the ongoing operation of the business.

Overview Items

1–1 COMMITMENT TO EXCELLENCE

The company has an obsession with excellence; there is dissatisfaction with the status quo. Executives provide the leadership necessary for change. They articulate the motivations for positive change and other core values, and communicate them widely throughout the organization—by actions as well as by words.

1–2 BUSINESS STRATEGY/VISION

There is an explicit written business strategy that includes a vision and/or mission statement. This strategy articulates the commitment to excellence and the overriding importance of customer satisfaction.

1–3 BENCHMARKING

The company continuously measures its products, services, and practices against the toughest competitors, within and outside the industry. This information is used to identify best practices and establish performance benchmarks.

1–4 SUSTAINABLE COMPETITIVE ADVANTAGE

The business strategies recognize the principle of sustainable competitive advantage: those items not directly under the company's control may not yield competitive advantage over the long run.

* Courtesy Oliver Wight Publications.

1–5 ONGOING FORMAL STRATEGIC PLANNING

There is an ongoing formal strategic planning process in place, in which all senior executives have active, visible leadership roles.

1–6 CONGRUENCE TO STRATEGY

Requests for capital expenditure are tested first for congruence to the business strategy and appropriate functional strategies (i.e., does this proposal fit the strategy?).

1–7 BUSINESS PLANNING

A business process is used to develop and communicate annual financial plans that incorporate input from all operating departments of the company.

1–8 GENERATION OF PRODUCT COSTS

Executives and managers believe the accounting system generates valid product costs, reflecting the true costs involved in producing and delivering the company's products. Activity-based costing as well as other costing methodologies are understood, and the most suitable costing method is used.

2. People/Team Processes

Qualitative Characteristics

Class A — Trust, teamwork, mutual respect, open communication, and a high degree of job security are hallmarks of the employee/company relationship. Employees are very pleased with the company and proud to be part of it.

Class B — Employees have confidence in the company's management and consider the company a good place to work. Effective use is made of small work groups.

Class C — Traditional employment practices are largely being used. Management considers the company's people to be an important, but not vital, resource of the business.

Class D — The employee/employer relationship is neutral at best, sometimes negative.

Overview Items

2–1 COMMITMENT TO EXCELLENCE

All levels of management have a commitment to treating people with trust, openness, and honesty. Teams are used to multiply the strength of the organization. People are empowered to take direct action, make decisions, and initiate changes.

2–2 CULTURE

A comprehensive culture exists to support and enhance effective people and team processes.

2–3 TRUST

Openness, honesty, and constructive feedback are highly valued and demonstrated organizational traits.

2–4 TEAMWORK

Clearly identifiable teams are used as the primary means to organize the work, as opposed to individual job functions or independent workstations.

2–5 EMPLOYMENT CONTINUITY

Employment continuity is an important company goal as long as the employee exceeds the minimum acceptable job requirements and the level of business is viable.

2–6 EDUCATION AND TRAINING

An active education and training process for all employees is in place and is focused on business and customer issues and improvements. Its objectives include continuous improvement, enhancing the empowered worker, flexibility, employment stability, and meeting future needs.

2–7 WORK DESIGN**

Jobs are designed to reinforce the company goal of a team-based, empowered workforce.

2–8 CONGRUENCE

People policies, organizational development, and educational and training maintain consistency with the company vision and business strategies.

3. Total Quality and Continuous Improvement Processes

Qualitative Characteristics

Class A Continuous improvement has become a way of life for employees, suppliers, and customers. Improved quality, reduced costs, and increased velocity contribute to a competitive advantage. There is a targeted strategy for innovation.

Class B Most departments participate in these processes; they have active involvement with suppliers and customers. Substantial improvements have been made in many areas.

Class C Processes are used in limited areas; some departmental improvements have been achieved.

Class D Processes are not established, or processes are established but static.

**Same item included in more than one business function.

Overview Items

3–1 COMMITMENT TO EXCELLENCE
There is a commitment to total quality in all areas of the business and to continuous improvements in customer satisfaction, employee development, delivery, and cost.

3–2 TOP MANAGEMENT LEADERSHIP FOR QUALITY AND CONTINUOUS IMPROVEMENT
Top executives are actively involved in establishing and communicating the organization's vision, goals, plans, and values for quality and continuous improvement.

3–3 FOCUS ON CUSTOMER
Various effective techniques are used to ensure that customer needs are identified, prioritized, and satisfied. Customers are identified both internally and externally, and all functions participate. External customers include users, other external links in the chain to the end user, shareholders, stakeholders, and the community.

3–4 CUSTOMER PARTNERSHIPS
Strong "partnership" relationships that are mutually beneficial are established with customers.

3–5 CONTINUOUS ELIMINATION OF WASTE
There is a companywide commitment to the continuous and relentless elimination of waste. A formal program is used to expose, prioritize, and stimulate the elimination of non-value-adding activities.

3–6 ROUTINE USE OF TOTAL QUALITY CONTROL TOOLS
Routine use of the basic tools of total quality control and the practice of mistake proofing has become a way of life in almost all areas of the company.

3–7 RESOURCES AND FACILITIES—FLEXIBILITY, COST, QUALITY
Resources and facilities required to receive, produce, and ship the product economically are continuously made more flexible, cost effective, and capable of producing higher quality.

3–8 PRODUCE TO CUSTOMER ORDERS
The time required to manufacture products has been reduced such that the planning and control system uses forecasts to project material and capacity needs, but production of finished products is based on actual customer orders or distribution demands (except where strategic or seasonal inventories are being built).

3–9 SUPPLIER PARTNERSHIPS
Strong "partnership" relationships that are mutually beneficial are being established with fewer but better suppliers to facilitate improvements in quality, cost, and overall responsiveness.

3–10 PROCUREMENT—QUALITY, RESPONSIVENESS, COST

The procurement process is continuously being improved and simplified to improve quality and responsiveness while simultaneously reducing the total procurement costs.

3–11 KANBAN

Kanban is used effectively to control production where its use will provide significant benefit.

3–12 VELOCITY

The velocity and linearity of flow is continuously measured and improved.

3–13 ACCOUNTING SIMPLIFICATION

Accounting procedures and paperwork are being simplified, eliminating non-value-adding activities, while at the same time providing the ability to generate product costs sufficiently accurate to use in decision making and satisfy audit requirements.

3–14 USE OF TOTAL QUALITY CONTROL AND JUST-IN-TIME

A minimum of 80 percent of the plant output is produced using the tools and techniques of TQC and JIT.

3–15 TEAMWORK

Clearly identifiable teams are used as the primary means to organize the work, as opposed to individual job functions or independent workstations.

3–16 EDUCATION AND TRAINING

An active education and training process for all employees is in place and is focused on business and customer issues and improvement. Its objectives include continuous improvement, enhancing the empowered worker, flexibility, employment stability, and meeting future needs.

3–17 WORK DESIGN

Jobs are designed to reinforce the company goal of a team-based, empowered workforce.

3–18 EMPLOYMENT CONTINUITY

Employment continuity is an important company goal as long as the employee exceeds the minimum acceptable job requirements and the level of business is viable.

3–19 COMPANY PERFORMANCE—QUALITY, DELIVERY, COST

Company performance measurements emphasize quality, delivery, and cost. Performance measures are communicated to all through visible displays that show progress and point the way to improvement (e.g., run charts coupled with Pareto charts).

3–20 SETTING AND ATTAINING QUALITY GOALS
Short- and long-term quality goals that cause the organization to stretch are established, regularly reviewed, and monitored. These goals are targeted on improvements in total cost, cycle time (or response time), and customer quality requirements.

4. New Product Development Processes

Qualitative Characteristics

Class A   All functions in the organization are involved with and actively support the product development process. Product requirements are derived from customer needs. Products are developed in significantly shorter time periods, meet these requirements, and require little or no support. Internal and external suppliers are involved and are an active part of the development process. The resulting revenue and margins satisfy the projections of the original business plan proposals.

Class B   Design engineering (or R&D) and other functions are involved in the development process. Product requirements are derived from customer needs. Product development times have been reduced. A low to medium level of support is required. Few design changes are required for products to meet the requirements.

Class C   The product development process is primarily an engineering or R&D activity. Products are introduced close to schedule but contain traditional problems in manufacturing and the marketplace. Products require significant support to meet performance, quality, or operating objectives. The manufacturing process is not optimized for internal or external suppliers. Some improvement in reducing development time has been achieved.

Class D   The products developed consistently do not meet schedule dates or performance, cost, quality, or reliability goals. They require high levels of support. There is little or no internal or external supplier involvement.

Overview Items

4–1 COMMITMENT TO EXCELLENCE
An intense commitment to excel in the innovation, effectiveness, and speed of new product development is broadly shared by all levels of management throughout the organization.

4–2 MULTIFUNCTIONAL PRODUCT DEVELOPMENT TEAMS
Multifunctional product development teams—including manufacturing, marketing, finance, quality assurance, purchasing, suppliers, and, where appropriate, customers—are used during the design process for new product development.

4–3 EARLY TEAM INVOLVEMENT

Design for manufacturability/concurrent engineering processes are used at the beginning and throughout the product development processes.

4–4 CUSTOMER REQUIREMENTS USED TO DEVELOP PRODUCT SPECIFICATIONS

Customer requirements are determined utilizing processes such as quality function deployment (QFD) and competitive benchmarking. These customer requirements are used to develop the specifications for the product.

4–5 DECREASE TIME-TO-MARKET

An ongoing effort to decrease the time-to market—the elapsed time between the start of the design and the first shipment of the product—is viewed as an important competitive weapon, as highly visible, and as generating improvements.

4–6 PREFERRED COMPONENTS, MATERIALS, AND PROCESSES

There is an agreed-upon product development and manufacturing policy on commonality and use of preferred components and preferred processes in new designs.

4–7 EDUCATION AND TRAINING

An active education and training process for all employees is in place and is focused on business and customer issues and improvements. Its objectives include continuous improvement, enhancing the empowered worker, flexibility, employment stability, and meeting future needs.

4–8 NEW PRODUCT DEVELOPMENT INTEGRATED WITH THE PLANNING AND CONTROL SYSTEM

All phases of new product development are integrated with the planning and control system.

4–9 PRODUCT DEVELOPMENT ACTIVITIES INTEGRATED WITH THE PLANNING AND CONTROL SYSTEM

Where applicable, product development activities in support of a customer order are integrated with the planning and control system.

4–10 CONTROLLING CHANGES

There is an effective process for evaluating, planning, and controlling changes to the existing products.

5. Planning and Control Processes

Qualitative Characteristics

Class A   Planning and control processes are effectively used company wide, from top to bottom. Their use generates significant improvements in customer service, productivity, inventory, and costs.

> Class B   These processes are supported by top management and used by middle management to achieve measurable company improvements.
>
> Class C   The planning and control system is operated primarily as a better method for ordering materials and contribution to better inventory management.
>
> Class D   Information provided by the planning and control system is inaccurate and poorly understood by users, providing little help in running the business.

Overview Items

5–1 COMMITMENT TO EXCELLENCE
There is a commitment by top management and throughout the company to use effective planning and control techniques—providing a single set of numbers used by all members of the organization. These numbers represent valid schedules that people believe and use to run the business.

5–2 SALES AND OPERATIONS PLANNING
There is a sales and operations planning process in place that maintains a valid, current operating plan in support of customer requirements and the business plan. This process includes a formal meeting each month run by the general manager and covers a planning horizon adequate to plan resources effectively.

5–3 FINANCIAL PLANNING, REPORTING, AND MEASUREMENT
There is a single set of numbers used by all functions within the operating system that provides the source data used for financial planning, reporting, and measurement.

5–4 WHAT-IF SIMULATIONS
What-if simulations are used to evaluate alternative operating plans and develop contingency plans for materials, people, equipment, and finances.

5–5 ACCOUNTABLE FORECASTING PROCESS
There is a process for forecasting all anticipated demands with sufficient detail and an adequate planning horizon to support business planning, sales and operations planning, and master production scheduling. Forecast accuracy is measured to improve the process continuously.

5–6 SALES PLANS
There is a formal sales planning process in place, with the sales force responsible and accountable for developing and executing the resulting sales plan. Differences between the sales plan and the forecast are reconciled.

5–7 INTEGRATED CUSTOMER ORDER ENTRY AND PROMISING
Customer order entry and promising are integrated with the master production scheduling system and inventory data. There are mechanisms for matching incoming orders to forecasts and for handling abnormal demands.

5–8 MASTER PRODUCTION SCHEDULING
The master production scheduling process is perpetually managed to ensure a balance of stability and responsiveness. The master production schedule is reconciled with the production plan resulting from the sales and operations planning process.

5–9 MATERIAL PLANNING AND CONTROL
There is a material planning process that maintains valid schedules and a material control process that communicates priorities through a manufacturing schedule, dispatch list, supplier schedule, and/or a kanban mechanism.

5–10 SUPPLIER PLANNING AND CONTROL
A supplier planning and scheduling process provides visibility for key items covering an adequate planning horizon.

5–11 CAPACITY PLANNING AND CONTROL
There is a capacity planning process using rough-cut capacity planning and, where applicable, capacity requirements planning in which planned capacity, based on demonstrated output, is balanced with required capacity. A capacity control process is used to measure and manage factory throughput and queues.

5–12 CUSTOMER SERVICE
An objective for on-time deliveries exists, and the customers are in agreement with it. Performance against the objective is measured.

5–13 SALES PLAN PERFORMANCE
Accountability for performance to the sales plan has been established, and the method of measurement and the goal has been agreed upon.

5–14 PRODUCTION PLAN PERFORMANCE
Accountability for production plan performance has been established, and the method of measurement and the goal has been agreed upon. Production plan performance is more than ±2 percent of the monthly plan, except in cases where mid-month changes have been authorized by top management.

5–15 MASTER PRODUCTION SCHEDULE PERFORMANCE
Accountability for master production schedule performance has been established, and the method of measurement and the goal has been agreed upon. Master production schedule performance is 95 to 100 percent of the plan.

5–16 MANUFACTURING SCHEDULE PERFORMANCE
Accountability for manufacturing schedule performance has been established, and the method of measurement and the goal has been agreed upon. Manufacturing schedule performance is 95 to 100 percent of the plan.

5–17 SUPPLIER DELIVERY PERFORMANCE

Accountability for supplier delivery performance has been established, and the method of measurement and the goal has been agreed upon. Supplier delivery performance is 95 to 100 percent of the plan.

5–18 BILL OF MATERIAL STRUCTURE AND ACCURACY

The planning and control process is supported by a properly structured, accurate, and integrated set of bills of material (formulas, recipes) and related data. Bill of material accuracy is in the 98 to 100 percent range.

5–19 INVENTORY RECORD ACCURACY

There is an inventory control process in place that provides accurate warehouse, stockroom, and work-in-process inventory data. At least 95 percent of all item inventory records match the physical counts, within the counting tolerance.

5–20 ROUTING ACCURACY

When routings are applicable, a development and maintenance process is in place that provides accurate routing information. Routing accuracy is in the 95 to 100 percent range.

5–21 EDUCATION AND TRAINING

An active education and training process for all employees is in place and is focused on business and customer issues and improvements. Its objectives include continuous improvement, enhancing the empowered worker, flexibility, employment stability, and meeting future needs.

5–22 DISTRIBUTION RESOURCE PLANNING (DRP)

Distribution resource planning, where applicable, is used to manage the logistics of distribution. DRP information is used for sales and operations planning, master production scheduling, supplier scheduling, transportation planning, and the scheduling of shipping.

Enterprise Resource Planning, and Beyond

In the late 1990s, the pace of change in manufacturing systems was increasing exponentially as information technology tools grew and advanced. Keeping up to date with all the improvements is not an easy task, but it is a mandatory one. The intention of this chapter is to highlight the technology being offered in the marketplace today. We want to convey to the reader a basic view of the expanded enterprise-wide optimization tools that build on the foundation of traditional MRP and MRP II. Consideration of the new enterprise resource planning tools will lead to an exciting review of the operational issues facing companies at the beginning of the century. If there was ever an area in manufacturing that will require lifelong learning, the study of the evolving enterprise control systems is it.

Enterprise resource planning is one of the newer system concepts that focuses on the integration of business systems. These integrated systems support all of the functional departments in the enterprise: sales and order entry, engineering, manufacturing, finance and accounting, distribution, order planning and execution, and the supply chain flows. It is a logical extension of traditional manufacturing resource planning (MRP II), which takes advantage of increased computer power and new client/server architectures to bring heavy duty data analysis tools to the user's desktop.

> Enterprise resources planning (ERP) system can be defined in one of two ways: (1) an accounting-oriented information system for identifying and planning the enterprise-wide resources needed to take, make, ship, and account for customer orders. An ERP system differs from the typical MRP II system in technical requirements such as graphical user interface; relational database; use of fourth-generation language, and computer-assisted software engineering tools in development, client/server architecture, and open-system portability; (2) more generally, a method for the effective planning and control of all resources needed to take, make, ship, and account for customer orders in a manufacturing, distribution, or service company. (*APICS Dictionary*, 9th Ed., 1998.)

Figure 9–1 traces the development of manufacturing software, beginning with the early computer-based systems for inventory control and moving toward the

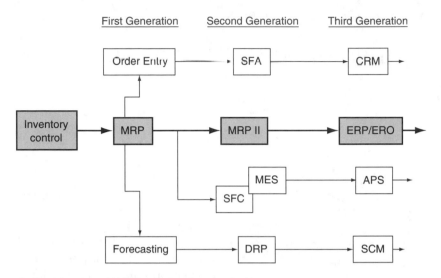

Figure 9–1 Manufacturing Software Development Path.
Source: SIBC Corp. Technology Course Series 1999.

modern systems popular at the start of the millenium. During this evolution, software companies have moved toward increasing the integration of the disconnected enterprise systems with the more mature manufacturing and financial systems. The primary path runs through the center of the figure, beginning at the left with inventory control and working across to ERP/ERO at the right. System capabilities were developed and expanded as computers and software became more powerful and cost efficient. System users came to appreciate and develop relationships with the functional areas that provided them with the critical data and information needed to operate manufacturing efficiently. For example, MRP is a more effective process if the higher level plans that feed into it are more accurate and timely. Improving the inputs to MRP required more participation and inputs from functional areas outside of the traditional manufacturing organization. The development of detailed plans for material resources became the logical driver of detailed capacity plans for physical resources and the evolution of closed-loop systems for manufacturing resource planning. The customer focus evident in the SME enterprise wheel provided an additional driving force to develop advanced systems that considered the resources of the enterprise and all the functional areas that it encompasses. The experience of the past twenty years has proven that manufacturing cannot function effectively by itself. The ERP architecture forces each functional area, including manufacturing, to join the program. Examples in the chapter will show that obtaining full cooperation is not an easy task. Besides ERP, several other system solutions have evolved to address specific planning problems. Descriptions of the other boxes in Figure 9–1 are explained in Appendix 9–1 at the end of this chapter.

9–1 MRP II: A DRIVER OF EFFECTIVE ERP SYSTEMS

MRP II is an essential part of ERP. Figure 9–2 depicts the core modules of the standard MRP II model and highlights the new modules offering expanded functioning for ERP. The core system logic enabling the MRP explosion process remains intact. Significant improvements have been made to planning and analysis modules such as finance and accounting, demand management, engineering, and global planning. These changes will be covered in detail in this chapter. At the other end of the model, the plant and supplier communication module and the schedule execution modules have been restructured through capabilities such as supply chain planning and optimized scheduling software. Again, these tools have been in place for many years, but they have been structured for broader applications across industries in ERP systems. The schedule execution modules have received significant attention as companies look toward just-in-time flows of material and short cycle manufacturing with kanban-type pull signals. The close-in schedule is much more focused on finite execution and broadcast capabilities than on a frozen master schedule with multiple time fences for accepting customer orders. Today's approach is to make to the demand schedule and keep the electronic order board open around the clock.

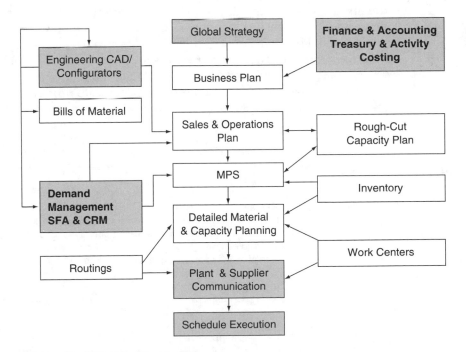

Figure 9–2 The MRP II Core Inside ERP.
Source: SIBC Corp. Technology Course Series 1999.

9–2 INFORMATION TECHNOLOGY

The synopsis presented in Figure 9–3 was prepared by a consultant familiar with the backroom operations taking place in the information technology (IT) department of many companies. His recent experiences depict the challenges IT teams face as they respond to various system issues. The IT team may have to address upgrades or replacement of software code, the implementation of new systems, conversion of data from legacy subsystems, the introduction of new architectures (for example, client/server), and debugging of code. Other issues may include the training of system users, process reengineering, documentation of new procedures, upgrading the staff skills of the department, and searching for new employees. All of this takes place while IT placates sales and marketing and engineering functions that are not getting critical software systems implemented in their areas. It is clear that companies will have to face many difficult systems issues and have to make many difficult systems and operations decisions in the short- and long-term.

Many schedule execution problems are caused by the slow transmission of critical data throughout the organization. Software companies around the world have responded to this problem and now offer several features in a transaction processing and control system known as enterprise resource planning (ERP) system. Such enterprise-wide information systems attempt to connect all business units in a company and facilitate data acquisition and sharing across all departments and functional areas. Companies using ERP can now share data much more efficiently and make major advances toward sharply decreasing data redundancy. Design engineering can now gain access to the customer information gathered by the sales and marketing departments to make product design decisions. Sales and marketing personnel can electronically communicate product configuration data, cost estimates, price quotations, and forecasting information with manufacturing

Figure 9–3   Today's Typical Systems Environment.
Source: SIBC Corp. Technology Course Series 1999.

- Everything is a "fire drill"
- Most technologies are only partially implemented or interfaced
- Budget constraints constantly cut back deployments
- Acquisitions & divestments force major steps backward
- Because of time constraints some ERP implementations were abandoned half way in favor of fixing legacy code
- Months of "Y2K" preparation drew blood from IT teams
- IT back room operations personnel are "winging it"
- Short term needs interfere with obtaining quality deliverables
- Sales/Engineering/Marketing/Purchasing/Field Service feel they have been abandoned
- Re-implementations have already started on new ERP systems
- The user/customer/client isn't well trained
- IT analysts and users need a long rest

and engineering. This ability provides a significant improvement in the ability to schedule the shop floor finitely. In this view, ERP is much like MRP II but with some new features that make it more attractive for all the functional areas.

Poorly integrated information technology infrastructures can make a company's information systems seem like "islands within the enterprise." Consultants from Grant Thornton report,

> The systems implemented to control vital manufacturing functions often cannot communicate with one another. Data and information does not flow where it is needed, but must be shared through manual processes and/or by rekeying data into multiple systems within the company. Poor integration can lead to critical omissions, erroneous data, production delays, and ultimately reduced profit.

Data Collection and Control Issues

ERP systems have helped solve many of the data acquisition and consistency problems encountered by medium-size to large manufacturers. Kurt Freimuth, president of BA Solutions, has found that problems may persist even after the implementation of an ERP system. Personnel often have to ask, "How did we make that part the last time? Sam always did it this way, but what do we do now?" The Deloitte & Touche consulting firm has diagnosed some of the problems with ERP implementations in their tool kit titled "Second-Wave." Their assessment of many ERP implementations is that they are only partially completed and lack the data accuracy and integration that helped to sell the systems to company officers in the first place. They now have a large contingent of consultants performing "second-wave" implementations and cleanup tasks at these ERP sites.

Manufacturing resource planning is an important component of ERP systems that provides critical information to all areas of an organization, from design engineering to sales and marketing, accounting, inventory management, and shipping. While important to the daily operations of a manufacturing concern, MRP is more or less an accounting and scheduling system oriented to ship on time, every time. In many midsize companies, it is an information system that runs daily or weekly and is not real time. Variations on the shop floor between the MRP system and actual practice may not be discovered or corrected until the next time the system is run. The biggest problem with using MRP alone is that it is based on two major assumptions: (1) the based data is accurate, and (2) the lead times are unchanging. The key to accurate data on the shop floor lies in the base files that support the entire planning hierarchy within an organization.

In practice, there are several fairly static base files used by MRP, such as the bill of materials, part/item master, inventory location master, work center file, routing files, vendor master, lead times, and the employee master file. These files constitute the foundation for all planning activity and provide the reference foundation for planning; however, they are not updated regularly in many systems. The information contained in the base files may be gathered only once every one to five years. The resulting information that supports the entire ERP system may be based largely on static historical data rather than dynamic current data.

The Information System Integration Nightmare

Companies often operate with systems that have evolved over many years. It is not uncommon to find companies operating several different software packages to satisfy the need for information and to help them plan and operate the business. Systems are often found to operate on different computer hardware platforms (mainframe, midrange, and more recently networked PCs). The information systems group in many companies has custom-designed and programmed elaborate and complex support systems to link the software programs, databases, and systems used in the different functional areas. Figure 9–4 provides a system map depicting various modules on different platforms with varying levels of interconnection at an actual case study company.

The challenge facing many companies with systems at all levels of integration and installed across various computer platforms is how to pull them together and how to keep them running. These two tasks are not easy for the information systems department, and they can become more difficult. The technology advances are relentless and powerful, yet many companies do not take advantage of

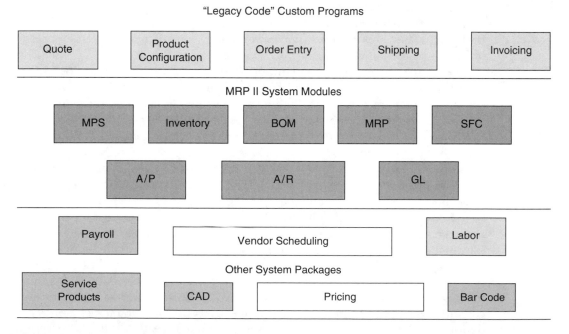

Figure 9–4   A Multiplatform System Environment Map.

Source: SIBC Corp. Technology Course Series 1999.

the new systems because of problems that plague their current operations. Companies have short-changed the investments that should have been made in platform consolidation and software upgrades. Company executives complain about their large investment in systems that do not show any improvement to the bottom line of the corporation. The risk to those progressive companies that have implemented ERP is that they too may not achieve the original payback projections if the systems are implemented only partially and lack the data accuracy and integration necessary to be classified as a Class A system.

9–3 THE DECISION TO IMPLEMENT AN ERP SYSTEM

Companies that choose to implement an ERP system face several critical decisions. There is much more to this process than selecting a software package. The rapid growth of information systems technology and the explosion of the Internet have made this process more complex. According to Patrick Delaney, president of SIBC Corporation, some of the key questions that must be addressed include:

What are the hardware and software requirements?

Will the system architecture be client/server?

Will there be an extranet Web site?

What bolt-on software modules might be required?

How will data be warehoused and mined?

What are the e-commerce applications of the system?

What business entity partitions will be needed in the system?

What network bandwidth will be needed to support a responsive system?

What is the migration path from where we are now to the future system?

Many of these questions are new to today's executives, but they must be answered correctly and with an eye toward the future. These are critical questions and the way they are answered can have a significant impact on the potential for success and survival of the company.

The decisions made in the past ten years related to systems and technology have a direct impact on companies now. Decisions made today will have far-reaching effects on the company well into the future. For many companies, the picture today is not optimistic. Bankrupt software companies have left system users with no support, system maintenance, or software upgrades. Complex interfaces written to link software packages on different systems often complicate migration to a new package. Decisions made in the past have limited the network capability to expand for e-business, Internet applications, and mobile access. Discontinued or obsolete hardware may make migration to a new system expensive; examples are reported weekly as companies move to decentralized computing models that require "fat clients" on the desktop. In many corporations, the information technology staff members do not have the skill set needed to work in the networked PC environment evolving rapidly toward full e-commerce capability.

Regardless of the difficulty and complexity, many companies will invest in new ERP systems. The systems can help companies sustain a competitive advantage in their industry. ERP systems can help companies manage information, a critical resource, more effectively. Figure 9–5 provides a list of some of the typical information needs of ERP systems. Notice that a lot of information in the list is outside the scope of the traditional manufacturing functions. The information requirements outlined in the figure address the enterprise, not just manufacturing.

Figure 9–5   Information Requirements of an ERP System.

- Sales, customer & order demand-related information
- Manufacturing resource data
- Inventory status data
- Manufacturing process information
- Internal control & security access tables for client/server
- Cost collection: standard-actual-activity costs
- Performance measurement extracts
- Customer information
- Customer satisfaction information
- Stockholder & treasury information
- Vendor & Supply chain detailed data
- Employee H-R data

Many of the new generation of manufacturing systems that were formerly known as MRP or MRP II systems now promote themselves as ERP systems. They still try to utilize a single database. New networking and open computer architectures have made the communication and integration of companywide systems possible. New links that were not possible just a few years ago can now be made to allow links to other systems through application interface programs. The continuing growth of the Windows™ operating systems and open system architectures have made communications between many of the systems used in manufacturing companies much easier. Interface programs are written to connect the ERP core software and the central database with detailed systems that support the operations throughout the company. This connection allows, for example, data collected by the sales force to be incorporated seamlessly into the central database and to be used for improving planning in ERP. The advancements in the available technology are making meaningful and efficient new links between the functional area programs and the central planning programs possible. Figure 9–6 provides an illustration of these relationships.

9–4 FEATURES OF MODERN MANUFACTURING, PLANNING, & CONTROL SYSTEMS

This section introduces an example of several of the important features of the modern ERP systems. The examples were originally developed by Joe Wilkinson, creator of the WinMan software system, and enhanced for presentation at the SME Education in Manufacturing Conference in 1998. WinMan is a Windows™ based package that takes advantage of relational database technology and gives a fresh

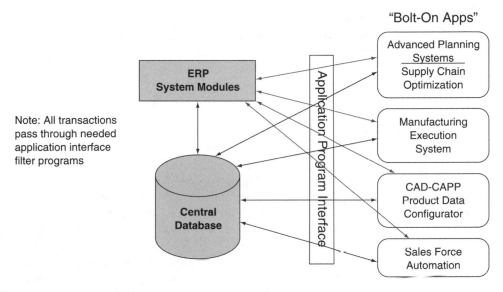

Figure 9–6 Advanced System Architectures.
Source: SIBC Corp. Technology Course Series 1999.

look at how a system can support manufacturing. Details of the example are explained in Appendix 9–3.

A central database for item-specific information is a common feature of modern MP&C systems. The creation of items in the system forces the system users to consider the type of item being created. These items include the item numbering system, the effective use of descriptive text, the unit of measure, the person responsible for planning and buying the item, the cost of the item, and procurement lead time information. Lots of questions need to be answered before the system can do any planning. All material items are not the same and must be carefully considered. Partial lists of the key material issues that must be addressed include:

- What is the unit of measure?
- What planning parameters do we want the system to use?
- How much inventory do we want to keep?
- What items may require safety stock?
- How will the warehouse be controlled?
- How will locations be named and identified?

Remember, this list is partial.

Current versions of business systems come with special controls and field edits to protect the integrity of the system data. New users must first sign on to the system and will be granted access to screens and transactions based on their assigned security level. Some of the information accumulated in the system is accessible to people throughout the organization. Users who have been given the appropriate authorization can access security-controlled information and data.

Typically there are several standard reports and screens that come ready to use. System users soon learn where to look in the system to find the data and answers needed to solve problems. Most systems now utilize online updates for most transactions. Additions of items to the system, reporting, and maintenance transactions are typically processed online. The processing of most transactions is completed quickly with little user waiting time.

There is a growing trend toward system designs that support a nearly paperless operation. Most often, the information needed to solve a problem is available for online inquiry and action. Occasionally, a printed report is required to find the needed information or for special documentation. It is recommended to view information on the screen whenever possible or to route report output to a screen window rather than to generate a printed page. Printed reports are available to users on a request basis. Report writers often provide filters that can be applied to limit the size of the print output or to focus on a particular issue. In addition to standard reports, a report writer function may be available to system users. Report writing features often allow more experienced users to select the fields they wish from the database and to develop customized reports. Print output can be directed to a screen for viewing or to printers on a network.

Describing Items and Products in a Formal System

The work with a formal system for manufacturing really begins with the entry of item-specific information into the database. Nothing can happen in the system until the specific information that defines an item has been entered and processed. Once the item is created in the system, the information that has been loaded can be viewed and used by people throughout the company. For example, the product designer/engineer may select the item for use in a product structure. The purchasing department may place orders for items and actually schedule delivery of items needed to complete production. Manufacturing may issue some of the item out of stock and use it to produce a product.

Modern systems try to make the entry of item information into the system as easy as possible. However, many decisions are required as part of the item setup process. The current generation of systems with graphical user interfaces often leads the user step-by-step through the data entry process. In many cases, systems will provide the user with a prompt in a dialog box or with a pull-down menu of choices. Often the data-entry process is as easy as pointing and clicking on a desired entry using the mouse. Selecting from the choices provided and minimizing the actual entry from the keyboard helps reduce data entry errors. Editing or correcting fields that were entered incorrectly is easier in the modern systems.

Accurate data entry is a critical part of the formal system process. The infamous "central database" is only as good as the data entered to it. People from all around the company will be using this data. The system cannot do a good job planning for you if the data you input is not the best you have. A sample view of the product setup screen for a new item is shown in Figure 9–7.

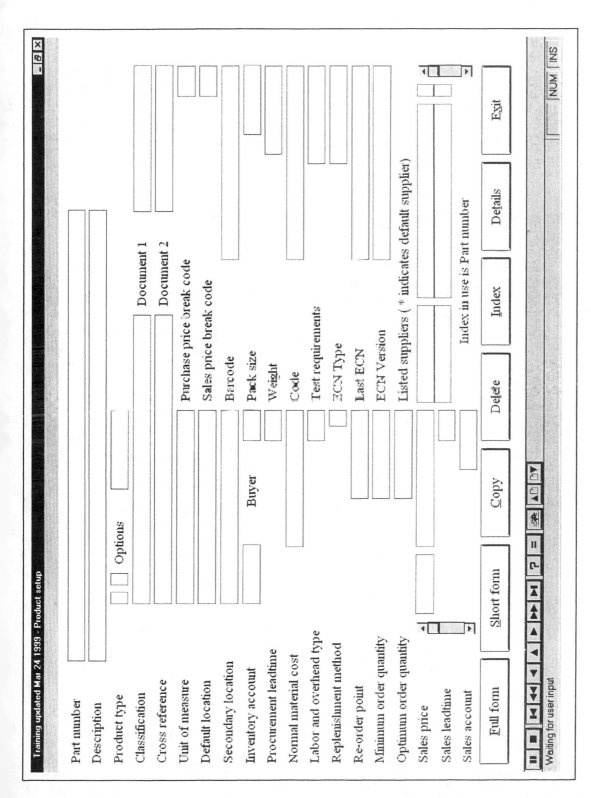

Figure 9-7 Product Setup Screen from WinMan Software by TTW, Inc.

Maintenance of the data is also important. Much of the data entered into the item master files is fairly static and will not change often. Several fields are more dynamic, such as the procurement lead time (in days) and the order policy information. The new generation of formal systems has made changing these fields as easy as clicking on the field and following the instructions in the resulting dialog box. Additional discussion of planning using WinMan is included in Appendix 9–3.

9–5 DEVELOPING TECHNOLOGIES: CONVERGING AND ENABLING

The advancements in technology and the development of modern pull systems for manufacturing are forcing several systems to be combined. Islands of technology that once separated the finance, engineering, manufacturing, and sales/marketing functions are being networked together. Figure 9–8 lists several of the primary system groups. No single system now covers all the areas and systems.

The ongoing development of computer operating systems and hardware platforms has made sharing data and improved communications possible. The Windows NT™ operating system, for example, allows programs to share data,

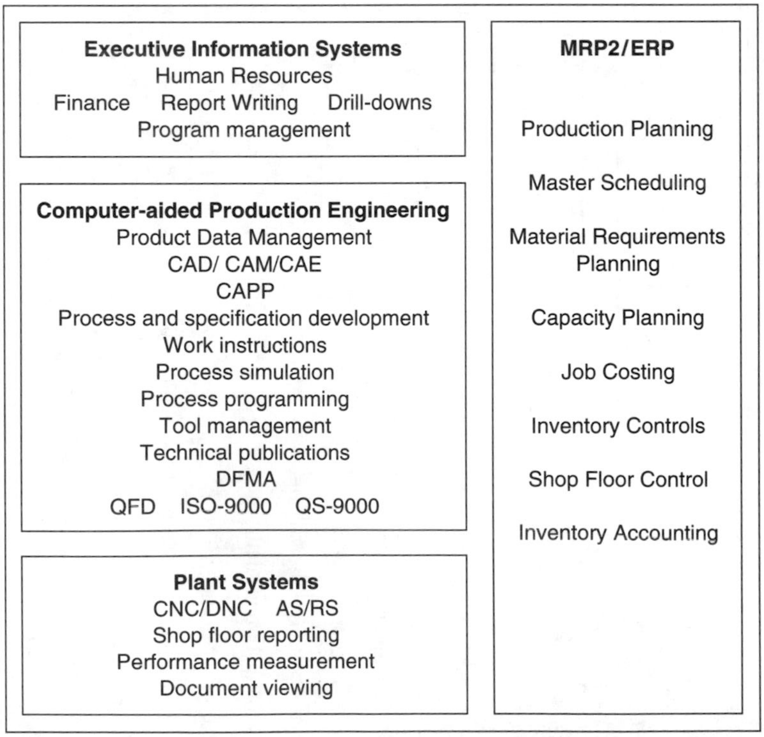

Executive Information Systems Human Resources Finance Report Writing Drill-downs Program management	**MRP2/ERP** Production Planning Master Scheduling
Computer-aided Production Engineering Product Data Management CAD/ CAM/CAE CAPP Process and specification development Work instructions Process simulation Process programming Tool management Technical publications DFMA QFD ISO-9000 QS-9000	Material Requirements Planning Capacity Planning Job Costing Inventory Controls Shop Floor Control Inventory Accounting
Plant Systems CNC/DNC AS/RS Shop floor reporting Performance measurement Document viewing	

Figure 9–8   Converging Systems for Manufacturing Planning and Control Systems.

pictures, text, spreadsheet and word processor documents, and even sound and video files. Report-writing features built into modern systems allow users to prepare database queries that in the past required special attention from someone in data processing. Drill-down features allow users to work through system fields to find the real drivers of schedules and costs. Production engineering functions related to design and documentation of parts and processes are much more efficient when data sharing is possible. Modern quality programs such as ISO-9000 (the international quality standard) and QS-9000 (the automotive industry quality standard) require extensive documentation and controls that are enhanced through the use of new computer-based systems. Systems are designed to take advantage of the emerging technologies to improve communication and break down barriers that cause confusion and add cost to the operation, and these systems help companies become more competitive.

Information will be a significant driver of the factory of the future. The newly linked programs, systems, and the central database will provide a framework for even more computer-based integration. The key systems issues will continue to address the improvement of the company's data accuracy and its ability to communicate internally and externally. The management and control of the company's data cannot be left to chance. The data and information contained in the company's systems are real assets that have value and must be protected. Figure 9–9 provides a

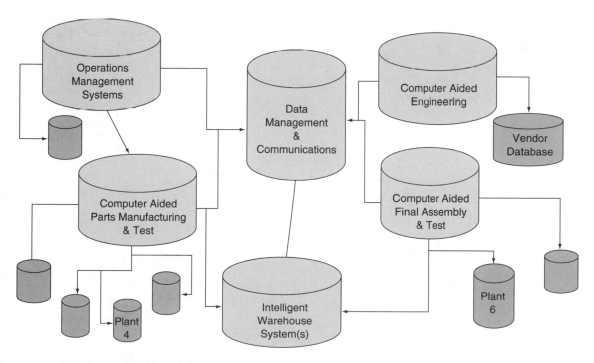

Figure 9–9 Factory of the Future Database Model.

Source: Adapted from Fundamentals of Computer Integrated Manufacturing, *Prentice Hall, 1991.*

factory of the future vision according to General Electric. Each site is seamlessly connected to the major databases that provide data to all of the plant sites. Data and information flow throughout the organization in a fashion similar to the way the human brain and nervous system send signals and commands throughout the body. The environment is paperless, intranet enabled, extranet dependent, client/server structured, online, and real-time enabled with fully staffed teams of "knowledge workers" maximizing the flow of the information pipeline.

Many companies have seen their systems grow and evolve over the past fifteen to twenty years. The result is often a messy system that has been patched, modified, bridged, enhanced, customized, etc. This type of system is extremely difficult for the information systems groups to work with and maintain. In these environments, the selection of a single new software package is also difficult. Software packages have often been developed to solve a particular issue or support a particular functional area. Many of the leading packages are recognized for their functioning in a particular area. In a growing effort to provide a total solution for a company, software packages have added features that fall outside their niche so they can round out their approach to the marketplace. The result is a package that is strong for some functions but very weak for others.

Some major corporations have elected to implement systems that utilize the best features from several different commercial packages. Figure 9–10 provides an

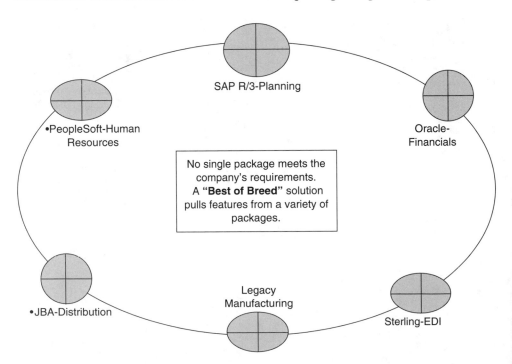

SAP R/3-Planning

•PeopleSoft-Human
Resources

Oracle-
Financials

No single package meets the
company's requirements.
A **"Best of Breed"** solution
pulls features from a variety of
packages.

•JBA-Distribution

Legacy
Manufacturing

Sterling-EDI

Figure 9–10   Single Corporation Selecting Diverse Applications.
Source: SIBC Corp. Technology Course Series 1999.

illustration of this approach. In the figure, the company could not find a single system that covered all its needs. It did recognize, however, that several packages are known for their strengths in particular areas and that by pulling in the best of the available systems, it could reach the best system. It will be interesting to see if this approach works effectively, or if the company has defaulted to a politically complex situation in the short term. Working with only one central package and its related interfaces is a huge challenge. Trade-offs must be considered. Companies must not lose sight of their competitive and strategic reasons for investing in information systems.

The system users in the functional areas may become closely attached to their systems. A change in how the work gets done, even if it represents an improvement, can turn people against it. The system implementation team will need to develop carefully a plan for communicating the need for change to the people involved. It will be critical to get the end users to accept the new system and thus ensure its success.

The systems that will drive the factory and enterprises of the future are continuing to be developed. Figure 9–11 presents some of the emerging applications that will become part of the integration challenge of the future. Nearly instantaneous communication from one point to any other around the world is becoming a reality. The growth in communication and the information it will make available to companies will highlight the challenge. Communications and systems will no longer have an internal company focus. Much of the growth will come as companies attempt to reach out to customers and suppliers along the length of the supply chain. The growing customer focus will drive companies to use the emerging technology for becoming more responsive and for making business easier to conduct. The amazing growth of the Internet has only begun for industrial applications. The specific applications are emerging slowly, but the

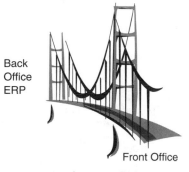

Back
Office
ERP

Front Office
Customer Relationship
Management

- Sales Force Automation (SFA)
- Laptop configurators
- QA-Remote Diagnostics
- Global Dealer Pipelines
- Web-based Procurement
- Supply Chain Management
- In-transit GPS Tracking
- Legacy Data Mining
- E-Commerce Ready Applications
- Electronic Catalogs
- Supply Chain Optimizers
- APS "Bolt-On" Applications
- Finite Execution Systems
- RF Data Collection
- Mobile Access

Figure 9–11   The Integration Challenge of the Future.
Source: SIBC Corp. Technology Course Series 1999.

direction toward doing more business online, for example, is clear. The technology is already allowing companies to do more and do it faster, with virtual plant tours of remote sites, transmission of complex engineering graphics, direct customer access to product configuration systems, and order tracking through each link in the supply chain.

Figure 9–12 highlights several of the critical issues related to e-commerce. E-commerce is already changing the way we communicate and do business outside and inside the firm. We may no longer have retail stores and displays if the value that they provide to the customer can be provided more economically using the Internet. The traditional distribution channels that linked suppliers with distributors, resellers, and the customers may be drastically shortened. A new computer-based warehouse and distribution system may emerge for many companies. The pressure to respond quickly in a competitive environment will compel companies to explore and implement new communications technologies. These technologies will directly affect each company's information and planning systems.

"Virtual manufacturing" may take on new meaning as companies link together to share information electronically and respond to manufacturing orders initiated over the Internet. Imagine that a company needs to have a particular part produced with the highest quality and at the lowest cost. The specifications and order details could be posted to the Internet at a site were approved suppliers can find the demand. Bids for producing this part could be accepted in a defined bid window. Proposals from suppliers could be presented and managed electronically. The parts made to the company's specifications could arrive at a specified assembly operation at a given time, and the finished products can be completed and shipped. Thus, traditional orders and paperwork could be replaced by interactive systems using the Internet.

Connecting the field sales and service personnel with the main corporate system is an exciting possibility with the advanced information systems and communications technologies. These systems can dramatically change the way a company does business. Imagine the power and potential competitive advantages that can

- **Distributors and retail stores may no longer add value**

- **Internet purchases and EC links between end customers and OEMs may eliminate the need for a mid-level buffer**

- **Combining "Quick Response Manufacturing" with Electronic Commerce may eliminate a major linkage in the supply chain, and reduce costs.**

Figure 9–12   E-Commerce Issues.
Source: SIBC Corp. Technology Course Series 1999.

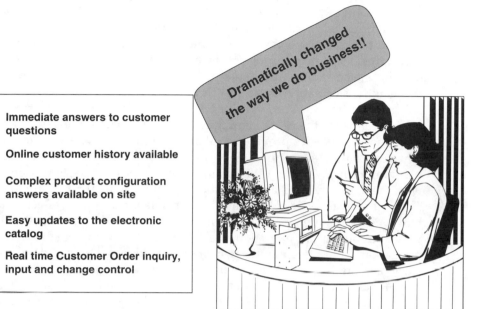

- **Immediate answers to customer questions**

- **Online customer history available**

- **Complex product configuration answers available on site**

- **Easy updates to the electronic catalog**

- **Real time Customer Order inquiry, input and change control**

Figure 9–13   Sales and Field Force Automation
Source: SIBC Corp. Technology Course Series 1999.

result from services listed in Figure 9–13. Customer questions can be handled in real time with real answers. Order-entry people can have access to a customer's order history, making it easier for them to reorder. Configuration issues and option selections can be managed better before production begins and errors are found on the shop floor. Printed catalogs that become outdated can be replaced by electronic versions maintained on the Internet. Changes to customer orders can be managed better and the implication of changes on the production process identified sooner. Companies with this improved communication capability will have the potential to become the preferred supplier and thus enhance their competitive position.

9–6 IDENTIFYING ERP SYSTEM SUPPLIERS

ERP systems are now available from several software suppliers. These systems run on various computer platforms. Costs for ERP systems vary greatly. Note that specific system implementation cost figures are not typically provided by software vendors. When comparing costs, make sure you know what is included and what is not. Interested readers can find some of the most current information about ERP and available cost estimates for ERP systems by visiting sites on the Internet. Appendix 9–2 describes several of these sources.

Modern ERP systems such as WinMan support the concepts of lean production. Lantech, Inc., a maker of shrink-wrapping equipment in Louisville, Kentucky, worked for nearly two years to become a lean manufacturing company. Lantech redesigned its manufacturing processes to eliminate waste and move to lean production methods. The traditional push-style MRP II system was in place before the lean transformation was replaced with the WinMan ERP system and an effective kanban system. The improved workflow and details of the operation have been described in the new planning system. Lantech used the data fields and planning parameter codes provided in WinMan to match the system to what is actually done on the shop floor. (See Appendix 9–3 for a more specific example of planning using WinMan.) WinMan has helped provide system users at Lantech with the data they need and helped to minimize manual reporting to the system. The cost savings realized by Lantech as a result of running an effective formal system have been significant. The work to streamline the processes and eliminate waste throughout the operation also helped Lantech complete the WinMan implementation in only two months, which saved additional time and money. The case story of Lantech can be found in the book *Lean Thinking*.

Refer again to Figure 9–1 and note that there is another three-letter abbreviation associated with ERP: ERO. This subtle change to enterprise resource optimization (ERO) is descriptive of the work being done to enhance and develop the concept. It is also a contributor to the jargon problem in manufacturing terminology that can often confuse matters. More abbreviations will be coming and the systems will continue to grow more complex, but do not let ERP/ERO confuse you.

And what can we predict about the future? We can be sure that our systems will continue to change. Computer systems will continue to increase in speed and become more cost effective. At the same time, the systems will become more complex and sophisticated. Patrick Delaney gives us a possible scenario for the future in Figure 9–14. ERP systems may grow to absorb the advanced production scheduling (APS) functions that are now considered "bolt-ons" to current packages. The functions now handled by third-party software packages will become part of the true functioning of the ERP system. These changes could take us into a systems environment with fewer software providers involved, and they should make the information systems function better able to manage and control the system. The traditional interface programs that have to be written and maintained by the IS function today will be absorbed into the primary ERP system. This move would really bring ERP up to the level that was intended for it.

With ERP functioning at the highest level and traditional bolt-on applications absorbed and working seamlessly across all the functional areas, it may be possible to move to a higher level of system use and performance. Traditional systems have provided tools for companies to make better decisions about their resources. The tools documented plans and scheduling changes, and the system tools helped companies determine how to change the plans in light of the new conditions. Perhaps we may see new system engines that will help companies optimize their operations by going beyond the current responses to change. Future systems should

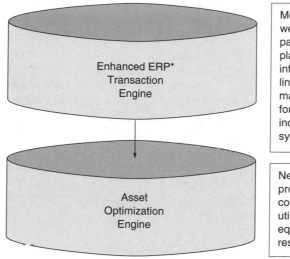

Module functions that formerly were achieved using "bolt on" packages such as advanced planning and execution systems, integrated design and process links, configuration management tools and sales force automation tools are incorporated into the ERP system.

New optimization software provides tools to help companies get the best utilization of material, equipment and human resources.

Figure 9–14   Asset Optimization Software—Possible by 2005?

make it easier for companies to test multiple scenarios and play the what-if" game more effectively. Future systems may provide users with more options to consider and do it in such a way that considering the options becomes a timely reality.

It is clear that computer-based systems are here to stay in manufacturing. They bring great opportunities but will place great demands on the people in the organizations who use them and maintain them. The next several years should be exciting in manufacturing.

9–7 SUMMARY

Systems for manufacturing planning and control can be expected to evolve and grow as new computer and information systems technologies develop. Enterprise resource planning systems represent the leading edge of applied technology at the start of the new century. The new generation of systems is helping companies realize the unfulfilled promises of earlier systems. Material requirements planning cannot be effective without reliable schedule inputs. The potential of manufacturing resource planning to unite the functional areas may have been lost due to its apparent manufacturing focus. The enterprise resource planning systems have the potential to integrate the functional areas of manufacturing and the supporting areas that make up the enterprise, as well as the customers who are served and the suppliers who provide the materials.

This chapter has given the reader a look at the leading edge of technology and provided some speculation about the future direction of computer-based systems for manufacturing. All of the systems presented in this chapter will soon be replaced by a newer, faster, and better set of systems. Companies will continue to access and use information about their operations to sustain their competitive advantage. Careful management and planning will be needed as the technology evolves and older systems are upgraded or replaced. The decisions made by the leaders of the operations and information systems areas will have a significant impact on the overall performance and profitability of the company. Manufacturing professionals will be challenged as they try to keep pace with the continuing flood of changes and enhancements to existing manufacturing planning and control systems.

REFERENCES

Cox, J. F., and J. H. BLACKSTONE, eds., *APICS Dictionary*, 9th Ed. Falls Church, VA: APICS—The Educational Society for Resource Management, 1998.

FOSTON, A. L., C. L. SMITH, and T. AU. *Fundamentals of Computer Integrated Manufacturing*. Upper Saddle River, NJ: Prentice Hall, 1991.

KRAEBBER, H. W. Teaching manufacturing systems. *Manufacturing Education for the 21st Century: Volume V. Manufacturing Education for Excellence in the Global Economy*. Dearborn, MI: Society of Manufacturing Engineers, 1998, pp. 259–264.

POND, K., A. BOND, and A. DAILEY, *On-line Research Note, 1998 EP Large User Magic Quadrant*. Boston MA: Gartner Group, 1998.

SIBC Corp., *Technology Course Series*. Evanston, IL, 1999.

TTW, Inc., *WinMan Version 5.26*. Reston, VA, August 1999.

WOMACK, J. P., and D. T. JONES, *Lean Thinking*. New York: Simon and Schuster, 1996.

PROJECTS

1. Enterprise resource planning and related planning and control systems are changing rapidly. You can find some of the most current information possible on the Internet. Complete a keyword search on the Internet for one or more of the new systems for enterprise planning and control. For example, select a search engine such as Alta Vista™ and search for information on sales force automation or advanced planning systems. You should find links to some of the most current publications available. Write a short summary about your findings.

2. Visit the Internet sites listed in Appendix 9–2. Write a summary outline of the key issues discussed and presented at the sites.

APPENDIX 9–1: IDENTIFYING THE ABBREVIATIONS INCLUDED IN FIGURE 9–1 *

APS—advanced planning and scheduling
CRM—customer relationship management
demand management
DRP—distribution requirements planning
ERP/ERO—enterprise resource planning/enterprise resource optimization
forecasting
inventory control
MES—manufacturing execution system
MPS—master production schedule
MRP—material requirements planning
MRPII (MRP2)—manufacturing resource planning
order entry
SCM—supply chain management
SFA—sales force automation
SFC—shop floor control

APPENDIX 9–2: IMPORTANT ERP-RELATED SITES ON THE INTERNET

The Internet has quickly become an important source of information about manufacturing systems and the companies that supply them. The sites listed below are expected to remain active; however, in the fast-paced world of the Internet, it is not uncommon for Web addresses to change. Finding current active sites may require the use of an Internet search engine that will list the addresses of sites of possible interest based on keywords you enter.

- APICS (The Educational Society for Resource Management): APICS has been a leader in developing the body of knowledge of manufacturing systems such as MRP and ERP.

  www.apics.org

- The Gartner Group: A special consulting company that studies trends in manufacturing and operations and makes predictions about the future. They are credited with the abbreviation ERP.

  www.gartnergroup.com

- Techra Corporation: Founded in 1995 to conduct research for the software industry. They provide a site where people can obtain important information about ERP vendors and products with minimal sales hype. They created the ERPSuperSite:

  www.erpsupersite.com

---

* Many of these terms are so new to the body of knowledge that they were not included in the *APICS Dictionary*, 9th Ed., published in 1998.

APPENDIX 9–3: AN ERP EXAMPLE USING WINMAN™

The product structure is a series of parent-item to component-item relationships. The way the product is structured can have a great impact on the operation of the planning system. The type of product, the structure of the product, and the planned routing of the product through its required operations each have a significant impact on the planning done by the WinMan system. Excessive product structure levels create unnecessary planning activities and reporting requirements for manufacturing. Decisions on the level of control required will also affect the number and complexity of reporting transactions. It is easy to make the product structure much more complicated than necessary; however, the ocean liner *Queen Mary* has been described in fourteen structure levels. The space shuttle is reported to have been described in only eight structure levels. Most manufactured products can be described in five levels or less. The use of fewer levels greatly simplifies manufacturing and the material planning process.

This example begins with a simple product. A graphical representation of the structure of product A and its related subassemblies and component items is shown in Figure 9–15. The structure diagram for product A shows how the required parts are put together in the finished product. Each step in the manufacturing process can take some time. The actual operating time to complete the work is critical for determining costs, but at this level, a formal planning system is looking for something a little different. The system must know how to plan for the total time needed to manufacture the item. This total time includes collecting the needed materials, the setup of the required workstations, the time to do the work, any quality control checks or inspections, and any material handling time. The expected total production time is the sum of all these inputs and any others that are relevant to this production work. The time is calculated and entered into the planning system to the nearest whole day. This time allowance is fairly coarse,

---

* This description of the operation of part of the WinMan™ ERP system has been adapted from documents prepared for TTW, Inc., and a paper presented at the 1998 SME Education in Manufacturing conference held in San Diego. Additional information about WinMan may be found at its Internet site: *http://www.winman.com.*

Figure 9–15 A Simple Product Structure "Tree Diagram" for Product A.

Source: Used by permission of the Society of Manufacturing Engineers. Copyright 1998.

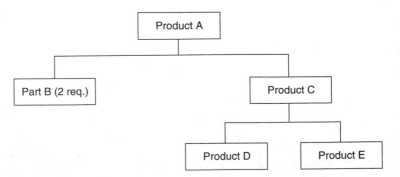

but for most material planning applications, it works quite well. The detailed capacity planning and shop-floor execution schedules require more detailed time standards.

The diagram in Figure 9–16 adds the dimension of the processing location (AS1 or AS2 in this example) to the given product structure information. It is assumed in this example that a special piece of equipment is needed to assemble parts D and E and that equipment is in assembly location AS2. A future improvement to this process might lead to the redesign of the two workstations into one. Product C could then be coded as a transient item, which would simplify the planning, assembly process, and reporting of work completed.

Requirements for product A may come from a sales order, the master production schedule, a specific job, a higher level product or assembly, a what-if order, or a reorder point signal. The system logic considers requirements for product A, the quantity needed, and the date it is required. The system works backward from the required date to determine the appropriate material plan for all the items in the product structure. Plans are developed by the system based on the planning parameters defined for each item. During the planning cycle, the demands for lower level items are driven from the top-level item in the structure. The planning process works through all the lower level assemblies and components that make up the manufactured item. The formal system uses the product structure, stated lead times, planning parameters, and current stock (inventory levels) of the required items to generate a time-phased material plan.

An MRP planning record for product A can now be considered. Product A is a top-level finished-goods item. The replenishment requirements for product A are a function of demand, current stocks, and the planning parameters. In Figure 9–17, the current stock on hand for product A is 2 units.

Some discussion of this record is needed. During the execution of the plan above, the quantities stated as planned order releases and planned order receipts will be reviewed and, if approved, will be replaced with firm scheduled receipts. For example, if the planned order release quantity of 40 on day 1 is confirmed, that

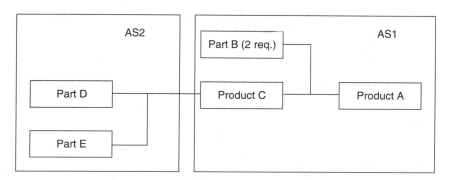

Figure 9–16   Linking the Product Structure with Workstations on the Shop Floor.
Source: Used by permission of the Society of Manufacturing Engineers. Copyright 1998.

Item: A

	Period Number											
	1	2	3	4	5	6	7	8	9	10	11	12
Gross Requirements					25					25		3
Scheduled Receipts												
Projected On-Hand	2	2	2	2	2	17	17	17	17	17	0	0	2
Planned Order Receipts					40						8		5
Planned Order Releases		40				8			5			

Lead time = 4   Order Policy = MRP   Optimum Order Quantity = 20

Minimum Stock = 0   Minimum Order Quantity = 5

Figure 9–17   MRP Planning Record for Product A.

Source: Used by permission of the Society of Manufacturing Engineers. Copyright 1998.

quantity will drop out of the planned order release and planned order receipt lines and appear as a scheduled receipt quantity of 40 on day 5. The scheduled receipt will be available at the start of day 5. In reality, the formal system must schedule receipt to occur by the end of work on day 4.

Note also how the system applies the planning parameters. The first new order release for product A is for a quantity of 40 units, or two times the stated optimum quantity. The second release is for the exact number of units required based on the MRP logic because the required quantity is between the stated minimum and optimum. The last scheduled release for 5 units was made because of the stated minimum requirement. These quantities may seem a little confusing at first, but they do provide planning flexibility for the system users.

The formal system will review and calculate the replenishment requirements for all the assemblies and component items that go into product A. Consider the proposed manufacturing order for a quantity of 8 units of product A scheduled for completion on day 10. Requirements for lower level items are driven by the planned order release for 8 units at the start of day 6 based on the four-day lead time for product A. The manufacturing order for 8 units of product A might look like Figure 9–18.

Material moving and handling is often one of the highest costs of manufacturing that does not add any value to the product. The developers of Win-Man have designed a system of locations and priority codes that allows users to tailor the system to fit their operation and reduce the associated costs of reporting and system transactions. WinMan encourages the definition and use of storage locations for materials at the location where the work is to be done. Parts will be issued automatically from the defined storage locations if the parts are shown by the system to be on hand. The material on hand the longest in

---

Information on the Manufacturing Order for Product A

Description: Assembled Product A   MO: 23987

Start date: x/ 6/xx   Due date: x/10/xx   Order Qty. = 8

Oprn # Operation description

0010   Assemble part D to part E as shown in the drawing "Product C/100"

0020   Test in accordance with Proc 456/89, then deliver to location AS1

0030   Assemble with two part B as shown in the drawing "Product A/001",
        then deliver to location "FINGDS"

Figure 9–18   Information on the Manufacturing Order for Product A.
Source: Used by permission of the Society of Manufacturing Engineers. Copyright 1998.

approved locations will be issued first, creating a dynamic first in, first out system. If the parts are not in the expected location when they are needed, the system will look for other available parts in approved locations before generating new planned orders for material. Manual issues can be made to an open manufacturing order at any time from any inventory holding. WinMan will issue the parts required on the planned issue date from the location designated in the product setup record.

Consider the issue of part B to workstation AS1. Sixteen of these parts are required to complete the order for product A. Stock of part B can be found in several locations, as shown in Figure 9–19. Following the priority scheme in WinMan, the four items in the designated manufacturing location AS1 will be issued first because the work is done in that location and material already there will be used

---

INVENTORY INFORMATION FOR "Part B":

Location	Qty.	Age	Availability
AS1	4	5	Y (only issue to manufacturing in AS1)
AS2	8	30	Y (only issue to manufacturing in AS2)
FINGDS	40	30	S (only shipped via a sales order)
QNTN	10	25	N (not to be issued or shipped)
MNST	10	25	B (both manufacturing or sales issues)
MNST	20	12	M (for manufacturing issue only)

Figure 9–19   Inventory Information for Part B.
Source: Used by permission of the Society of Manufacturing Engineers. Copyright 1998.

first. The system then looks for material in other locations to complete the issue of the remaining 12 units of part B. WinMan will look at each available batch and issue the material on hand the longest first, which means that 10 units from the 25-day-old batch in the main stockroom (MNST) and 2 units from the 12-day-old batch in MNST will be issued to the order. Note that parts in a location such as the main stockroom have an additional code letter that allows more control of material issues. Parts with the B code may be used for sales or manufacturing, while the M code allows parts to be used only for manufacturing.

Note that some older pieces of part B in stock were not issued. The parts in assembly area 2 (AS2) are not issued because they are expected to be issued to manufacturing orders assigned to location AS2. Parts in the finished goods warehouse location (FINGDS) that are coded with an S are issued only to sales orders. Parts in the quarantine location (QNTN) may not be issued or shipped without approval. In many companies, a material review board will evaluate quarantined parts and make a decision on their status. Only authorized individuals with access to the system may change the availability indicator of a batch of material.

Enabling Processes and Systems for Modern Manufacturing

After you have completed Part 4, it will be clear to you that:

- Properly designed manufacturing processes, machines, and systems are the CIM elements found on the shop floor.
- Industrial robots, automated material handling devices, machine control computers, and programmable controllers are the automation foundation of the factory floor.
- The technology required for automated control of manufacturing cells and production areas is available and within reach of manufacturing operations implementing CIM.
- Systems for effectively controlling machines and systems are elements of the CIM system.
- An enterprise-wide data and information network is a necessity for successful implementation of CIM.
- Quality is a major element in the CIM system.
- Successful CIM depends on the people of the organization.

Production Process
Machines and Systems

Production machines and systems are the fundamental building blocks on the shop floor. The concepts presented in this chapter provide a broad overview of the types of machines used for the production of finish goods. The many types of production systems and machine configurations are too numerous to cover in detail in a CIM introductory text; however, the production processes and machines commonly used are identified and described in detail. In this chapter we focus on the materials processing part of factory automation; the remaining components are addressed in Chapter 11.

Material and parts are actively processed only a small percentage of the time that they are in production. For example, the spindle in Figure 5–1 could be machined from bar stock in approximately 3 minutes of value-added time. In many manufacturing operations, however, the bar stock could spend 3 or more hours of non-value-added time at the machine for setup, which includes setting up the production tooling, verifying production programs, and checking for the first good part. During this period, metal is not being removed from production parts. If additional work at other work centers is required after this operation, the ratio of value added to non-value-added time is repeated and the wasted time increases. This ratio is illustrated graphically in Figure 10–1. The figure indicates that the machine is initially inactive during setup, and then the majority of the production time is divided among setup, moving, waiting, and inspection. Reducing move time, shortening travel distance, reducing setup time, and decreasing the need for inspection are the goals of world-class manufacturing.

Finally, we cover the linking of different production machines together to satisfy a production problem. Figure 10–2 illustrates the types of production machine or system typically used in the five manufacturing systems categories introduced in Chapter 2. Comparison of the manufacturing systems in Figure 2–3 with the graph in Figure 10–2 indicates that the project has the lowest volume output and continuous processes have the highest. The graph shows that *manual* and *programmable machines* are adequate for low-volume production, but that *fixed automation* systems are required for the large volumes associated with the continuous production systems found, for example, in the petroleum and chemical industry. A special type of fixed automation system, called a *transfer line*, is often used for high-volume

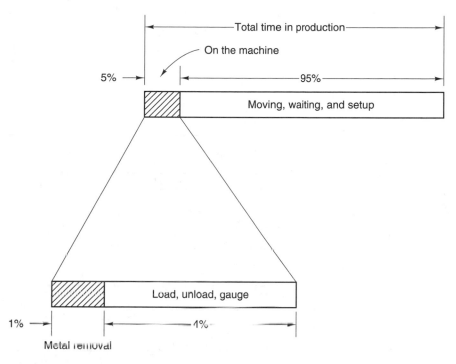

Figure 10–1 Production Efficiency.

production of a complex machined part or family of parts. The production of components such as air-conditioning compressors for the automotive industry typically uses a repetitive manufacturing system. Therefore, the manufacturing floor would have different production systems, ranging from *programmable machines* to *flexible automation* systems. The description of production machines and systems starts with the basic processes used to change raw material into finished parts and products.

10–1 MATERIAL AND MACHINE PROCESSES

A review of the production process chart in Figure 3–17 is a logical starting point for this study of machine processes. Study the production processes in the figure until the names and basic operations are again familiar.

Process Operations

The process operations transform raw material into a finished product without adding other materials or components. The shape, physical properties, and surface of the material are altered in the manufacturing process by the application of energy. For example, the mechanical energy expended by a cutoff saw when it cuts the end from a 2-inch-diameter steel bar changes a steel bar into a cylinder 2 inches

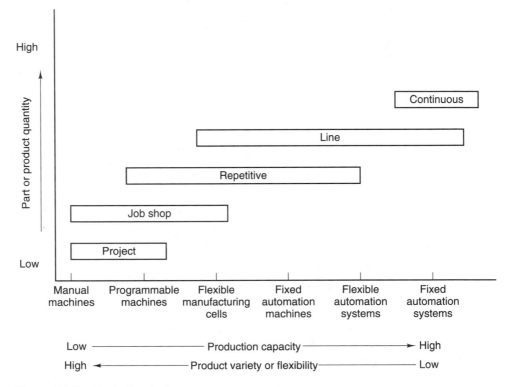

Figure 10–2 Production Systems.

across and 5 inches in height. The process operations performed in manufacturing are classified into one of four categories:

1. Primary operations.
2. Secondary operations.
3. Physical properties operations.
4. Finishing operations.

A review of these four operation areas in manufacturing is provided in the following sections.

Primary Operations

A *primary operation* converts raw material into the basic geometry required for the finished product. Metal casting is a good example of a primary operation. The raw material is metal in a molten state, and a void in the mold cavity is the rough geometry of the product. When the molten metal is poured into the mold and cools, the primary operation is complete, and the raw casting has the general shape of the finished product. Typical primary operations include *casting, forming, oxyfuel and arc cutting,* and *sawing.*

Casting.   Casting is a manufacturing process that uses molten metal and molds to produce parts with a shape close to that of the finished product. The molten metal is poured or forced into a mold with a cavity shaped like the part desired. After the metal fills the cavity and returns to a solid state, the mold is removed and the casting is ready for secondary processing. The green-sand molding process is used for 75 percent of the 23 million tons of castings produced in the United States annually. A cross-sectional view of a typical green-sand mold is provided in Figure 10–3.

The process starts with a pattern of the desired part, usually constructed from wood. The lower half of the mold, called the *drag,* is filled with green-sand, a sandlike material that sets up or hardens. The pattern is placed on the surface of the lower mold half (Figure 10–3), and the upper half of the mold box, called the *cope,* is placed on top and filled with additional green-sand. After the green-sand sets, the upper and lower half of the molds are separated at the *parting line,* and the pattern of the casting is removed. The mold halves are put back together and the molten metal is poured into the pouring basin and enters the cavity through the *sprue* and *runners.* The molten metal fills the cavity and forms the part. The green-sand method just described is one of several traditional casting processes. Other traditional casting techniques include *sand-mold, dry-sand, shell molding, full-mold, cement mold,* and *vacuum mold.* Casting operations vary from totally automated systems that produce over 300 castings per hour to manual productions. The type of metal used in casting includes most of the ferrous metals, along with aluminum, brass, bronze, and other nonferrous materials.

Several casting processes that do not rely on sand to produce the mold have been developed in the last thirty years. These contemporary casting and molding processes are generally used for smaller geometry parts and often do not require

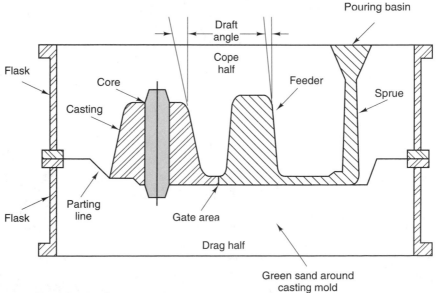

Figure 10–3   Cored Casting Cross-section.

additional processing or finishing on the surfaces with noncritical dimensions. The most frequently used process, called *high-pressure die casting,* uses metal molds. The basic die casting machine has four components: (1) a two-part die, (2) a die mounting and closing mechanism, (3) a molten metal injection device, and (4) a source for the molten metal. The sequence of operation for the process is illustrated in Figure 10–4. Note that in (a) the two-part die with two movable cores is closed and metal is being poured into the cylinder. In (b) the molten metal is forced into the mold by the ram, and the metal cools. The die separates in (c), and when the die is open and the cores are withdrawn, the part is ready for removal (d).

Die-cast machines can utilize either a *hot-chamber* or a *cold-chamber* process. The older, hot-chamber process incorporates a furnace in the machine to keep a reservoir of molten metal ready for the injection device. Machines of this type can typically produce 100 castings an hour weighing up to 50 pounds (23 kilograms) or 1000 parts in the 1-ounce range. The hot-chamber process uses zinc alloys primarily, because aluminum and copper alloys chemically attack the submerged metal injection device.

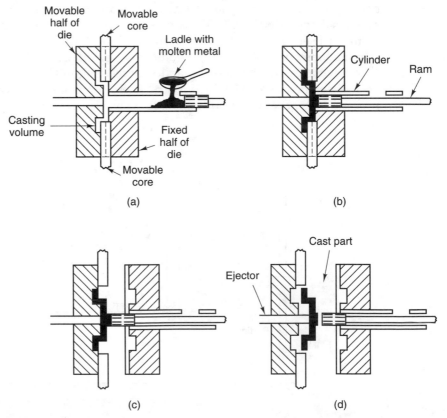

Figure 10–4   Sequence of Operation for a Cold-chamber Die Casting Machine.

Cold-chamber machines have the molten metal introduced from an external source, like the operation shown in Figure 10-4. A complete cold-chamber machine has a *traveling plate* and *front plate*. Half of the die is attached to the travel plate and the other half to the front plate. A closing cylinder moves the travel plate with one-half of the die attached so that the die is closed. Molten metal is introduced by the shot cylinder, and a part is produced. Other contemporary casting processes include *permanent-mold, centrifugal, plaster molding, investment,* and *solid-ceramic.*

Forming.   In the forming process, the shape of raw material is changed by the application of force. For example, in the *extrusion* process shown in Figure 10–5, pressure from the pressing stem forces a hot round billet of metal into a die, and the extruded part that results takes the shape of the hole in the die. A cookie press operates in a similar fashion to produce cookies with different shapes. Other forming processes frequently used include forging (hot and cold), roll and *spin forming, shearing, punching, drawing, upsetting,* and *swaging.*

Oxyfuel and Arc Cutting.   Oxyfuel gas cutting constitutes a group of four cutting processes (oxyacetylene, oxyhydrogen, air–acetylene, and pressure gas) in which the base metal is heated to an elevated temperature and then cut by the chemical reaction of oxygen with the base metal. In some cases, a chemical flux or metal powder is used to cut oxidation-resistant metals.

Plasma arc cutting is an arc cutting process that uses the difference in the voltage between the cutting electrode and the material to cut through the metal (Figure 10–6). The area where the arc penetrates the metal causes localized melting, and the molten metal is removed by the high-velocity ionized gas directed by

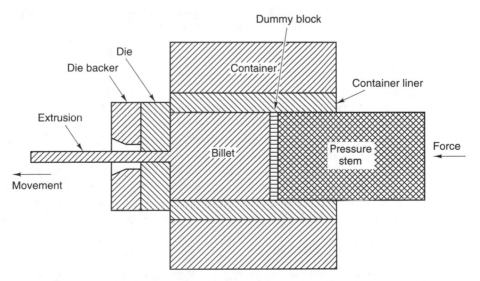

Figure 10–5   Cross-section of a Forward Extrusion Press.

Figure 10–6 Gas-shielded Plasma Arc Cutting.

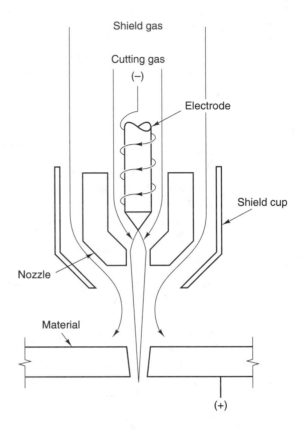

the nozzle. The process is used to cut any material that conducts electricity. Other arc-cutting methods include *oxygen arc, gas metal arc, gas tungsten arc, shielded metal arc,* and *carbon arc.*

Sawing. Every manufacturing operation uses raw material in the form of wire, bars, tubes, pipes, extrusions, sheets, plates, castings, or forgings. In many cases, the material must be cut to the required length to prepare for secondary operations. *Sawing* is a primary machining process using straight, band, or circular blades to cut the raw material to the rough shape required for the secondary machining operations. The cutoff operation, sometimes called *slugging,* is performed most often on power hacksaws, bandsaws, and circular saws. Hacksaw machines are available in both manual and numerical control models. Bandsaws come in the same range of control from manual to fully automated, like the saws in Figures 10–7 and 10–8. Cut-off is also performed on lathe-type machines using single-point and parting tools.

Secondary Operations

Secondary operations are performed to give the raw material the final shape required for the finished part or product. For example, after iron is cast into the general

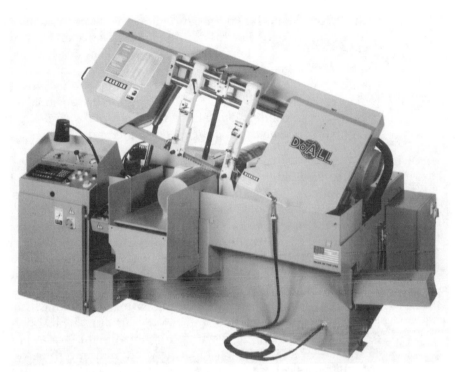

Figure 10–7 High Production Power Saw with 10 HP Drive Motor and NC Control.
Source: Courtesy DoALL Company, Des Plaines, Illinois.

Figure 10–8 Tilt-frame Bandsaw with NC Controls.
Source: Courtesy DoALL Company, Des Plaines, Illinois.

shape of an automotive engine, the cylinder walls are machined so that the pistons fit smoothly inside. As in this example, secondary operations often follow primary manufacturing operations. The primary difference between the two types of operations is that secondary operations take the material to the final geometry required for use: Also, many secondary operations, using different types of machine tools, are required to achieve proper shape and dimensional accuracy in the finish part. Typical secondary operations include *turning, boring, milling, drilling, reaming, grinding,* and *nontraditional machining processes.*

Turning.   Turning is performed on a machine tool called a lathe and can be manual or fully automated with sophisticated numerical controls (NC) or computer numerical controls (CNC). In turning operations, the material or workpiece is held in a spindle and rotated about the longitudinal axis while a cutting tool is fed into the rotating workpiece to remove material and produce the desired shape. The workpiece is held in the lathe spindle by a chuck, collet, fixture or faceplate, or between centers. In a turning operation, the material can be removed from either the external surface (OD) of the workpiece or from the internal part (ID) of the material in the spindle. The OD operations include facing, chamfering, grooving, knurling, skiving, threading, and cutoff (parting). The ID operations are called recessing, drilling, reaming, boring, and threading.

The NC and CNC machines generally fall into three categories: (1) center-type machines, where primarily OD cutting is performed with the workpiece held between centers; (2) chucking-type machines, with larger and lower speed spindles, with cutting performed on both the inside and outside diameters; and (3) universal machines, which have characteristics of both preceding types. The workpiece and cutting tool turret are visible on the CNC Cincinnati Milacron machine in Figure 10–9.

Boring.   Boring is a machining process for producing precision internal cylindrical surfaces using either manual and NC/CNC lathes or special-purpose boring machines. In the lathes, the workpiece rotates and the single-point or multi-edge cutting tool is stationary. However, in special-purpose boring machines (Figure 10–10), the tool is rotated and fed into the stationary fixed workpiece. Common applications include the precision finishing of cored, pierced, or drilled holes and contoured internal surfaces. In addition to the precision operation, the massive rigid construction of boring machines is ideal for precision heavy cutting operations.

Milling.   Milling is the process of removing material using a rotating cutter with multiple cutting edges. The primary characteristic of milling is that each cutter tooth takes a small individual chip of material as the cutter and workpiece move relative to each other. Cutting is achieved by feeding the workpiece into a stationary rotating cutter, feeding a rotating cutter into a stationary workpiece, or a combination

(a)

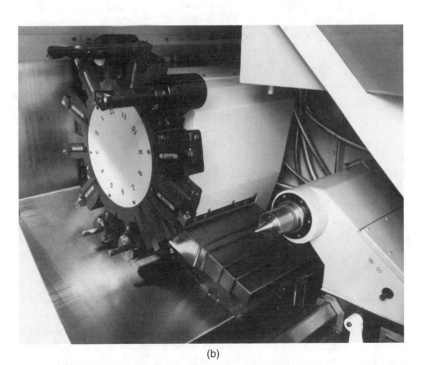

(b)

Figure 10–9 (a) CNC Turning Center; (b) Indexing Tool Turret.
Source: Courtesy Cincinnati Milacron, Inc.

Figure 10–10   CNC Boring Machine.
Source: Courtesy Cincinnati Gilbert, Inc.

of the two. In most applications, the workpiece is mounted on the machine table, and the table is advanced at a relatively slow rate past a cutter turning at relatively high speed.

The number of machine axes refers to the different number of directions the workpiece can be moved simultaneously past the cutter. A simple single-axis machine can move the table only in the X direction, while a six-axis machine could move the table in the X, Y, Z directions and a three-axis move of the workpiece. Multi-axis machines of this type are used to mill the complex shapes required in wing spars and ribs required by the aircraft industry. Milling is one of the most widely used secondary processes and includes a large number of different types of machines. Most of the machines are grouped into two categories: horizontal and vertical. The orientation, horizontal or vertical, identifies the direction of the center axis of the rotating tool. Figures 10–11 and 10–12 show examples of horizontal and vertical CNC mills. Milling methods are generally grouped in two major categories: *peripheral* and *face*. A selection of different types of milling cutters is shown in Figure 10–13. While milling is one of the most widely used secondary processes, it is also one of the most complex machining methods.

Drilling.   Drilling is a process used to produce or enlarge a hole in a workpiece using one of the following drilling methods: conventional, deep-hole, or small-hole. The tool used in the drilling process; called a twist drill, is a rotary end-cutting tool with one or more cutting lips and flutes. Figure 10–14 shows a twist drill with all the significant parts. Study the figure until all the parts of the drill are identified. Twist drills are classified using the following characteristics: drill material, shank type, number of flutes, direction of rotation for cutting, length, diameter, and point configuration. When a drilling operation is performed on a vertical CNC

Figure 10–11   Vertical CNC Mill.

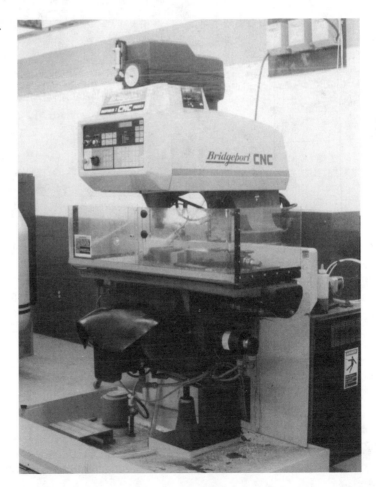

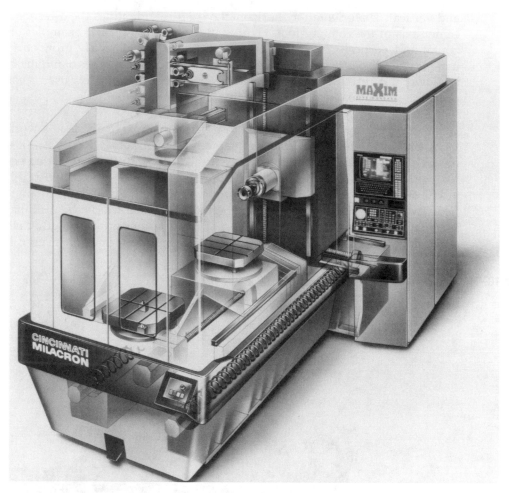

Figure 10–12   Horizontal CNC Mill.
Source: Courtesy Cincinnati Milacron, Inc.

mill, the workpiece is fixed, and the rotating drill is fed into the part. Figure 10–15 shows a drill mounted on the tool turret of a turning center. In this type of machine, the part is rotated and the drill is stationary.

Reaming.   The drilling operation just described is not considered a precision process; rather, a drilling operation is selected because it is fast and economical. If hole precision is necessary, the drilling operation is followed by a *reaming* process. Reaming is used for enlarging, smoothing, and accurate sizing of existing holes.

Grinding.   Grinding ranks first according to the number of machine tools in use, exceeding processes such as turning and drilling. Grinding is an abrasive process that removes material by means of an abrasive grain. Thousands of types of grinding

Figure 10–13   Milling Cutters.

wheels and surfaces are available to the production engineer determining the optimum production method. In addition to the many varied shapes, grinding wheels differ based on five elements: (1) type of abrasive agent used; (2) size of the grain particle used; (3) type of bonding material used to hold the grains in operational shape; (4) grade, which determines the hardness or strength of the wheel; and (5) structure, which sets the proportion and configuration of the grains and bond. A wheel's porosity is determined by a combination of structure and grade. A manual surface grinder is illustrated in Figure 10–16; the wheel, magnetic chuck surface to hold the part, and a small block ready for grinding are visible.

Grinding machines are available in a wide range of sizes with manual and CNC controls. In addition, machines are designed to grind cylindrical stock or rectangular material. Grinding of cylindrical surfaces is performed with the part held between centers, like in a lathe operation, or using a centerless grinding process. Grinding of rectangular parts is performed on a surface grinder (Figure 10–16) or jig grinder. The grinding process produces smoother surface finishes than those achieved through either milling or turning. As a result, grinding is frequently the last operation in a surface-finishing production sequence.

A standard part of the grinding operation includes *trueing* and *dressing* of grinding wheels. The trueing process adjusts the surfaces, outside diameter, and sides of the grinding wheel so that they are at the correct angle and distance relative to the drive shaft. The dressing process removes metal particles produced from previously ground parts that collected on the wheel and restores the wheel to the original geometry.

Metal-cutting machines such as lathes, mills, and grinders use cutting fluids to lubricate and cool the cutting tools and grinding wheels during the production process. However, grinding exhibits characteristics that separate it from other

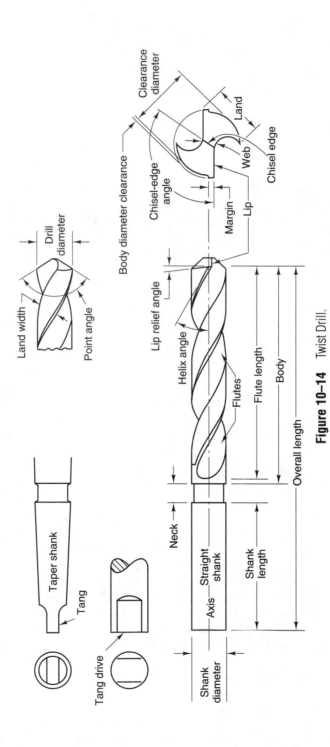

Figure 10-14 Twist Drill.

Figure 10–15   Drill in Tool Turret of Turning Center.
Source: Courtesy Cincinnati Milacron, Inc.

Figure 10–16 Manual Surface Grinder.

types of material removal. As a result, grinding different material may require a specific type of cutting fluid to ensure minimum friction, long wheel life, minimum heat buildup, and optimum finish on the part.

Nontraditional Machining Processes. The category of nontraditional machining includes a large number of processes developed after 1940 that use mechanical, electrical, thermal, or chemical energy to remove material. Many of the processes remain experimental; however, several have become the mainstay of specialty machining areas.

The most frequently used mechanical process in the nontraditional area is *hydrodynamic* machining. Hydrodynamic machining is used primarily for slitting and contour cutting of nonmetallic materials such as wood, paper, asbestos, plastic, gypsum, leather, felt, rubber, nylon, and fiberglass. A narrow cutting width or *kerf* is produced as workpiece material is removed by a high-pressure (usually 60 psi), high-velocity stream of water or a water-based cutting fluid. Frequently, the motion of the cutting heads are computer numerically controlled so that complex contours are cut with a high degree of repeatability. A second mechanical process in the nontraditional machining area is *ultrasonic* machining. In this process, the tool is vibrated at ultrasonic frequency to enhance the conventional cutting action in the workpiece.

The nontraditional machining *electrical* process, called electrochemical or electrolytic processing, uses chemical electrolytes in combination with electrical energy to create a cutting action. As Figure 10–17 illustrates, the cutting action is the result of a reverse plating action between the tool and the workpiece. The chemical electrolyte is forced down through the tool into a gap between the tool and workpiece. The high-current dc power source and the electrolyte force electrons from the workpiece material, and the metallic bonds of the workpiece atoms are broken at the surface. The workpiece atoms enter the electrolyte solution as metal ions and are washed away by the constant flow of electrolyte. Hard metals with the ability to conduct an electric current are frequently processed with this method. In a similar process, called *electrochemical discharge grinding,* an ac or pulsating dc source of electrical energy is used. Again, most of the metal removal results from the electrochemical process; however, sparking between the graphite grinding wheel and the workpiece removes the oxide developed by the electrochemical process. This grinding process is used routinely to grind and sharpen carbide tools.

The nontraditional machining *thermal processes* use thermal energy to remove metal from the workpiece. The four methods used to generate the necessary thermal energy for machining include *electrical discharge, electron beam, laser beam,* and *plasma beam.* The first method is widely used to machine intricate parts using two methods: *electric discharge machining (EDM)* and *electric discharge wire cutting (EDWC).* The latter process, EDWC, is often called wire EDM. The components of an EDM machine are illustrated in Figure 10–18. Metal is removed from the workpiece as a result of rapid electrical discharges (arcing) between the tool electrode and the workpiece immersed in a liquid dielectric. Small, hollow metal chips are produced as the workpiece material is removed by melting and vaporization. The resulting shape of the workpiece matches the contour of the tool electrode.

The second popular electrical discharge method, wire EDM, uses a fine metal wire as the cutting electrode (Figure 10–19) to cut metal sheets and plates into intricate shapes. Note the fine vertical wire visible in Figure 10–19. The parts in Figure 10–20 are examples of wire EDM work performed on the CNC wire

Figure 10–17 Electrochemical Machining.

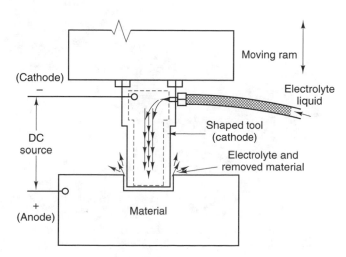

Figure 10–18 Components of an Electrical Discharge Machine.

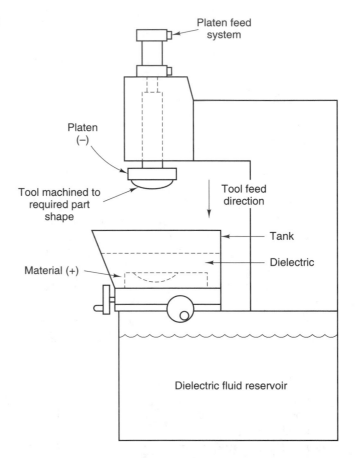

Platen feed system

Platen (−)

Tool machined to required part shape

Tool feed direction

Tank

Dielectric

Material (+)

Dielectric fluid reservoir

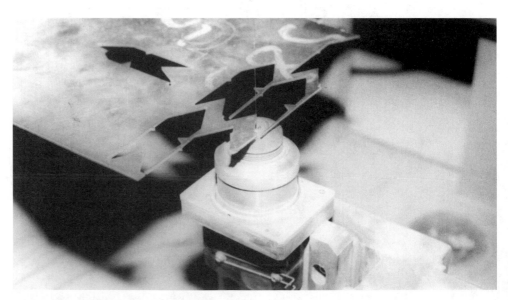

Figure 10–19 Wire EDM Cutting Wire.

Figure 10–20   Parts Produced with
Wire EDM.

EDM pictured in Figure 10–21. This machining method is frequently used in the production of stamping and extrusion dies.

In the chemical nontraditional machining process, chemical energy is used as the primary source for metal removal. Although this process was used for many years, it was after World War II that a process was developed by the North American aviation industry to use chemical milling in volume production. The two major chemical processes used in industry are *chemical milling* and *photochemical machining*. Chemical milling is an etching process where metal is removed by a chemical. The amount of metal removed or the depth of the etch is related to the chemical used and the immersion time in the solution. Photochemical machining uses a photographic process to place a chemical-resistant image on the surface of a metal sheet. The metal not protected by the resistant coating is then removed through a chemical milling process. Both of these processes are used in a wide variety of applications where removal of a relatively small amount of material is required.

The primary and secondary operations produce the finished geometry for the production part. In many applications, the processing phase of manufacturing ends with the last secondary operation; however, it is often necessary to change the properties of the material or add a finish to the exterior of the product.

Physical Properties Operations

The process that changes the physical properties of the part does not change the part geometry significantly. The most common physical property change performed

Figure 10–21   Wire EDM Machine with Fanuc Controller.

in manufacturing is *heat treating.* Heat treating of ferrous and nonferrous metals is an operation or combination of operations involving the heating and cooling of solid metals and alloys to change the microstructure of the material. The changes in the material are generally grouped into one of two categories:

1. Processes that increase the strength, hardness, and toughness of metals through either hardening of the entire part or hardening of the surface metal on the part.
2. Processes that decrease the hardness of metals through annealing and normalizing of the part to improve homogeneity, and machinability, and formability or to relieve stresses.

Setting up heat treating processes requires a mixture of science and practice. The science of heat treating includes a complex analysis of the chemical and metallurgical properties of metals and metal alloys. Although the science of metal structure and behavior is an important part of the process, much of the success in heat-treating operations results from experience gained in heat treating of parts over many years.

Finishing Operations

Like physical properties, finishing operations do not significantly change the geometry of the finished parts. In most cases, finishing operations add a thin layer to the surface of the part to improve the operational life and serviceability. The finishing operation frequently prevents surface oxidation of the metal used to produce the part, especially in parts produced from steel. The most frequently used finishing operation to reduce oxidation of parts is painting. Other finishing operations commonly used include (1) plating of the parts using either hot dips or electrolysis, and (2) acid etching or pickling of the surface.

The four basic material process operations, described in the preceding four sections, use general-purpose production machines to support manufacturing requirements. In many manufacturing situations, however, specialized machines are required for high-volume production. A discussion of these machines and manufacturing systems follows.

10–2 FLEXIBLE MANUFACTURING

Rapid response and production flexibility in manufacturing were identified in Chapter 1 as a world-class measurement standard for productivity and an important *order-winning criterion*. Take a few minutes to review Figures 1–4 and 1–6. In Figure 1–6, "flexibility" refers to the number of different parts that a workstation can produce under normal production conditions. The values presented indicate that a world-class company exceeds the U.S. average by a factor of 10. Flexibility in manufacturing is often described as (1) the ability to adapt easily to engineering changes in the part, (2) the increase in the number of similar parts produced on the system, (3) the ability to accommodate routing changes that allow a part to be produced on different types of machines, and (4) the ability to change the system setup rapidly from one type of production to another. In each case, the critical point is that properly designed production areas make production planning and scheduling easier and result in quicker response to customer needs. The production systems illustration in Figure 10–2 shows how selection of the production system affects manufacturing flexibility.

The list of *order-winning* criteria in Figure 1–4 includes flexibility and lead time. Both of these criteria, necessary to maintain or increase market share, are affected by the design of the production area. Group technology, covered in Chapter 5, focused on the design of production cells to handle a family of parts with common production characteristics. Further evidence that integrated production cells are critical elements of CIM is provided by the movement toward self-directed work teams. This important human resource development activity, described in Chapter 13, emphasizes the grouping of machines into production cells that are managed by a team of workers. Realizing the benefits of CIM and the integrated enterprise in many product areas requires implementation of *flexible manufacturing cells* (FMCs) or *flexible manufacturing systems* (FMSs) on the shop floor.

Flexible Manufacturing Systems

Global market pressures in the early 1980s demanded higher production efficiency, lower cost, and faster response; as a result, manufacturers installed flexible manufacturing systems (FMSs) for mid- and low-volume production products. An FMS is defined by the *Automation Encyclopedia* (Graham, 1988) as follows:

> *A flexible manufacturing system is one manufacturing machine, or multiple machines that are integrated by an automated material handling system, whose operation is managed by a computerized control system. An FMS can be reconfigured by computer control to manufacture various products.*

The first FMS, designed in the mid-1960s by a British firm, was called *System 24.* Due to insufficient control and computer technology, the system was never completely installed. The most notable early installation of an FMS in the United States was at Caterpillar Inc. by Kearney & Trecker. The goals set for the FMS were met or exceeded because the system goals were specific and addressed specialized applications. The FMS did not exhibit the flexibility described by the preceding definition; however, it did satisfy the Kearney & Trecker definition, which follows:

> *An FMS is a group of NC machine tools that can randomly process a group of parts, having automatic material handling and central control to balance resource utilization dynamically so that the system can adapt automatically to changes in parts production, mixes, and levels of output.*

A review of the two definitions indicates some common elements, such as *numerical control* (NC) or smart production machine tools, *automatic material handling, central computer control, production of parts in random order,* and *data integration.* As the definitions indicate, an FMS is a collection of hardware linked together by computer software. The process hardware frequently includes NC and CNC machines like many of those described in Section 10–1. In addition, an FMS has production tooling and setup systems, part cleaning and deburring stations, raw material and finished-parts automatic storage and retrieval systems (ASRS), and coordinate measuring machines (CMM). The systems are linked with automated material handling, which ranges from less sophisticated belt conveyors to highly sophisticated robots and automatic guided vehicles. A typical FMS layout is illustrated in Figure 10–22. Study the figure until you locate all the elements in the FMS.

   The FMS shown in Figure 10–22 is designed to produce a family of machined metal parts that can be manufactured with three-axis vertical machining centers. Five machining centers are required to meet the production demands; locate them in the figure. Raw material for the parts is delivered to the automatic work changers (number 5 in the figure) and is loaded onto pallets or fixtures that will hold the material for one or more milling operations. Frequently, a partially finished part will be removed from one fixture and placed on another in a different orientation for additional machining. The pallets or fixtures are delivered to the correct machining center cell by the computer-controlled cart or automatic guided vehicle (AGV). The vehicles have no person on board for navigation and use a current-carrying wire embedded into the floor and electronics on the cart for path, direction, and speed control. The material handling AGVs are also used to deliver tools

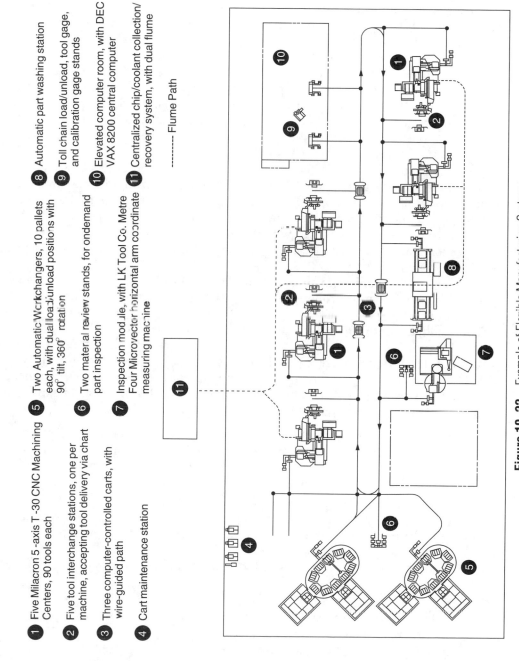

1. Five Milacron 5-axis T-30 CNC Machining Centers, 90 tools each

2. Five tool interchange stations, one per machine, accepting tool delivery via chart

3. Three computer-controlled carts, with wire-guided path

4. Cart maintenance station

5. Two Automatic Workchangers, 10 pallets each, with dual load/unload positions with 90° tilt, 360° rotation

6. Two material review stands, for ondemand part inspection

7. Inspection module, with LK Tool Co. Metre Four Microvector horizontal arm coordinate measuring machine

8. Automatic part washing station

9. Toll chain load/unload, tool gage, and calibration gage stands

10. Elevated computer room, with DEC VAX 8200 central computer

11. Centralized chip/coolant collection/recovery system, with dual flume

------- Flume Path

Figure 10–22 Example of Flexible Manufacturing System.
Source: Courtesy Cincinnati Milacron, Inc.

369

from the setup and calibration area (9 in the figure) to the tool interchange stations (2 in the figure). From the tool interchange stations, tools are changed automatically on the machining centers to match the requirements of the parts to be produced. Finished parts still mounted to the pallets are delivered by the AGV system to the parts washing station (8 in the figure) prior to inspection and shipping. To track the quality of finished parts, the AGV delivers machined parts to the coordinate measuring machine (7 in the figure) for automatic inspection or to the manual inspection stations (6 in the figure). A centralized chip and coolant recovery system collects metal removed in machining and filters the cutting fluid used in the cells. The cells in the FMS are under area computer control from a central computer (10 in the figure).

The hardware in the FMS is interfaced to computer controllers at several levels: (1) the sensors, gauges, switches, and controls supporting the operation of a machine are linked directly to the computer in the machine through the discrete signal input/output interface (Figure 10–23a) or the devices are connected to a cell control computer that interfaces with the machine computer (Figure 10–23b), (2) computer-controlled machines located in a manufacturing cell are linked together by a cell computer controller (Figure 10–23b), (3) the automated cells form an FMS when the cells are linked to an area controller (Figure 10–24); and (4) the FMS is linked to the system controller or enterprise computer and the intranet/Internet (Figure 10–24). Sophisticated computer software is required in this complex FMS with multiple control levels to handle the high degree of variability associated with the production of many different parts. The degree of complexity is evident when the levels of hardware and software present in the FMS are identified. Typically, a minimum of five technology levels are present in an FMS. For example, the five levels of technology present in the FMS in Figure 10–22 are listed below (the numbers refer to the equipment in the figure):

- *Enterprise level:* scheduling production requirements for the FMS, preparation of computer programs and code for the production system and machines, generating purchase orders for raw material, and creation of shipping documents for finished goods.
- *System level:* centralized coolant and chip collection system (11); control and scheduling of computer-controlled carts (10); downloading of computer code for production machines (10); synchronization of all cell operations (10); central calibration and setup of tools and tooling for the machines (9); and tracking of tooling, raw materials, and finished-goods inventory (10).
- *Cell level:* machining cells (1), tool gauge and calibration station (10), material load and unload stations (5), testing and quality control cell (7), and parts washing cell (8).
- *Machine level:* CNC machining centers (1), manual operations (6), automatic wire guided carts (3), work holders and changers (5), quality testing machines (7), automatic parts washing machine (8), and tool interchange stations (2).

Figure 10–23 FMS Computer
Control Options.

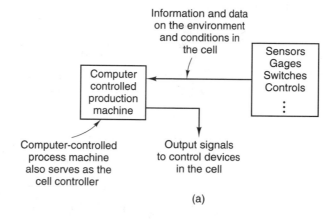

(a)

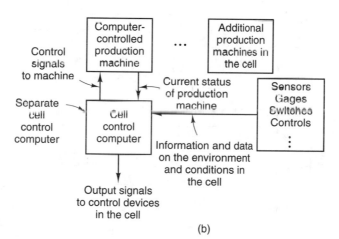

(b)

- *Device level:* sensors, ac and dc motors, pneumatic and hydraulic components, tools, fixtures, electrical components, connectors, wire, and fiber optics.

The level of complexity is magnified by the number of different hardware and software vendors required to assemble an FMS. Some of the most successful FMSs have been built and integrated by machine tool manufacturers. A review of the operation of the typical system in Figure 10–22 and the technology levels just identified indicates that the complexity associated with an FMS is the major disadvantage and a serious obstacle in implementation. However, successfully implemented flexible manufacturing systems offer several tangible advantages:

- *Inventory reduction.* In some implementations the work-in-process inventory and finished-goods inventory are reduced by over 80 percent. If the FMS permits production of parts in lot sizes of one, a just-in-time operation is possible, and finished-goods inventory for that operation in near zero.

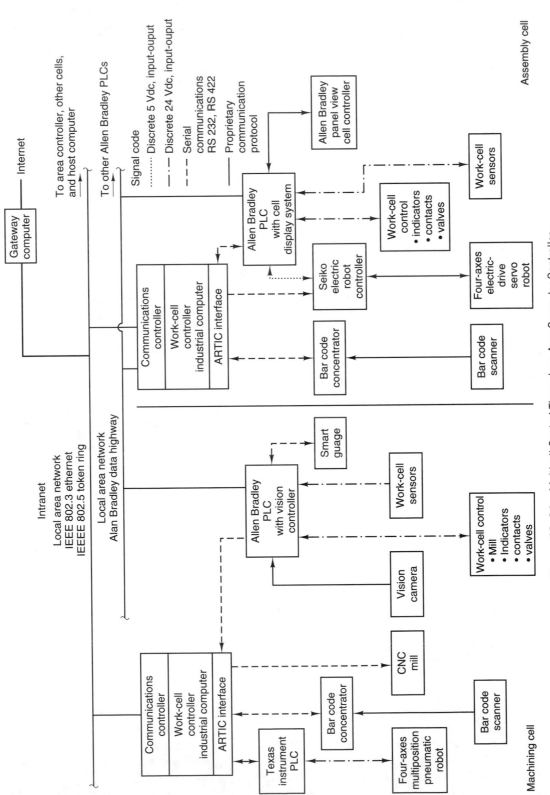

Figure 10–24 Multicell Control Through an Area Computer Controller.

- *Direct labor cost reduction.* The automation associated with FMS implementations permits a large production system to operate over three shifts with fewer workers than would be necessary when manual moving and loading of raw materials into production machines is practiced. The manual direct labor component for the cost of goods sold (COGS) is typically only about 10 percent, so the reduction in direct labor due to the automation is not as significant an advantage.

- *Machine utilization increase.* Another benefit of automation is the ability to operate expensive production machinery over three shifts and seven days a week. This continual operation causes improved capital utilization and increased capacity without a significant increase in labor cost. As a result of this increase in capacity, fewer machines are required and the need for production floor space decreases.

- *Supports world-class standards.* The standards used by world-class enterprises to measure manufacturing effectiveness were described in Chapter 1 and illustrated in Figure 1–6. A properly designed FMS improves each of these standards: low setup time, high quality, manufacturing space a large percentage of total factory space, low in process and finished-goods inventory, flexibility, low travel distance for material through production, and good machine up-time.

- *Supports order-winning criteria.* When properly designed and implemented, an FMS directly supports the *order-winning criteria* discussed in Chapter 1 and listed in Figure 1–4: low price, high quality, short lead times, high percentage of on-time delivery, and good flexibility in manufacturing a product family.

The advantages associated with a successful FMS are impressive, and several systems have been installed around the world that prove FMS technology can work. However, the cost, complexity, and the level of technology required to implement an FMS limits FMS solutions to the very large manufacturers. As a result, a greater emphasis in automation design is currently placed on *flexible manufacturing cells* (FMCs).

FMC Versus FMS

A flexible manufacturing cell is defined in the *Automation Encyclopedia* (Graham, 1988) as follows:

> *An FMC is a group of related machines that perform a particular process or step in a larger manufacturing process.*

Based on this definition, the production building blocks used to assemble an FMS are just flexible manufacturing cells. The manufacturing root of the term *cell* is difficult to trace; however, in most cases, the term referred to a production area that had one or more NC or CNC machines. Today, production hardware grouped in the formation of cells is segregated for reasons that include raw material needs, operator requirements, manufacturing cycle times, and group technology. In general, cells can be grouped into two classifications:

1. The traditional stand-alone production cell.
2. The automated and integrated production cell.

In both cases the cells can have more than one production machine and operation. In addition, the production machines may be a combination of manual and computer controlled. For example, a traditional, stand-alone system may have two CNC machines, a manual punch and brake, and an assembly station so that a complete subassembly is produced from several parts made in the cell. The machines shown in Figures 10–9, 10–10 and 10–12 are good examples of CNC machines used in traditional production cells. Since automatic tool change capability is not always present, it is important to have an operator available to change tools as required by the machining sequence. The traditional cell usually relies on human operators to load raw material and run the machines. If the number of operators is equal to the number of machines, it is a *one-to-one* operation. Frequently, one operator runs two CNC machines, and the process is called a *two-to-one* operation. Two-to-one operation works especially well on production jobs with long machine cycle times.

As the name implies, the automated and integrated production cell usually has automated material handling hardware to load and unload one or more computer-controlled machines in the production cell. In cases where a milling operation is placed into a fully automated cell, the machine tools have tool carousels that can hold over 100 tools. As a result, a large variety of milling cutters and drill sizes are available to the programmer as the part cutting geometry and tooling are specified. However, human operators are not excluded; for example, in some automated cells, operators work with industrial robots in the machine load and unload cycle. The welding operation in Figure 10–25 is a good example of this type of operation. Study the picture until you can identify the robot and the circular welding table. The robot in Figure 10–25 is performing metal inert gas (MIG) welding. The welding table that holds the parts is on the left of the picture and has a plastic screen that separates the robot side of the table from the operator side. The circular table has two fixtures to hold the parts, one on each side of the screen. The operator loads the parts that need to be welded into the fixture and then presses a switch that rotates the circular table to put parts in front of the robot. The rotation of the table delivers to the operator the parts just welded by the robot. While the robot welds the new parts, the operator removes the finished parts and loads up another set to be welded. When the robot finishes the welding process, the table is again rotated by the operator and the process is repeated. The screen allows the worker to see the welding process to verify that everything is correct but protects the operator from the sparks, flash, and the vision-damaging ultraviolet light from the welding process. The operator and robot are performing as a team.

In other cases where workers are in an automated cell, operators perform intricate secondary operations, such as deburring or inspection, on production parts after a robot unloads the part from the machine. The automated cell is also characterized by multiple levels of computer control and a complex control algorithm.

Figure 10–25   Welding Robot Work Cell.

Integrating the automated cells into the CIM computer network and data base is usually easier than the task of bringing an entire FMS up to speed. The cells that form the FMS in Figure 10–22 are good examples of automated and integrated production cells.

It is difficult to determine when a large FMC becomes an FMS. A distinguishing factor is the level of control above the cell control hierarchy for the synchronization of the cells. If area controller software is used to control the sequence of operation at the cell level, the system is crossing from an FMC to an FMS. Other factors that differentiate FMCs from FMSs are tool capacity and flexibility in manufacturing. As the software used to control the production environment improves, the number of automated production systems that are true flexible manufacturing systems will increase. The control strategies used for FMCs and FMSs are described in Chapter 11.

10–3 FIXED HIGH-VOLUME AUTOMATION

A manufacturing system capable of large volumes of discrete parts is necessary for the economic production of some products. A good example of this type of product is the disposable razor or lighter. Although the production rate is well below that of disposable razors, discrete-parts production in the automotive industry for items such as engine crankshafts also falls into this category of manufacturing. Manufacturing systems capable of satisfying this type of production are called *transfer machines or transfer lines.* The many different systems used for large volume discrete-parts production are collectively called *Detroit-type automation.*

In-Line Fixed Automation

In-line fixed automation systems, similar to the system illustrated in Figure 10–26, have the following characteristics:

- A series of closely spaced production stations are linked by material-handling devices to move the parts from one machine to the next.
- There is a sequential production process, with each station performing one of the process steps and the cycle time at each machine usually about equal.
- Raw material enters at one end and finished parts exit at the other.
- The number of stations in the system is dictated by the complexity of the production process implemented.
- The production stations used in the system could include process machines such as mills, drills, and lathes; automated assembly stations; and automated inspection equipment.

Although most machines of this type are fully automated, it is possible to include manual stations along the line to handle operations that are either not economical or are difficult to automate. Figure 10–27 shows an in-line fixed automation production system that performs a series of manufacturing and assembly operations on small products. Frequently, buffer storage is placed in the in-line production system to smooth out the work flow due to irregularities and to continue to supply parts when maintenance is necessary on an upstream station.

Rotary Fixed Automation

Rotary fixed automation locates the production stations around a circular table or dial. As a result, this type of production system is called an *indexing machine* or a *dial index machine.* The system pictured in Figure 10–28 has seven production stations around the machine. In both this system and the in-line system in Figure 10–27, automatic parts feeders supply the pneumatic robots with components for the final assembly. One type of parts feeder, a *vibrating bowl* feeder, is pictured in Figure 10–29. A rotary fixed automation system is limited to smaller workpieces and fewer production stations than the in-line type.

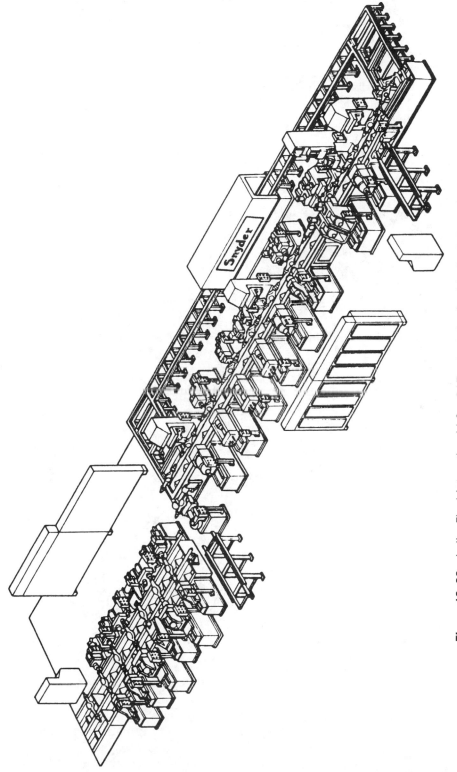

Figure 10–26 In-line Fixed Automation: (a) Special Twenty-station Combination Pallet and Free Transfer Machining System for Truck Rear-axle Housing; (b) Truck Rear Axle Produced on the System.

Source: Courtesy Synder Corp.

377

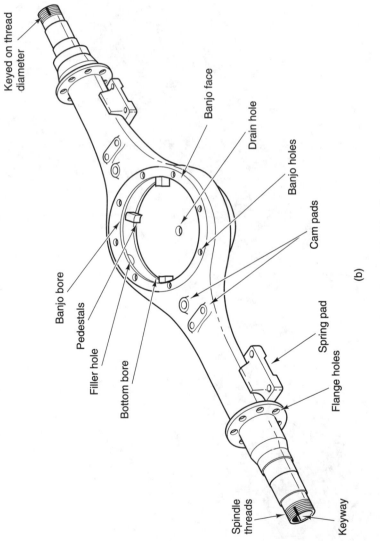

Keyed on thread diameter

Banjo face

Drain hole

Banjo holes

Cam pads

Banjo bore

Pedestals

Filler hole

Bottom bore

Spring pad

Flange holes

Spindle threads

Keyway

(b)

Figure 10–26 Continued.

Figure 10–27   In-line Fixed Automation System for Small Parts.
Source: Courtesy of Metro-fer, Inc., Pittsburgh, PA.

Selection Criteria for Fixed Automation Systems

Automation of this type generally requires a product with a high degree of stability and long life. For example, the production of a specific type of automotive engine crankshaft could have a five-year production life because the same engine is used in several different car models. Even different engine models could use the same crankshaft. In addition, products made on these machines have high production rates that are driven by high product demand. Also, this type of automation is selected when the labor content would be excessive if the part were produced manually. This type of automation improves on the following world-class metrics from Figure 1–6: low setup time, improved quality, small floor space area requirement, reduced in-process inventory, and minimum distance traveled in production. In addition, *order-winning* criteria from Figure 1–4 that are supported include lower product cost, better quality, lower lead time, and improved delivery performance.

10–4 SUMMARY

The processes in manufacturing are grouped into four operation categories: primary operations, secondary operations, physical properties, and finishing. Typical

Figure 10–28   Indexing or Dial Index Production System.
Source: Courtesy of Metro-fer, Inc., Pittsburgh, PA.

primary operations include casting, forming, oxyfuel and arc cutting, and sawing. Typical secondary operations include turning, boring, milling, drilling, reaming, grinding, and nontraditional machining. The physical properties of the part are often changed with a heat-treating process where the material is hardened in some cases and in other cases the hardness is reduced. The final processing group is finishing, where painting, plating, and pickling are used to change the surface of the part.

To achieve the flexibility and rapid response required by customers, many manufacturers are turning to flexible manufacturing systems (FMSs). An FMS is one or more machines integrated with automated material handling and managed by a computer control system. Typically five levels of technology are present in an FMS: enterprise, system, cell, machine, and device. Successful implementation of FMS systems provides the following benefits: inventory reduction, direct labor cost reduction, machine utilization increase, and support for world-class standards and order-winning criteria.

A building block of the FMS is the flexible manufacturing cell (FMC). The FMC is a group of related machines that perform a particular process or step in a

Figure 10–29   Vibratory Bowl
Feeder.

*Source: Courtesy of Metro-fer, Inc.,
Pittsburgh, PA.*

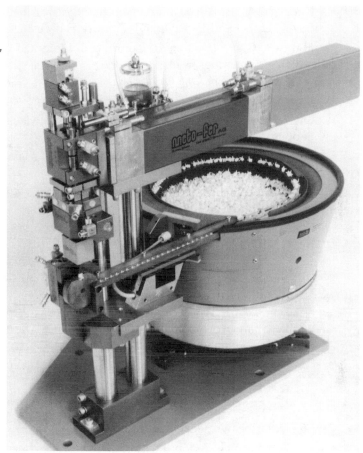

larger manufacturing process. The FMC operation falls into one of the following
categories: (1) traditional stand-alone production cell, or (2) automated and integrated production cell.

   Fixed automation is a manufacturing system focused on high-volume production of very specific parts or part mixes. Machines capable of this type of
production are often called transfer lines or Detroit-type automation. Fixed automation systems are generally configured using either in-line or rotary geometries.

REFERENCES

GOETSCH, D. L., *Modern Manufacturing Processes.* New York: Delmar Publishers,
1991.

GRAHAM, G. A., *Automation Encyclopedia.* Dearborn, MI: Society of Manufacturing
Engineers, 1988.

GROOVER, M. P., *Automation, Production Systems, and Computer-Integrated Manufacturing.* Upper Saddle River, NJ: Prentice Hall, Inc., 1987.

HAGGEN, G. L., "History of Computer Numerical Control." *Industrial Education,* August/September 1990, pp. 14–16.

LUGGEN, W. W., *Flexible Manufacturing Cells and Systems.* Upper Saddle River, NJ: Prentice Hall, 1991.

OWEN, J. V., "Flexible Justification for Flexible Cells." *Manufacturing Engineering,* September 1990.

PALFRAMAN, D., "FMS: Too Much, Too Soon." *Manufacturing Engineering,* March 1987.

REHG, J. A., *Introduction to Robotics in CIM Systems,* 4th ed., Upper Saddle River, NJ: Prentice Hall, 2000.

SOBCZAK, T. V., *A Glossary of Terms for Computer Integrated Manufacturing.* Dearborn, MI: CASA of SME, 1984.

QUESTIONS

1. What are process operations?
2. Name the four categories of process operations.
3. What is the function of the primary operations?
4. Describe two primary operations.
5. What is a secondary operation?
6. Describe two secondary operations.
7. What is included in the physical properties operations?
8. Describe the two categories of physical property changes.
9. What are finishing operations?
10. What are the attributes of flexible manufacturing?
11. Compare and contrast flexible manufacturing cells and flexible manufacturing systems.
12. Define an FMS in your own words.
13. Describe the five technology levels that may be present in an FMS.
14. What advantages do FMSs offer?
15. What is the definition of an FMC?
16. How are cells typically grouped in an FMC?
17. What distinguishes a large FMC from an FMS?
18. What is Detroit-type automation?
19. What are the characteristics of in-line fixed automation?
20. Describe the selection criteria for fixed automation systems.

PROJECTS

1. Using the list of companies developed in Project 1 in Chapter 1, develop a matrix to show the process operations used at the companies. Organize the operations in the primary, secondary, physical properties, and finishing groups.

2. Using the list of companies developed in Project 1 in Chapter 1, determine which companies are using flexible manufacturing cells or flexible manufacturing systems.

3. Select one of the companies in Project 2 that uses either an FMC or FMS and develop a case study on the implementation.

4. Select one of the companies in Project 1 that uses either an in-line fixed automation system or rotary fixed automation and develop a case study based on the implementation.

APPENDIX 10–1: HISTORY OF COMPUTER-CONTROLLED MACHINES

Computer-integrated manufacturing is a process, adopted by a manufacturer of goods and/or a provider of services, to increase productivity and serve customers more effectively. The definition of CIM in Chapter 1 emphasizes the integration of systems and use of data communications to achieve these increased productivity and customer service goals. Although the term *computer* is absent from the definition, the implementation of any CIM system requires the use of computer-driven devices and machines. Therefore, an overview of the development of computer-controlled machines provides a perspective on the development of CIM technologies and some insight on the direction of future technology initiatives.

Computers were included in production machines to speed the processing of data, perform arithmetic calculations, and make decisions based on production conditions. One of the first examples of a counting machine was the *abacus*, used over 1500 years ago. Many centuries later, in 1620, William Aughtred and others invented the *slide rule* to handle calculations more rapidly. In that same century, Blaise Pascal, a French philosopher and mathematician, and Gottfied Wilhelm developed wheel-and-cog devices that performed the four basic mathematical operations.

It was not until the eighteenth century that production machines were interfaced with crude control devices. In 1804, Joseph M. Jacquard built an automated loom capable of complex weaving designs using hole patterns in wooden boards as a program. Later, in 1833, an Englishman, Charles Babbage, designed two devices: the *difference engine* and *analytical engine.* Although not built successfully by Babbage, these two devices were the forerunners of mechanical adding machines and cash registers. In 1816, another British mathematician, George Boole, developed what was to be the universal language for computers. His *boolean algebra* used statements that were either true or false and involved three basic operations: *AND, OR,* and *NOT.* The counting of the U.S. census data was automated in 1890 by placing the data on punched cards and reading the data from the cards with an automated counter. The previous census processing time was reduced from 7 $\frac{1}{2}$ years to 6 weeks for the tabulation and 2 $\frac{1}{2}$ years for the statistical analysis. As a result of the success of this project, the developer of this punched-card process, Herman Hollerith, formed the Tabulating Machine Company, which later became the International Business Machine Corporation (IBM).

The first application of Boole's universal language to the solution of computational problems occurred in 1936, when Claude Shannon, a 21-year-old graduate student, published a paper entitled "A Mathematical Theory of Communications." Over the next four years, George Stibitz made three significant contributions to computing: (1) the first electromechanical circuit using Boolean algebra, (2) the first working information network in the United States, and (3) remote computing through the use of a terminal. Following this early work, a series of fully functional computers were built for the first time. The *Mark I*, an electromechanical computer built by IBM in 1944, was 51 feet long and had 750,000 parts connected with 500 miles of wire. These early machines were designed for computational solutions; applications in machine control for manufacturing was only a dream.

The development of the transistor in the early 1950s and the integrated circuit in the early 1960s began the first revolution in machine control. For the first time, computer technology could be applied practically to the problem of machine control in manufacturing. Some of the first machines to benefit from computer technology were the machine tool and the robot. In 1952, a project at the Massachusetts Institute of Technology (MIT) sponsored by the Air Force produced the first numerically controlled (NC) milling machine. Each machine movement was assigned a code. The code, along with the distance to travel, was fed into the machine using a 1-inch-wide paper tape on which the location of punched holes provided the program information. In the late 1950s, the first NC programming language, called *automated programmed tools* (APT), was developed by MIT. The computer simplified the manufacturing automation problem by converting the English language–like statements in APT into the elaborate codes needed to cut complex parts on a machine.

In the early 1950s, George Duval, the inventor of the industrial robot, developed a drum programming unit that controlled the motion of his early robot designs. The programming drum had adjustable protrusions on the surface that triggered motion in the robot arm as the drum rotated. Borrowing technology from the developing computer industry, this mechanical programming device was later replaced by electronic programming units. As in machine tools, mechanical programming gave way to high-level computer languages used to program robots, in which the computer translated the language statements into robot motion.

The development of the microprocessor in the early 1970s started a second revolution in machine control for manufacturing automation. *Direct numerical control* (DNC), which connected machine tools directly to host computers, was replaced by *computer numerical control* (CNC). A CNC machine tool has a dedicated computer integrated into the machine controller and is capable of storing programs to cut complex parts inside the machine tool. DNC operation is still used, but now the host is only a storage repository for part programs that are downloaded to the

CNC machine's computer for execution. The development of the microprocessor and the microcomputer throughout the 1970s and 1980s permitted computer control to be added to almost every manufacturing device and system on the shop floor. The capability of current machine controllers exceeds the demands of most current applications. For example, controllers such as the BostoMatic, G.M. Fanuc, and Rolo-Con-II have the following features:

- Spindle speeds as high as 40,000 revolutions per minute.
- Simultaneous 200-inch/minute, simultaneous five-axis contouring.
- Analog and discrete cell control.
- Cell control of twenty axes and over 200 discrete inputs/outputs.
- Full-color 3-D operator interfaces with real-time display of cutting and control sequences.
- Storage of up to 999 part programs in the machine.
- Interface capability to the enterprise's computer networks.
- Software to support the manufacturing database and management information system.
- A full 32-bit computer architecture for speed and accuracy.

Prior to 1980, applications for computer control in manufacturing were limited by the ability of the technology to support the application. Since that date, the only limiting factor has been the ability of manufacturers to apply technology tools present in the marketplace.

Production Support Machines and Systems

Chapter

11

The production machinery and systems described in Chapter 10 add value to the product when the process is acting on the raw material. However, in many manufacturing operations, partially finished parts and raw material spend days or even months in the shop being moved, waiting for an available production machine, being loaded and unloaded from production machines, and being inspected. Reducing or eliminating these non-value-added operations is the goal for world-class companies; however, completely eliminating these cost-added operations is often not possible. As a result, production hardware such as robots, material-handling systems, and storage systems are necessary. Therefore, analysis of automation systems to minimize the impact of these cost-added operations is important and starts with a description of industrial robots.

11–1 INDUSTRIAL ROBOTS

Industrial robots are given a separate section in this chapter because they are unique production machines. Although several different production machines may appear to have the capacity to work like a robot, on closer inspection the robot is distinctly different. Without a definition, it would be difficult to differentiate industrial robots from the millions of automated machines so that they can be studied. The Information Standards Organization (ISO) defines an *industrial robot* in standard ISO/TR/8373-2.3 as follows:

> *A robot is an automatically controlled, reprogrammable, multipurpose, manipulating machine with several reprogrammable axes, which may be either fixed in place or mobile for use in industrial automation applications.*

The key words are *reprogrammable* and *multipurpose* because most single-purpose machines do not meet these two requirements. *Reprogrammable* implies two characteristics: (1) the robot's motion is controlled by a written program, and (2) the program can be modified to change the motion of the robot arm significantly. Programming flexibility is demonstrated, for example, when the pickup point for a part is located by a vision system, and the location is sent to the robot while the robot is moving to the part. *Multipurpose* emphasizes the fact that a robot can perform many different functions, depending on the program and tooling currently in use. For example, a robot could be tooled and programmed in one company to do welding, and in a second company the same type of robot could be used to stack boxes on a palletizer.

The Basic Robot System

A *robot system* is defined by the Robotic Institute Association as follows:

> *An industrial robot system includes the robot(s) (hardware and software), consisting of the manipulator, power supply, and controller; the end-effector(s); any equipment, devices, and sensors with which the robot is directly interfacing; any equipment, devices, and sensors required for the robot to perform its task; and any communications interface that is operating and monitoring the robot, equipment, and sensors.*

A basic robot system is illustrated in Figure 11–1. The system includes a *mechanical arm* to which the *end-of-arm tooling* is mounted, a *computer-based controller* with attached *teach station, work cell interface,* and *program storage device.* In addition, a source for pneumatic and/or hydraulic power is part of the basic system. The work cell controller, a part of the manufacturing system, connects to the robot through the robot work-cell interface.

Mechanical Arm

The arm is a mechanical device driven by *electric drive motors, pneumatic devices,* or *hydraulic actuators.* The basic drive elements will be either *linear* or *rotary* actuators. The combination of motions included in the arm will determine the type of arm geometry that is present. The basic geometries include *rectangular, cylindrical, spherical,* and *jointed spherical.* Figure 11–2 shows a jointed-spherical mechanical arm with all the motions indicated.

The six motions are divided into two groups. The first group, called the position motions, includes the *arm sweep, shoulder swivel,* and *elbow extension.* With a combination of these three motions, the arm can move to any position required within the work area. The second group of motions, associated with the wrist at the end of the arm, includes the *pitch, yaw,* and *roll.* A combination of these three motions, called *orientation,* permits the wrist to orient the tool plate and tool with respect to the work.

Robot Arm Geometry Classification

In general, the basic mechanical configurations of the robot manipulator are categorized as *cartesian, cylindrical, spherical,* and *articulated.* The cartesian is divided into two groups, *traverse axes* and *gantry,* and the articulated is divided into the *horizontal* and *vertical* groups. A description of these configurations follows.

Cartesian Geometry. A robot with a cartesian geometry can move its gripper to any position within the cube or rectangle defined as its working volume. Two

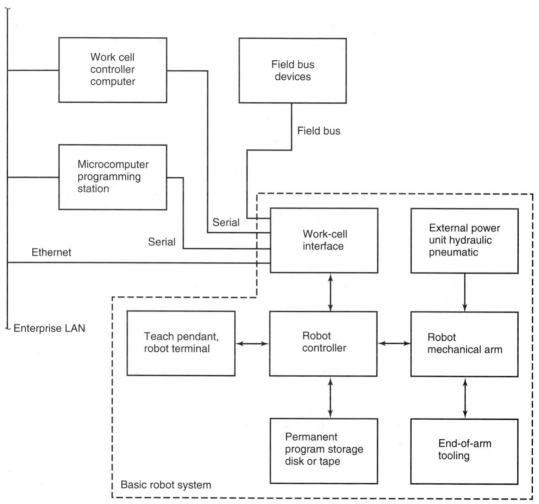

Figure 11–1   Basic Robot System.

Source: James A. Rehg Introduction to Robotics in CIM Systems, *Fourth Edition, © 2000, p. 24. Reprinted by permission of Prentice Hall, Upper Saddle River, New Jersey.*

configurations form this geometry, *traverse* and *gantry* machines. Figure 11–3 is an example of a gantry machine that can move in the X, Y, and Z directions as indicated. The tooling is attached to the tool plate, so work is performed from above. The rectangular work envelope of this type of robot is often used to move parts from conveyor systems into production machines. In the gantry illustration, the three degrees of freedom for positioning are indicated by arrows to show movement in the X, Y, and Z directions. Three degrees of freedom, A, B, and C, are provided on the wrist to orient the tool mounted on the tool plate. The second type of cartesian geometry, traverse, is pictured in Figure 11–4.

Figure 11–2 Jointed-Spherical
Mechanical Arm.

*Source: Courtesy of ABB
Robotics, Inc.*

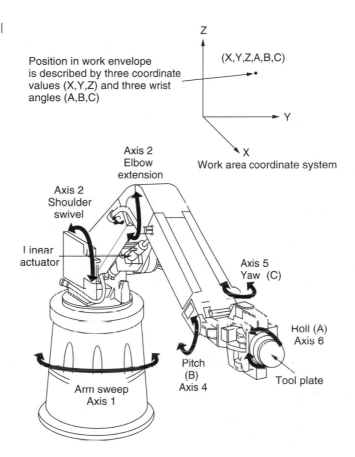

Cylindrical Geometry. Cylindrical coordinate robot systems, like the one illustrated in Figure 11–5, can move the gripper within a volume that is described by a cylinder. The cylindrical coordinate robot system is positioned in the work area by two linear movements in the X and Y directions and one angular rotation about the Z axis. The axes on cylindrical coordinate robots are driven pneumatically, hydraulically or electrically. A small pneumatic cylindrical geometry robot is pictured in Figure 11–6.

Spherical Geometry. The spherical geometry arm, sometimes called *polar*, is drawn in Figure 11–7. Spherical arm geometry robots position the wrist through two rotations and one linear actuation. As in the previous cases, the orientation of the tool plate is achieved through three rotations in the wrist (A, roll; B, pitch; C, yaw). In theory the rotation about the Y axis could be 180 degrees or greater, and the wrist rotation about the Z axis could be 360 degrees. Then if R, robot reach, went from the retracted to the fully extended position, the volume of operating space defined would be two concentric half-spheres. The work envelope for an early robot from Unimation that used spherical geometry machines is illustrated

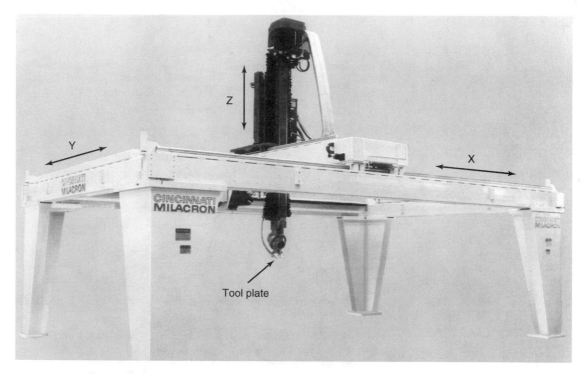

Figure 11–3 Gantry-type Robot.

in Figure 11–8. This type of robot was frequently used in early industrial applications. Note that the actual working volume is much less than the theoretical volume of the machine in Figure 11–7. Again, this results from mechanical design constraints. Spherical geometry machines use either *hydraulic* or *electric* drives as the prime movers on the six axes, with pneumatic actuation used to open and close the gripper.

Articulated Geometry. *Articulated* industrial robots, often called *jointed arm, revolute,* or *anthropomorphic* machines, have an irregular work envelope. This type of robot has two main variants, *vertically articulated* and *horizontally articulated.* The vertically articulated robot (Figure 11–9) has three major angular movements consisting of a base rotation (axis 1), shoulder (axis 2), and forearm (axis 3) joint. The irregular work envelope is illustrated in Figure 11–10. As in the previous arm designs, the orientation of the tool plate is provided by the three rotations in the wrist. Electric drives with feedback control systems are used on most machines. An example of a jointed-spherical configuration is illustrated in Figure 11–11. Note the ball-shaped *three-roll wrist* that generates the orientation motion in all three wrist axes for the spot-welding tooling attached to the tool plate.

The *horizontally articulated* robot arm has two angular movements for positioning the tooling (arm and forearm rotation), and one position linear movement that is a vertical motion. Horizontally articulated arms are implemented with two

Figure 11–4   Cartesian Geometry Robot.
Source: Courtesy of Metro-fer, Inc., Pittsburgh, PA.

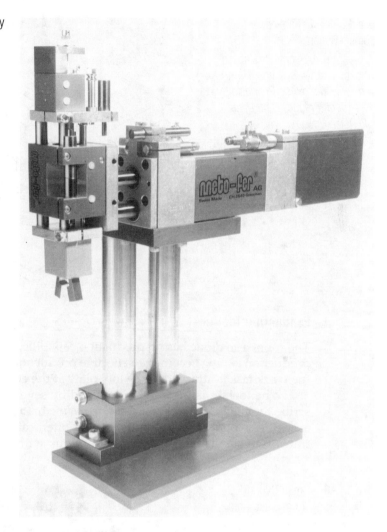

mechanical configurations: (1) the *SCARA (selective compliance articulated robot arm)* illustrated in Figure 11–12, and (2) the *horizontally based jointed* arm.

The SCARA machine (Figure 11–12) has two horizontally jointed arm segments fixed to a rigid vertical member. Positions within the cylindrical work envelope are achieved through changes in axis numbers 1 and 2. Vertical movement of the gripper plate results from the Z axis located at the end of the arm. SCARA machines usually have only one wrist axis, rotation. This arm geometry is frequently used in electronic circuit board assembly applications because this geometry is particularly good at vertical part insertion.

The horizontally based jointed arm uses the same construction as the SCARA with one exception; the vertical Z axis is located between the rigid vertical base and the shoulder joint of the upper part of the arm. With this configuration, the wrist and gripper mounting plate are located at the end of the forearm. Robots produced by the Reis Company made this cylindrical geometry popular.

Figure 11–5 Cylindrical Coordinate Robot.

Source: James A. Rehg, Introduction to Robotics in CIM Systems, *Fourth Edition, © 2000, p. 49. Reprinted by permission of Prentice Hall, Upper Saddle River, New Jersey.*

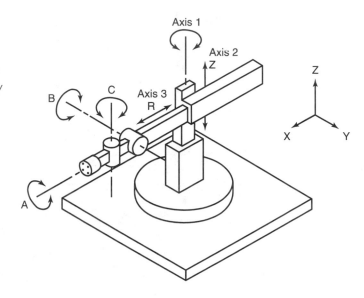

Production Tooling

The robot arm alone has no production capability, but the robot arm interfaced to production tooling becomes an effective production system. The tooling to perform the work task is attached to the tool plate at the end of the arm.

The tooling is frequently identified by several names. The term used to describe tooling in general is *end-of-arm tooling* or *end effector*. Tooling with an open-and-close motion to grasp parts is most often called a *gripper*. Figure 11–13 illustrates two standard gripper configurations with angular and parallel gripping

Figure 11–6 Small Cylindrical Coordinate Robot from Seiko Corp.

Source: James A. Rehg, Introduction to Robotics in CIM Systems, *Fourth Edition, © 2000, p. 53. Reprinted by permission of Prentice Hall, Upper Saddle River, New Jersey.*

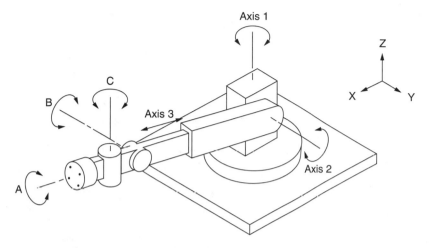

Figure 11–7 Spherical Coordinate Robot Geometry.
Source: James A. Rehg, Introduction to Robotics in CIM Systems, *Fourth Edition, © 2000, p. 53. Reprinted by permission of Prentice Hall, Upper Saddle River, New Jersey.*

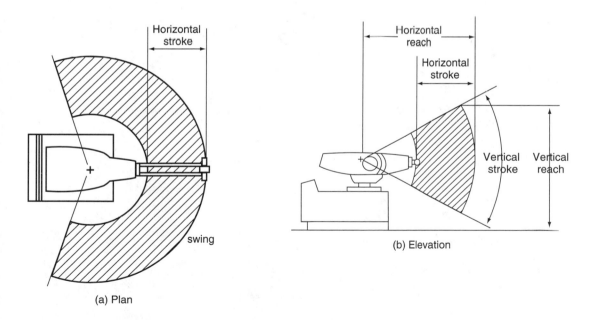

(a) Plan

(b) Elevation

Figure 11–8 Westinghouse/Unimation 2000 Work Envelope.
Source: James A. Rehg, Introduction to Robotics in CIM Systems, *Fourth Edition, © 2000, p. 54. Reprinted by permission of Prentice Hall, Upper Saddle River, New Jersey.*

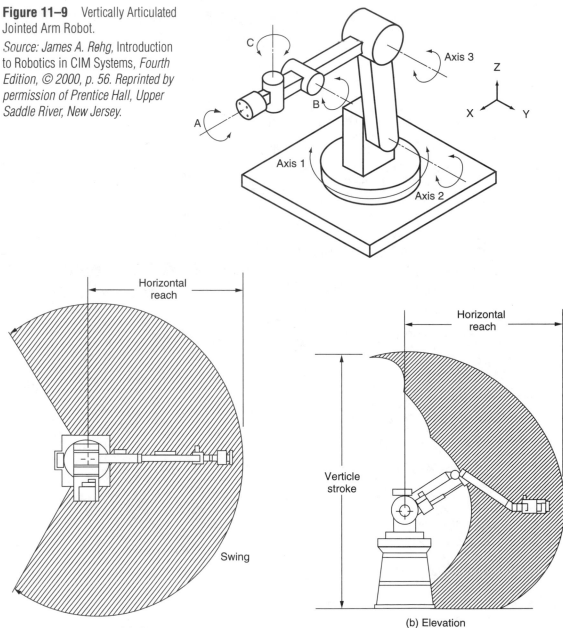

Figure 11–9 Vertically Articulated Jointed Arm Robot.

Source: James A. Rehg, Introduction to Robotics in CIM Systems, *Fourth Edition, © 2000, p. 56. Reprinted by permission of Prentice Hall, Upper Saddle River, New Jersey.*

(a) Plan

(b) Elevation

Figure 11–10 Work Envelope of a Vertically Articulated Robot.

Source: James A. Rehg, Introduction to Robotics in CIM Systems, *Fourth Edition, © 2000, p. 57. Reprinted by permission of Prentice Hall, Upper Saddle River, New Jersey.*

Figure 11–11 Spot Welding Body Assemblies with a Jointed-Arm Robot.

motions. The end-of-arm tooling used on current robots can be classified in the following three ways:

1. According to the method used to hold the part in the gripper.
2. Based on the special-purpose tools incorporated in the final gripper design.
3. Based on the multiple-function capability of the gripper.

The first category of gripping mechanisms includes standard mechanical pressure grippers. (Figure 11–14), tooling utilizing vacuum for holding or lifting

Figure 11–12 SCARA Robot Arm.
Source: Courtesy Adept Technology, Inc.

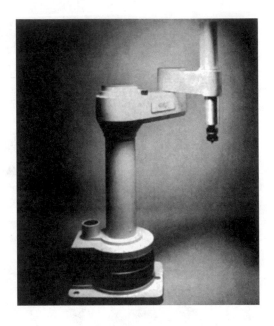

(Figure 11–15), and magnetic devices. Figure 11–14 shows one-half of a dual parallel gripper picking up a steel cylinder from a conveyor for loading into a turning center. The other parallel gripper is used to remove the finished part from the turning center before the steel cylinder blank is loaded. Figure 11–15 shows four vacuum grippers on a specially designed end-of-arm tool placing a window into a car on an automotive assembly line.

The second classification of tooling includes drills, welding guns (Figure 11–11) and torches (Figure 11–16), paint sprayers, and grinders. Figure 11–11

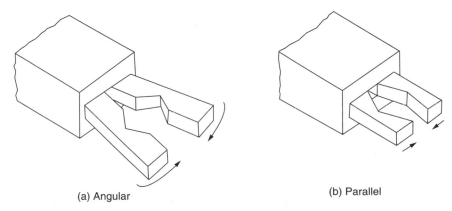

(a) Angular

(b) Parallel

Figure 11–13 Standard and Angular Grippers.
Source: James A. Rehg, Introduction to Robotics in CIM Systems, *Fourth Edition, © 2000, p. 82. Reprinted by permission of Prentice Hall, Upper Saddle River, New Jersey.*

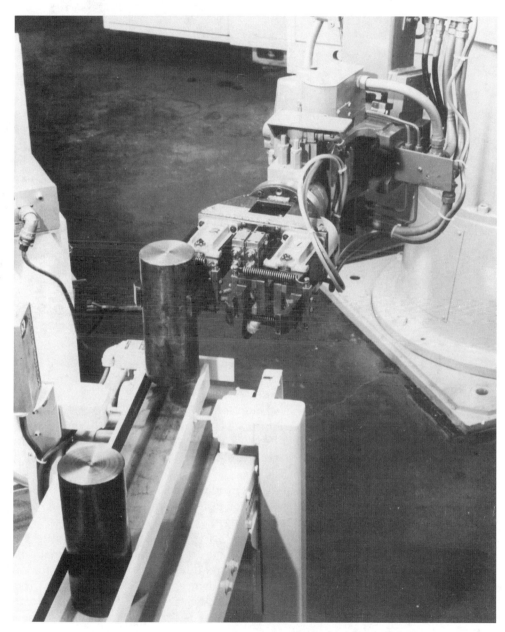

Figure 11–14   Two Parallel Grippers Loading a Turning Center with Steel Cylinders.

Figure 11–15 Vacuum Grippers Used to Install the Front Windshield in Cars.

shows a robot performing spot welding on a car body as the car moves past the robot work cell, and in Figure 11-16 a robot uses MIG welding tooling and a weld vision system.

The third type of gripper tooling includes special-purpose grippers that are designed for a specific task. For example, a robot with special tooling could lift a folded box and open it for filling.

In setting up any manufacturing system, some criteria must be used in selection of the equipment. When robots are included in the automation, an analysis of each sequence will establish the requirements of the gripper for each step of the manufacturing process. Every sequence in the manufacturing process should be examined and the relative difficulty established. The robot system capability, including the end-of-arm tooling, must be equal to or greater than the most demanding sequence in the process.

Robot Controller

The controller is the most complex part of the robot system illustrated in Figure 11–17. Figure 11–17 is a basic block diagram of a typical controller used on electric

Figure 11–16   MIG Welding Tool and Welding Vision System.

Source: James A. Rehg, Introduction to Robotics in CIM Systems, *Fourth Edition, © 2000, p. 310. Reprinted by permission of Prentice Hall, Upper Saddle River, New Jersey.*

robots. The controller, basically a special-purpose computer, has all the elements commonly found in computers, such as a central processing unit (CPU), memory, and input and output devices. The microprocessors that form the CPU in the controller have the primary responsibility for controlling the robot arm and the work cell in which it is operating. The controller can also communicate with external devices through the *input/output interface* (Figure 11–17).

Teach Stations

Teach stations on robots may consist of teach pendants, teach terminals, or controller front panels. Some robots permit a combination of programming devices to be used in programming the robot system, while others provide only one system programming unit. In addition, some robots use IBM PS/2s or PC-compatible microcomputers as programming stations. Teach stations support three activities: (1) robot *power-up* and preparation for programming, (2) *entry* and *editing* of programs, and (3) execution of programs in the work cell. The development of a program includes keying in or selecting menu commands in the controller language, moving the robot arm to the desired position in the work cell, and recording the position in the program.

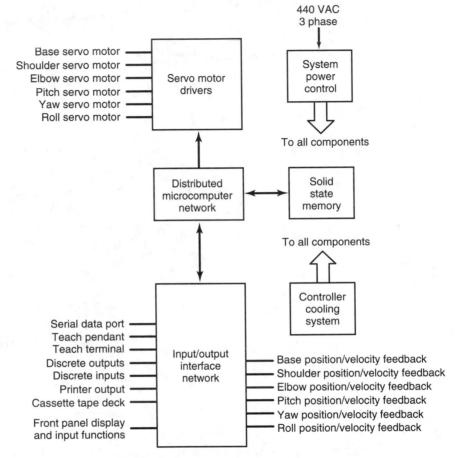

Figure 11–17   Robot Controller Block Diagram for Electric Robots.

Source: James A. Rehg, Introduction to Robotics in CIM Systems, *Fourth Edition, © 2000, p. 28. Reprinted by permission of Prentice Hall; Upper Saddle River, New Jersey.*

Robot Applications

The robot is a very special type of production tool; as a result, the applications in which robots are used are quite broad. These applications can be grouped into the following categories:

- Material processing.
- Material handling.
- Assembly and fabrication.

In the first group, robots use tooling to *process* the raw material. For example, the robot tooling could include a drill and the robot would be performing drilling operations on raw material. The robot in Figure 11–18 is performing a grinding operation to remove burrs from a raw casting.

Material handling, the second application group, is the most common application for robots. In many cells, robots replace human operators in loading and unloading production machines. The robot in the foreground of Figure 11–19 is loading and unloading two turning centers in a flexible manufacturing system making printing press rollers.

Assembly is another large application area for using robotics. Robots are used to assemble a wide range of products from cars on assembly lines (Figure 11–15) to the boxes of candy in Figure 11–20.

Selecting and Justifying Robot Applications

Integrating an industrial robot into an existing production station requires a detailed design process. The cost of a robot automation project depends on the complexity of the cell, the quantity and quality of existing production equipment, and the type of robot selected. The following three processes were adapted from Rehg (1992). The three basic steps in the robot cell development process are: (1) pick the best manufacturing situation for the implementation, (2) pick the best robot for the specific job identified in step 1, and (3) build the best work cell possible around the robot selected. Although overly simplified, the three-step process emphasizes an important point: *the first step is to identify the best manufacturing application for the robot project.* Identification of the best application area uses a two-step process: (1) implement a process to identify all the production areas where the possibility of success is between 90 and 100 percent; and (2) study in detail the production cells selected and choose the one with the highest return on investment.

1. Form an automation team (three to five people) to perform the initial plant survey. It is important that the team members have some training in robotics and automation. In addition, all shop floor employees should know why the survey is being performed and know that the written company policy

Figure 11–18 Robot in a Grinding Application.

Source: Courtesy ABB Robotics, Inc.

Figure 11–19   Robot Serving as a Material Handler in a Flexible Manufacturing Cell.

for automation-related job loss will be followed. To start the process, team members make independent visits to every production area and complete a survey form (Figure 11–21) for each. After all production cells are surveyed, the production areas that received six *yes* answers from every team member are identified as the areas for additional analysis. All the cells selected that meet the following criteria are identified.

■ Production areas where maintaining safe working conditions is difficult or costly.

■ Production areas where excessive protective equipment and clothing must be used by the operators.

■ Production areas with a history of worker opposition and worker discontent.

■ Production operations where material savings are possible with automation.

2. All the production areas identified by the last set of criteria have a high implementation success factor. A work area is selected for robot automation from this group of work cells, and a detailed design study is performed. The study must include a detailed job analysis that gathers data in three areas: (1) technical considerations, (2) economic considerations, and (3) human factors.

Figure 11–20   Robots Used to Assemble Candy Boxes. *Source: Courtesy Adept Technology, Inc.*

The study includes the selection of a robot and all the necessary work-cell hardware to support the automated production process.

As computers become less expensive and more sophisticated and as more automation is added to the production area, industrial robots will play a larger role in the production of manufactured goods.

11–2 AUTOMATED MATERIAL HANDLING

In most production systems, the part or raw material is either in transit, waiting in a queue, being processed, or in inspection. Except for the processing time, all of the other operations add cost, not value, to the product. As a result, the material-handling process for parts and raw material should be automated only after every unnecessary inch of material transport distance has been removed from the production process.

The work simplification and analysis process that precedes the design and selection of material-handling automation starts with a diagram of the production flow. Process flow analysis uses symbols, such as those illustrated in Figure 11–22, to diagram the production system. Starting with the symbolical representation of the production flow (Figure 11–22), the analysis of part movement and queues leads to the elimination of any unnecessary elements. With the distance traveled

---

Robot applications—initial plant survey

Answer the following questions for each workstation in the plant survey:

1. Can inspection by operators be eliminated from this workstation?   Yes   No
   It is difficult and expensive to include parts inspection in a robot work cell.

2. Is the shortest machine cycle 3 seconds or longer?   Yes   No
   Robot speed is limited. Human operators can work faster than robots when demanded by the process.

3. Can the robot displace one or two people for three shifts?   Yes   No
   If the average robot project costs $100,000, then it will be necessary to save the cost of one or two operators for three shifts to get a one- or two-year payback.

4. Can the parts be delivered in an oriented manner?   Yes   No
   Picking parts from a tote bin is easy for humans but difficult for robots. If the parts can come oriented for easy robot pickup, then robot automation is possible.

5. Can a maximum of six degrees of freedom do the job?   Yes   No
   Robots have one arm that moves through a restricted work space compared with the two-armed human. A single-armed robot must be able to do the job.

6. Can a standard gripper be used or modified to lift the part or parts?   Yes   No
   The tooling is a major part of the work-cell expense. The simpler the tooling, the greater the likelihood of a successful project. Also the weight of the part plus the gripper must be consistent with the robot's capability.

If all the answers are yes, then you have a prime candidate for a robot application.

Figure 11–21   Robot Survey Form.

Source: James A. Rehg, Introduction to Robotics in CIM Systems, *Fourth Edition, © 2000, p. 381. Reprinted by permission of Prentice Hall, Upper Saddle River, New Jersey.*

by parts and raw material reduced to the minimum possible, the selection of effective automation to transport and handle parts is possible.

The transfer mechanism used to move parts between work cells and stations has two basic functions: (1) move the part in the most appropriate manner between production machines, and (2) orient and position the part with sufficient accuracy at the machine to maximize productivity and maintain quality standards.

Automated Transfer Systems

The many mechanisms used to achieve these two goals are grouped into three categories: *continuous transfer, intermittent transfer,* and *asynchronous transfer.*

Continuous Transfer.   In a continuous transfer operation, the parts or material moves through the production sequence at a constant speed. The work does not

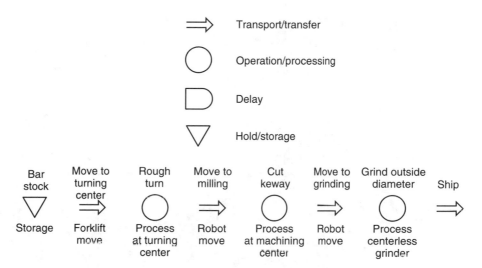

Figure 11–22   Symbols Used to Diagram a Production Flow.

stop at a workstation for a production operation to be performed. A good example of this type of transfer is in an automated car wash. The car is pulled through the car wash at a constant speed, and every washing mechanism, such as the brushes that clean the wheels, must perform their operation synchronized with the moving vehicle. In the manufacturing area, a prominent example is the assembly of cars on an automotive assembly line. The cars move at a constant rate through 95 percent of the assembly process, with the automated assembly machines and manual operations performed on a moving vehicle. The picture in Figure 11–23 shows a spot-welding operation performed on a moving car by a robot. In this case the robot has the capability to move the spot-welding tooling connected to the end of the robot's arm at a rate synchronized with the material-handling system carrying the car. The material-handling system moves each car through the same line past the robot. The robot senses the speed of the car and controls the motion of the welding gun with sufficient accuracy that the welds are placed with a repeatability of ¼ inch. For the continuous transfer system to function, the workstations must support part or material movement during the work process.

The types of mechanisms used to achieve continuous motion in manufacturing include:

- *Overhead monorail.* This mechanism is a continuous series of interconnected hooks attached to free-turning rollers that ride in overhead tracks. Parts are hung from the hooks to transport material between workstations or through a production operation. This type of system is frequently used to bring parts past human or robot paint sprayers to complete a painting or coating operation.

- *Monorail tow systems.* This system is the same as described above except that wheeled carts are attached to the hooks, and the overhead system pulls the carts from one destination to another.

Figure 11–23 Spot-Welding Operation on Moving Car Body.

- *In-floor tow systems.* This operation is similar to the monorail tow except that the towing mechanism is a continuous chain riding in a channel placed below the surface of the floor. The chain has hooks placed at regular intervals that pull a wheeled carrier through production. This method is the most popular for towing car carriers through the main automotive assembly process.

The second type of material movement system is called *intermittent* or *synchronized* transfer.

Intermittent Transfer. The intermittent or synchronized transfer system has the following characteristics:

- Workstations are fixed in place.

- The motion of the transfer device is intermittent or discontinuous so that parts cycle between being in motion and being stationary.
- The motion of the parts is synchronized so that all parts move at the same time and then are motionless during the same time frame.

This type of material and parts transfer system is usually found in machining applications, especially progressive die operations.

The most common type of mechanisms used to achieve synchronized intermittent motion is the *walking beam transfer system* illustrated in Figure 11–24. Study the figure until you are familiar with the names of the various parts. The transfer system has a *fixed rail* that has part carriers resting in notches. The *transfer rail* has a similar set of notches and is free to move. The synchronized motion occurs when the transfer rail picks up all the parts and moves them forward by one set of notches. Notice that in Figure 11–24a, the six parts are in the first six sets of notches on the left side of the fixed rail. After one cycle of the transfer rail, all parts are located one notch to the right of their previous position (Figure 11–24d).

The third type of material movement system is called *asynchronous transfer*.

Asynchronous Transfer.   The term *asynchronous* means *not synchronized*; therefore, an asynchronous transfer system is one where each part moves independently

Figure 11–24   Walking Beam Mechanism.

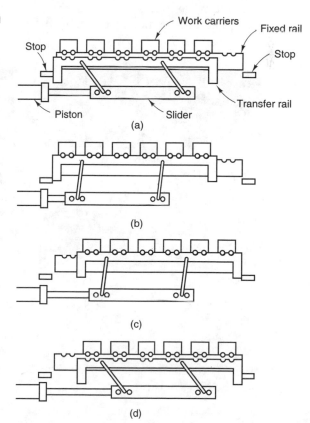

of other parts. The asynchronous transfer is often called a *power-and-free system* to indicate that parts can be either free from the transfer mechanism or powered by the transfer device. Several advantages of this type of transfer are evident in the shuttle pallet type of asynchronous system pictured Figure 11–25. Two pallets that hold parts for machining are held at a stop while the twin-belt conveyor continues to move under the stationary pallets. A third pallet is carried by the conveyor to a part loading station. In-process work storage is provided by asynchronous transfer with some pallets in production and others ready to move into production as soon as the previous part is machined. The process permits finished parts to move to the next production operation, while still other parts are in production.

Other advantages of this type of automated transfer include the ability to aid in line balancing by using parallel stations for operations with long cycle times and single production stations where the cycle time is shorter. This type of transfer is also used where manual production stations are integrated with fully automated operations. The singular disadvantage is that the cycle rates for asynchronous transfer systems are usually longer than the two previous types of material transfer.

The most common type of mechanism used to achieve asynchronized transfer motion is the *power conveyor.* The driven surface on the conveyor is frequently a belt made from steel, impregnated fabric, slats, or interlocking plates. In many applications, pallets, pulled by the frictional force between the pallet and the

Figure 11–25   Pallet Shuttle Conveyor System.

conveyor belt (Figure 11–25), move with the conveyor and hold a single workpiece or a family of parts.

The transfer system just described performs over 60 percent of all material-handling functions in automated production systems.

11–3 AUTOMATIC GUIDED VEHICLES

Automatic guided vehicles (AGVs) are defined by the Material Handling Institute as follows:

> *An AGV is a vehicle equipped with automatic guidance equipment, either electro-magnetic or optical. Such a vehicle is capable of following prescribed guide paths and may be equipped for vehicle programming and stop selection, blocking, and any other special functions required by the system.*

As the definition indicates, an AGV is a driverless vehicle capable of performing all the operations formerly available only with forklift trucks and other types of human-operated delivery vehicles. Like the vehicle pictured in Figure 11–26, AGVs are usually powered by electric motors that receive energy from electric batteries on the vehicle. AGV technology dates back to the early 1950s.

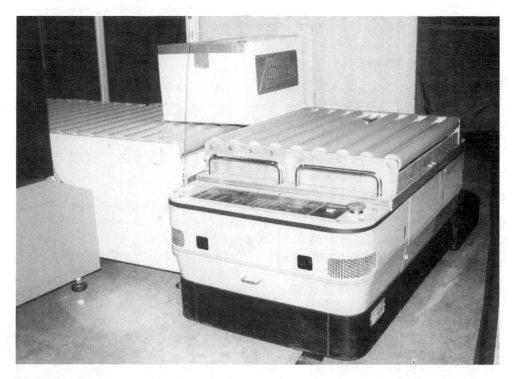

Figure 11–26   Automatic Guided Vehicle.

History of AGVs

The automatic guided vehicle concept was developed in the United States in the 1950s by Barret Electronics. The early technology used towing vehicles to pull a series of trailers, predominantly in the warehousing environment, to predefined locations. The negative climate for productivity automation prevented wide acceptance and installations of the system in the United States. However, the European manufacturing community aggressively adopted the concept and agreed on one standard pallet size (800 by 1200 mm), called the *Euro pallet.* With most of Europe using this common skid-type pallet without bottom boards, the development of driverless material-handling devices was enhanced in the European community.

In the 1970s, several factors caused widespread use of AGV technology in the United States. Competition from foreign markets forced a reduction in direct labor cost and a new interest in promoting productivity through automation. In addition, advances in the computer, especially the microprocessor, permitted the development of sophisticated AGV systems. Using a "land-based" AGV computer and microprocessor computers on-board, AGVs could provide bidirectional operation between points in manufacturing, operate on open- and closed-loop paths using FM radio signals for communication, handle traffic control and queuing in multiple-vehicle systems, and perform material tracking on all parts moved throughout the plant. Today, AGVs perform a valuable material-handling service on part transfers that are necessary for efficient production.

Types of AGVs

There are six basic types of AGVs: *towing, unit load, pallet truck, fork truck, light load,* and *assembly line.* The outline of each type is provided in Figure 11–27; review each outline until you are familiar with the design. The design incorporated into each type makes the AGV ideal for a specific manufacturing application. Four types of AGVs are pictured in Figure 11–28. The key features of each type are described below.

Towing Vehicles.   This type was the first introduced and continues to be used in many applications. The primary function is to tow trailers (Figure 11–28) with capacities from 5000 to 50,000 pounds at speeds up to 3 miles per hour. The number and variety of trailers that are pulled in these applications is widely varied, but ten is usually the limit. The primary application for this type of AGV is bulk movement of product into and out of warehouse areas.

Unit Load Vehicles.   Unit load vehicles are equipped with a deck that holds a pallet of material and generally have some type of automatic transfer device to deliver and pick up material. The primary application is in warehouse and distribution systems, with the AGVs transferring a high volume of material and parts to conveyors and work cells. Figure 11–29 shows a unit load type of AGV.

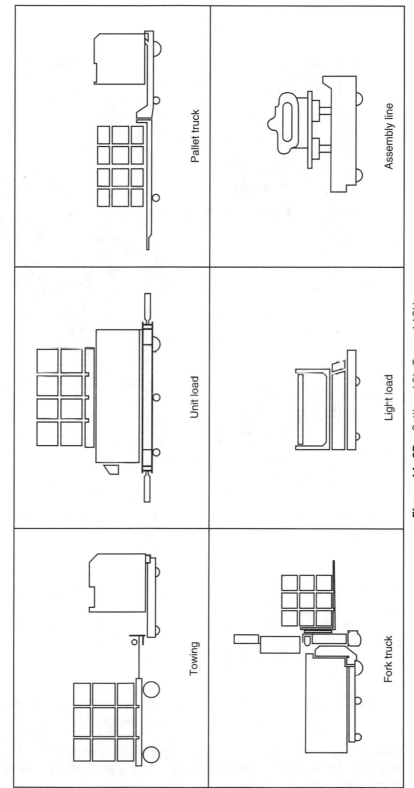

Figure 11–27 Outline of Six Types of AGVs.

Figure 11–28   Picture of AGV Types: Front to Back Unit Load, Unit Load, Assembly Line, Towing, Fork Truck.

Source: Courtesy Eaton-Kenway, Inc., Salt Lake City, UT.

Pallet Truck Vehicles.   The application is primarily in distribution and movement of palletized loads at floor level. The AGV is frequently loaded by an operator who boards the vehicle and backs it into a loaded pallet. With the pallet on board, the operator drives the AGV back to the guide path, inputs a destination code into the AGV, gets off the vehicle, and then uses a start command to send the AGV to the programmed destination.

Fork Truck Vehicles.   Fork truck AGVs (Figure 11–30) emulate the operation of a manual fork truck with the flexibility to pick up and deliver loads at different heights at each stop. This flexibility makes this type of AGV the most expensive to install and requires greater attention to the path location and accurate positioning of loads. Unlike the pallet truck AGV, the guided fork trucks operate without human intervention.

Figure 11–29   Unit Load AGV.
Source: Courtesy Eaton-Kenway, Inc., Salt Lake City, UT.

Light-Load Vehicles.   As the name indicates, the light-load AGV has only a several-hundred-pound-capacity. This type of AGV is ideal for moving small parts trays or bins from small parts storage areas to individual workstations where operators perform light assembly. Applications that justify this type of AGV include electronic assembly and small product assembly.

Assembly Line Vehicles.   The assembly line AGV (Figure 11–31) carries subassemblies, such as automotive engines, through a serial assembly line process until the assembly is complete. In some applications, the AGV passes through a parts staging area where trays of components required for assembly are loaded onto the AGV. This process is repeated until the subassembly is complete and ready for delivery to the main assembly line.

   The six types of AGVs are designed to satisfy different manufacturing requirements. However, each of the different types shares the same operational characteristics.

AGV Systems

AGV systems must perform five functions: *guidance, routing, traffic management, load transfer,* and *system management.* The significance of each of these functions on the effective operation of an AGV system is described in the following sections.

Guidance.   The vehicle must follow a predetermined path that is optimized for a given material flow pattern. The path is most often fixed by embedding a current-carrying wire just below the surface of the floor. Wire-guided AGV systems use

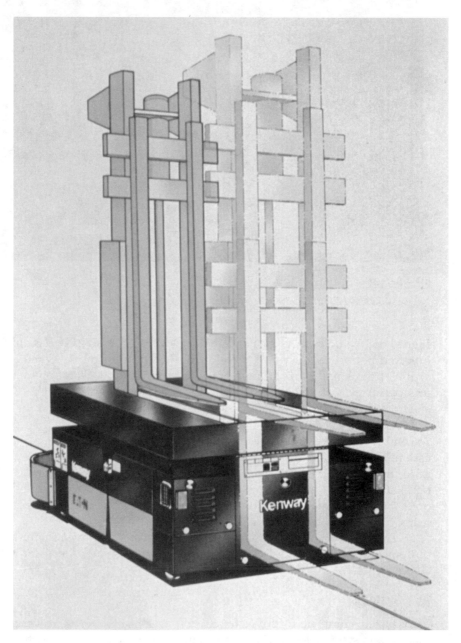

Figure 11–30   Fork Truck AGV.
Source: Courtesy Eaton-Kenway, Inc., Salt Lake City, UT.

Figure 11–31   Assembly-Line in AGV
Source: Courtesy Eaton-Kenway, Inc., Salt Lake City, UT.

on-board sensors to detect the magnetic field surrounding the current-carrying wire to keep the vehicle on the path. The wires are placed in the floor by cutting a narrow slot about ½ inch deep, laying the wire into the slot, and then sealing the cut with epoxy for a smooth finish. Figure 11–32 illustrates the relationship between the buried wire in the floor and the magnetic sensors on board the AGV.

In other applications, optical sensors on the AGV track a line painted on the floor to mark the path. The optical technique, called a chemical guide path, uses an invisible fluorescent dye approximately 1 inch wide. In clean room environments, the chemical guide path can be used; however, in heavy industrial manufacturing, the wire-guided systems are preferred.

Another classification of guidance techniques is called *off-wire technology*. Off-wire systems use various techniques to give the AGV path location information along the route. The chemical guide path is an example of this type of path control. Other techniques used in specialized applications include *optical triangulation, dead reckoning, laser guidance, inertial guidance, position reference beacons, and ultrasonic imaging.*

Routing.   The AGV must be capable of changing the route to a destination based on current conditions and needs in manufacturing. Study the simple AGV guide path layout in Figure 11–33. If an AGV is at *stop 1* and must proceed to *stop 4* over the shortest distance, the vehicle must pass two *decision points* at A and B. In addition, the vehicle must past through one location, called a *convergence*, where two guide paths merge into a single path. AGV routing uses two techniques, *frequency select* and *path switch select*, to direct the vehicle to the correct path at a decision point.

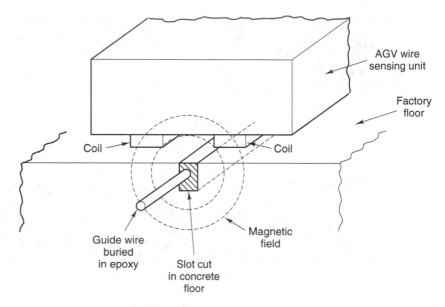

Figure 11–32   Operation of a Buried-Wire AGV System.

An AGV is alerted that a decision point has been reached by a passive marker in the floor. The marker is usually an embedded permanent magnet, metal plate, or some other coding device. In the frequency select method, the current in each guide wire leading away from a decision point is changing at a different alternating-current (ac) frequency. The AGV is programmed to follow the guide wire with the frequency that will cause the vehicle to reach the next stop over the shortest path. Study Figure 11-33 and note that, in going from stop 1 to stop 4, the AGV would choose frequency 2 at *decision* point A and continue to use that frequency at the second decision point. At the *convergence* point, the AGV would again track the guide wire with frequency 1.

The second routing method, *path switch select,* indicates the correct path for the AGV by switching the current *off* for the incorrect direction at the decision point. The AGV always follows the path at a decision point where the guide-wire current is *on*. For example, as the AGV approaches decision point A in Figure 11–33, the ac guide-wire current for the straight route would be switched off and the current in the guide wire heading toward decision point B would remain on. At the decision point, the AGV would sense the path with current and turn right toward decision point B.

Traffic Management.   The system must maximize material flow through the production system while minimizing interference and collisions with other AGV vehicles. Three traffic management techniques are generally used: *zone control, forward sensing,* and *combination control.* Zone control is the most frequently used traffic management technique. Study the path in Figure 11–33 until you have identified all

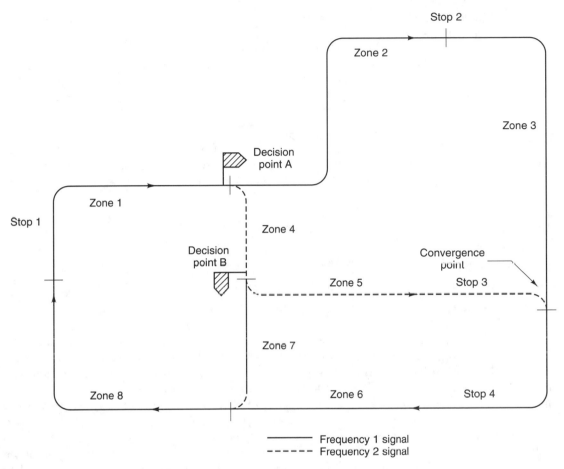

Figure 11–33 Typical AGV Path.

eight zones. When a vehicle occupies a zone, a trailing AGV cannot enter the zone; therefore, AGVs are separated by a minimum of one zone to prevent collisions. Three methods are used to implement zone control: *distributed zone control, central zone control,* and *on-board zone control.* In the first method, distributed zone control, *zone control boxes* sense the presence of an AGV in a zone and activate a *zone hold beacon* to stop entry into the occupied zone. The second technique, central zone control, achieves the same results but performs the control from one central control box. In the last method, on-board zone control, each AGV has the intelligence to recognize the zone it currently occupies and transmit that information to all the other AGVs on the guide wire. The first method is cost-effective on small systems and the second is used on larger systems. On-board zone control is becoming popular for any size system because it eliminates the interface wiring between the zones.

In addition to zone control, AGVs use *forward sensing* to perform traffic management. In this technique, each AGV has on-board sensors and electronics to detect that another AGV is on the guide wire in front of an approaching AGV. Three types of sensors are used to detect the vehicle in front: (1) *sonic*, which works like radar; (2) *optical*, which uses reflected infrared light; and (3) *bumper*, which uses contact. Forward sensing works well when the AGV system has several straight-line paths.

As the name indicates, the *combination control* is a combination of the other two types of control systems. Forward sensing is used for long straight sections of the path, and zone control is used for the curved sections.

Load Transfer.   The delivery and removal of material and parts from a terminal point must be performed without disturbance to the other production systems. The transfer of the load to and from the AGV is performed using one of five methods:

- Manual labor.
- Automatic coupling and uncoupling.
- Power rollers and belts.
- Horizontal power lifts.
- External power push and pull.

The type of method used is a function of the type of AGV in use.

System Management.   The operation of the AGVs throughout the site must be managed by the AGV system in a cost-effective manner. The system management function has two components: vehicle dispatch techniques and system monitoring methods. The dispatch techniques include *on-board dispatch, off-board call system, remote terminal, central computer,* and *combination.* The on-board dispatch system uses an operator interface panel on the AGV to input destination information. The off-board call system uses an operator interface external to the AGV to direct the vehicle to path locations. The remote terminal is an extension of the off-board control with the addition of a monitor screen or AGV locator panel to display current vehicle locations. Computer control, the highest level of AGV control, uses a central computer to control the movement of all the AGVs in the system. The last type, combination, uses a combination of control techniques in the same system environment.

System monitoring focuses on visual displays of current AGV positions and the manufacturing environment surrounding the AGV movement. The design of production systems that use an automatic guided vehicle requires consideration of each of these functions during the design of the AGV material-handling system.

Justification of AGV Systems

The primary advantages of using AGV systems include a net reduction in direct labor, floor space requirements, maintenance, peripheral material-handling equipment, and product damage resulting from material handling. In addition, AGVs also provide better control of material flow and inventory, higher throughput, and improvements in safety records.

The disadvantages, similar for any new technology, include the complexity of the hardware and software, which places new demands on training and increased skill levels for operators and maintenance personnel. Despite the problems associated with the adoption of high-technology equipment, the implementation of AGV systems is easily justified when specific conditions exist in manufacturing.

The justification of an AGV investment is based on current and future material-handling requirements. The presence of five or more of the following conditions, adapted from Miller's (1987) text on AGVs, warrants consideration of AGV technology.

- There is lost or late delivery of material and parts to work centers more than 5 percent of the time using manual material delivery systems.
- Ten or more pickup and delivery locations exist in the facility.
- Total material movement activity at all work centers falls between 35 and 200 loads per hour.
- Three or four forklifts are currently required on a minimum of two shifts.
- Over 300 feet of roller or power-and-free conveyor is currently in use.
- The production activity control or production control software requires on line real-time material tracking.
- The level of in-process inventory on the shop floor makes increased productivity and throughput impossible.
- Automation introduced in the work centers or in parts and material storage requires an automated material movement system.
- Damage during material handling using manual systems creates significant quality problems.

The use of current AGV technology in CIM implementations will continue to increase. However, as computer technology expands and intelligent AGV systems permit autonomous navigation without guide wires, applications for driverless material-handling systems will expand in all aspects of information, material, and parts distribution.

11–4 AUTOMATED STORAGE AND RETRIEVAL

In a manufacturing facility, the raw material is in one of two places: storage or production. When the raw material is in production, four substates are possible: in transit, waiting in a queue, being processed, or being inspected. Unsold finished products are categorized as finished goods inventory. The storage of raw material, the holding of production parts in a queue, and the warehousing of finished goods all add cost to the end product. In addition to these three major inventory items, several other types of inventory are normally carried in a manufacturing facility: (1) purchased parts and subassemblies used in the assembly of products, (2) rework and scrap that result from production operations, (3) tooling used in production, (4) spare parts for repair of production machines and the facility, and

(5) general office supplies. The elimination of all inventory is neither possible nor in many cases desirable; however, the application of modern storage technology minimizes the added costs associated with this necessary inventory.

The term used to describe modern mass storage technology, *automatic storage and retrieval systems* (AS/RS), is defined by the Material Handling Institute as follows:

> *AS/RS is a combination of equipment and controls that handles, stores, and retrieves materials with precision, accuracy, and speed under a defined degree of automation.*

Systems that satisfy this definition are frequently used in CIM systems to store and retrieve raw materials, production tools and fixtures, purchased parts and subassemblies, in-process inventory, and finished products. The systems are often unique because every manufacturer has distinctive production system require- ments and products. The AS/RS has four basic components: *a storage structure*, a *storage and retrieval machine, unit modules,* and *transfer stations.* Each of these com- ponents is described in the following sections.

AS/RS Components

The primary function of the AS/RS structure is to support the material placed in the storage system. In many cases a secondary function, roof and building sup- port, is present when the AS/RS structure is integrated into the structural system for the storage or manufacturing building. An AS/RS furnishing structural sup- port for the building roof and providing a large storage capability is pictured in Figure 11–34. An AS/RS consists of a series of storage aisles running from the floor to the ceiling. Each aisle separates large storage walls with numerous com- partments or bins to hold the inventory items listed in the preceding section. The AS/RS is often a substantial structure because the full weight of a fully loaded system plus roof loading must be supported safely.

Material is stored and retrieved from the AS/RS compartments by a *storage/ retrieval* (S/R) machine or crane that moves horizontally and vertically on guide rails in the aisles between the storage walls. The S/R machine performs two oper- ations: (1) the vertical and horizontal movement permits the carriage on the S/R machine to reach any compartment for delivery or pickup of material, and (2) mechanisms inside the carriage pull material from the storage compartment into the carriage or push material from the carriage into the storage bin. Also, the S/R machine must maintain a high degree of position accuracy between the car- riage and the storage compartment over the entire range of travel from one end of the AS/RS to the other. The carriage in some systems travels at horizontal and ver- tical speeds of 500 and 100 feet per second, respectively.

The material moved into and out of the storage compartments is placed on *storage modules.* The storage modules have a standard base size so that storage in any compartment is possible. A unit load system is shown in Figure 11–35. In ad- dition, the modules are usually held in the compartments by interlocking tracks or rails. The storage modules are usually configured using one of the following: pal- lets, wire baskets, tote pans, or special drawers. The carriage moves the storage

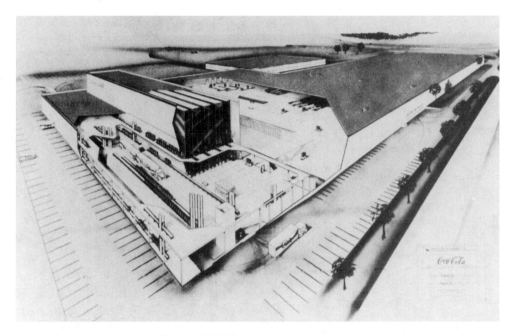

Figure 11–34 Integrated AGV and AS/RS System
Source: Courtesy Eaton-Kenway, Inc., Salt Lake City, UT.

modules between a *single delivery point,* called the *pickup and deposit station* (P&D), and the storage compartment in the AS/RS.

The pickup and deposit (P&D) stations are located at one or both ends of the AS/RS and aligned with the S/R machine traveling in the aisle (Figure 11–35). The material-handling capability present in the P&D station must be compatible with the S/R machine and the material-handling system used to bring storage modules to the AS/RS. Load transfer to the P&D stations from manufacturing is performed using manual load/unload, forklift vehicles, gravity and power conveyors, and AGVs.

Types of AS/R Systems

Automatic storage and retrieval systems are designed in various configurations. For example, some are totally automated and operate on a unit load transfer basis; others, called *human-on-board AS/RS,* have a human riding in the carriage to facilitate the picking of specific parts from the compartments. Other systems, called *miniload* and *automatic item retrieval AS/RS,* are designed to permit single-item retrieval from the AS/RS. In the miniload type, for example, a bin is retrieved from the storage location in the AS/RS and delivered to the P&D area, an operator takes out the necessary parts, and the bin is returned to storage. The second type, automatic item retrieval, permits individual items to be retrieved from a storage location without the need for a human operator.

Figure 11–35 AS/RS System.
Source: Courtesy Eaton-Kenway, Inc., Salt Lake City, UT.

Regardless of the type of configuration, the performance of any AS/RS is evaluated on four criteria: *storage capacity, material throughput, percentage of time used,* and *reliability.*

Carousel Storage Systems

Carousel storage systems offer a lower cost alternative in applications requiring a miniload type of AS/RS. Operation of the carousel system is exactly opposite that of the AS/RS. In the carousel system, the storage and retrieval (S/R) machine is fixed in place at the pickup and deposit (P&D) station, and the storage compartments rotate on a track that passes by the P&D station. A typical carousel storage system is illustrated in Figure 11–36. Carousel systems are frequently used in the following types of applications:

- *Small parts assembly.* Storage of a large variety of small parts used in mechanical, electrical, and electronic assembly is a typical application for a carousel system. A parts kit containing all the components required for an assembly is built by selecting parts from different bins in the carousel system.
- *Assembly transport.* In this application, the manual and automated assembly stations are located around the periphery of the carousel. At each assembly station, the partially completed assembly and parts for the product are removed from the carousel. The parts are added to the assembly and the partially finished product is placed back into the carousel. At

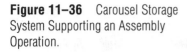

Figure 11–36 Carousel Storage System Supporting an Assembly Operation.

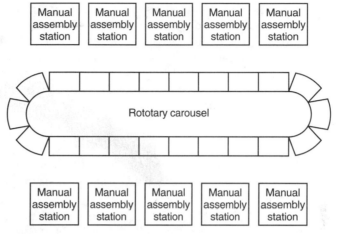

the next assembly station, the partially finished assembly is retrieved along with the additional components that must be added. At the end of the process, the assembly is complete.

In world-class enterprises, the level of inventory in storage is reduced to as low a level as possible. However, when large levels of inventory are necessary to support make-to-stock production operations, it is important to minimize the added cost to the product. Automatic storage and retrieval is one technique used to minimize the cost-added nature of inventory.

11–5 SUMMARY

Industrial robots are part of most automated production systems, where they perform specialized material handling or some production process. The basic robot system includes a robot arm in one of four geometries, production end-of-arm tooling, controller electronics, and a teach mechanism. The numerous robot applications are grouped into three categories: material processing, material handling, and assembly and fabrication.

Material-handling automation is a necessity for an automated CIM facility. The many mechanisms used to automate material movement are grouped into three categories: continuous transfer, intermittent transfer, and asynchronous transfer. The automatic guided vehicle (AGV) is a special-purpose material movement system. The types of AGVs in common use include towing, unit load, pallet truck, fork truck, light load, and assembly line. Some systems require no human operator intervention; others use operators to maximize production efficiency. AGV systems perform five functions: guidance, routing, traffic management, load transfer, and system management.

Automatic storage and retrieval systems (AS/RS) automate the placement and removal of the necessary inventory in a manufacturing facility. The AS/RS has four basic components: a storage structure, a storage and retrieval machine, unit modules, and transfer stations. The types of AS/R systems include human onboard, miniload, and automatic item retrieval. Another type of AS/RS, called carousel storage, provides a lower cost alternative to the storage and retrieval of parts, especially small parts assembly and applications where a partially assembled product must be transported through assembly.

REFERENCES

GOETSCH, D. L., *Modern Manufacturing Processes.* New York: Delmar Publishers, 1991.

GRAHAM, G. A., *Automation Encyclopedia.* Dearborn, MI: Society of Manufacturing Engineers, 1988.

GROOVER, M. P., *Automation, Production Systems, and Computer-Integrated Manufacturing.* Upper Saddle River, NJ: Prentice Hall, 1987.

LUGGEN, W. W., *Flexible Manufacturing Cells and Systems.* Upper Saddle River, NJ: Prentice Hall, 1991.

MATERIAL HANDLING INSTITUTE, *Consideration for Planning and Installing an Automated Storage/Retrieval System.* Pittsburgh, PA: MHI, 1977.

MILLER, R. K., *Automated Guided Vehicles and Automated Manufacturing.* Dearborn, MI: Society of Manufacturing Engineers, 1987.

REHG, J. A., *Introduction to Robotics in CIM Systems.* Upper Saddle River, NJ: Prentice Hall, 1992.

SOBCZAK, T. V., *A Glossary of Terms for Computer Integrated Manufacturing.* Dearborn, MI: CASA of SME, 1984.

WIERSMA, C. H., *Material Handling and Storage Systems: Planning to Implementation.* Mansfield, OH: Self-published, 1984.

QUESTIONS

1. What distinguishes a robot from other types of automation?
2. Draw a basic robot system diagram.
3. Describe the four basic arm geometries used for robots.
4. Compare and contrast position movement and orientation movement.
5. Describe the two types of cartesian geometries.
6. Describe the two types of articulated geometries.
7. What are the terms used to describe the production tooling used on robots?
8. Describe the three classifications used for robot tooling.
9. Describe the basic components in a robot controller.
10. Describe the three functions served by robot teach stations.
11. Describe the three categories of robot applications and give industrial examples of each.

12. Describe the selection and justification process used to select a possible robot implementation.
13. How does fixed high-volume automation differ from flexible automation?
14. What are the characteristics of fixed in-line automation?
15. Compare rotary automation to in-line automation.
16. Describe the selection process used for fixed automation systems.
17. Why is automated material handling a critical technology for a company that wishes to become a world-class manufacturer?
18. What are the two basic functions of material and part transfer mechanisms?
19. Describe the three types of transfers.
20. What are the three types of mechanisms used to achieve continuous motion transfer?
21. Describe a walking beam transfer system.
22. What is the definition of an automatic guided vehicle?
23. Describe the five basic types of AGVs.
24. Describe the five functions that AGV systems must perform.
25. What is off-wire AGV technology?
26. Compare switching and signal frequency routing.
27. Describe how wire-guided AGVs operate.
28. Describe the three traffic management techniques used for AGVs.
29. What are the five methods used to transfer the load between the AGV and a fixed site?
30. Why is system management of AGVs more difficult than the system management of robots?
31. What are the basic elements present in most AS/R systems?
32. Why is the study of AS/R systems important if inventory is an unwanted quantity?
33. What are the four metrics used to evaluate the effectiveness of an AS/RS?
34. Describe the different types of AS/R systems.

PROJECTS

1. Using the list of companies developed in Project 1 in Chapter 1, determine which companies are using robotics, fixed automation systems, automatic guided vehicles, and automatic storage and retrieval systems.
2. Select one of the companies in Project 1 that uses robots and classify the systems by arm geometry, control techniques, and type of tooling used.
3. Select one of the companies in Project 1 that uses AGVs and describe the types, guidance, route control, traffic control, and management used in the system.

CASE STUDY: AGV APPLICATIONS AT GENERAL MOTORS

The number of automatic guided vehicles (AGVs) used worldwide exceeds 13,000 in a wide variety of machine types and applications. AGV installations span every conceivable production industry and include several service industries. Automotive manufacturing worldwide is a major user of this technology. The following case information is drawn from experiences in AGV applications at General Motors and is adapted from case studies in *Automated Guided Vehicles and Automated Manufacturing* by R. K. Miller (see the References).

Oldsmobile Division

The General Motors Oldsmobile Division in Lansing, Michigan, installed the first AGV system for automotive assembly. The 185-vehicle AGV system had over 10,000 feet of guide wire and used Eaton-Kenway unit load carriers in several areas. The engine dress system area used 95 AGVs on 7500 feet of guide wire, the chassis system used 65 AGVs in two areas, and the engine stuff system used 25 AGVs in two areas. The AGVs supported the production of front-wheel-drive Oldsmobiles, Buicks, and Pontiacs by carrying engines through assembly until the engines were merged with the car bodies.

The AGVs offered the following advantages over previous assembly methods:

- The AGV carriers permit each engine to remain stationary at a workstation until the assembler is sure that the assembly operation was performed correctly. Therefore, engines released to final assembly had fewer quality problems.
- Engines held in the stationary position during assembly support the application of automated assembly technology, like robots.
- Assembly efficiency was enhanced because the AGV permitted the height of the engine from the floor to be adjusted to the needs of each assembler and assembly operation.

Assembly Plant

The General Motors Assembly plant at Orion, Michigan, used a 22-vehicle AGV system to move 70 percent of the incoming stock on a just-in-time basis to 69 drop zones in two departments. The system used 24,000 feet of guidepath and Conco-Tellus vehicles with various containers from 30 to 54 inches wide. The sophisticated vehicles had automatic alignment capability for pickup and drop-off of material using photo cells in the ends of the lift forks. In addition, the AGVs could go off-wire under some conditions using an FM radio communications system of data exchange with a central computer controller. All the vehicles were "smart devices" because they carried on-board microcomputers. Recharging of AGV batteries was automatic. When the battery life dropped below 20 percent, the AGV took itself out of service and reported to a recharge station for a full charge that would provide another 16 hours of continuous service.

Machine and System Control

The design and analysis systems and software introduced in this book, and the production control and process technology discussed earlier, must operate in harmony using a single product database in the CIM enterprise. The control mechanisms for this diverse set of technologies are spread across the entire enterprise and are organized in a hierarchical structure. On one extreme, this hierarchical structure links simple mechanical limit switches that sense the presence of parts on a conveyor for control of a machine. At another point in the hierarchy, mainframe computers are interfaced to external communication networks, like the Internet, to pass production orders electronically to computers at vendor locations around the world. Developing and implementing this complex mix of hardware and software technologies is an evolutionary process. Review the automation time line illustrated in Figure 1–1. Note on the time line that computer control in each of the three major functions described by the SME CIM wheel started many years ago. However, only in the last twenty years has the control of the individual elements matured sufficiently to allow integration of the many diverse enterprise functions. The technologies used to integrate the CIM enterprise and to establish the hierarchical control are described in this chapter.

12–1 SYSTEM OVERVIEW

A convenient starting point for the description of the control hierarchy used by the CIM enterprise is a review of two figures used in previous chapters and illustrated again in Figures 12–1 and 12–2. Study these drawings and see if you can find the interface that links the two figures.

Figure 12–1 shows typical enterprise functions linked by local area (computer) networks (LANs). Remember, a LAN is a nonpublic communications system that allows devices connected to the network to exchange data and information electronically over large distances. Each of the area LANs is bridged to the backbone so that everyone has equal access to the central database and links to external resources and systems.

The system illustrated in Figure 12–2 is a more detailed description of the level 1 elements in Figure 12–1. The interface between the two illustrations occurs at the top

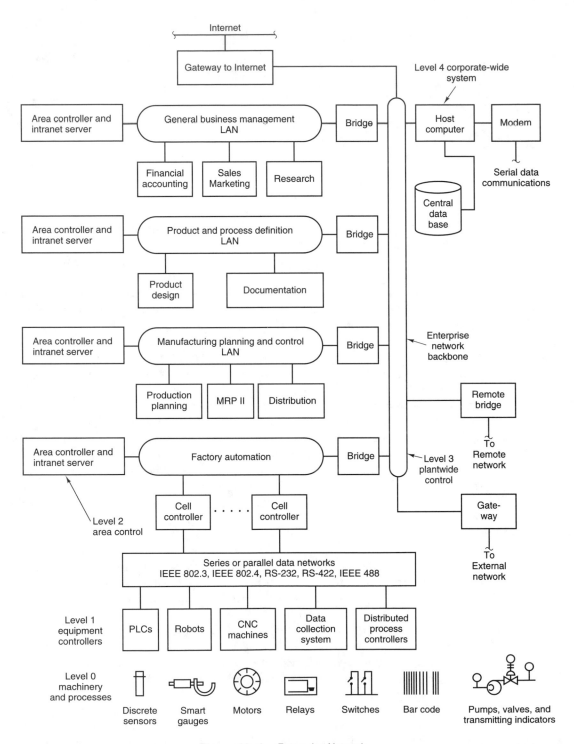

Figure 12–1 Enterprise Network.

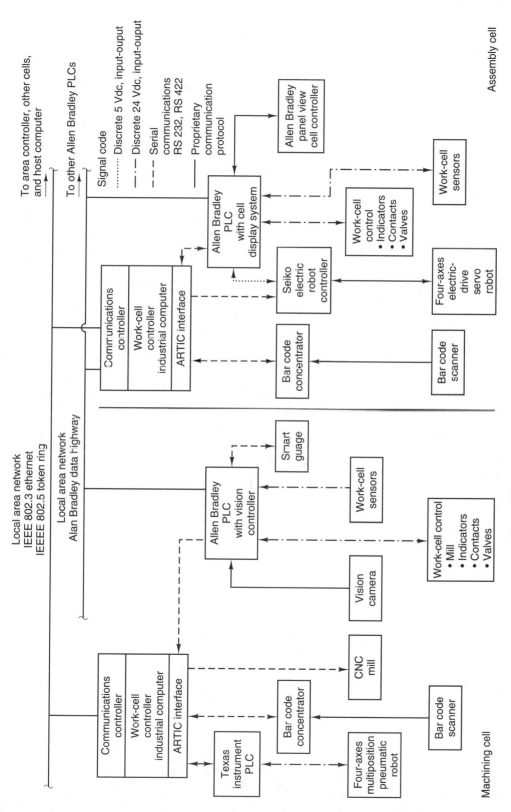

Figure 12-2 Level 1 Machine and Cell Controller.

of Figure 12–2, where the network has the reference. "To area computer, other cells, and host computer." Therefore, the *local area network* drawn at the top of Figure 12–2 is part of the *factory automation LAN* connected to the level 2 area controller in Figure 12–1. Note also that the work cell controllers are present in both figures.

Some key characteristics differentiate the cell control from the higher levels in this typical hierarchy. Figure 12–3 illustrates these differences in five different categories: controller hardening, response time, control versus information processing, number of users, and communications interface. The hardening column indicates the percentage of applications where a system must be able to operate in a harsh environment in which extreme changes in temperature, shock and vibration, power fluctuations, airborne contaminants, and moisture exist. The response time indicates how fast the system must be capable of generating an output based on input conditions. The control versus information processing addresses the percentage of control problems versus the percentage of information processing problems. The last two parameters indicate the typical number of users and the type of communication system. The system needs at the cell and device levels stress very fast response times with a heavy emphasis on the control of a small number of devices. At this low level, networks are replaced by point-to-point communications.

The control hierarchy illustrated in Figure 12–2 establishes the interface between production machines, production machine controllers, and the enterprise computer network. The cell controller links computer-controlled machines in the cell with the other enterprise departments. A good starting point for a description of CIM hierarchical control is at the lowest level, the production cell.

12–2 CELL CONTROL

The hardware in the work cells falls into two general categories: intelligent and nonintelligent. Like the programmable logic controllers (PLCs) in Figure 12–2, intelligent equipment has some type of internal computer control. In some applications,

Enterprise level	Harden applications	Response time	Control versus information	Number of users	Communication system
Plant host	0–3%	Days to seconds	0–5% 95–100%	Hundreds	LAN
Area controllers	8–12%	Minutes to seconds	5–10% 90–95%	Tens	LAN
Cell controllers	80–90%	Seconds to milliseconds	10–20% 80–90%	1–12	LAN and point-to-point
Device controllers	90–95%	Milliseconds to microseconds	90–95% 5–10%	1–2	Point-to-point

Figure 12–3   System Characteristics at Various Enterprise Levels.

intelligent machines are called "smart devices." Like the sensors and bar code readers in Figure 12–2, nonintelligent hardware often uses solid-state electronics for signal conditioning and conversion but it is not reprogrammable. Automated cells utilize an infinite variety of intelligent and nonintelligent hardware, control software, and every conceivable process machine. While production cells are rarely the same, the concepts learned from a study of the two cells in Figure 12–2 will transfer to the operation of other production cells. Therefore, an analysis of the operation of these two cells is a convenient starting point for the study of cell control techniques.

Cell Controllers

The cell controllers in the machining and assembly cells in Figure 12–2 are usually either industrial computers using Intel microprocessor chips or minicomputers with proprietary processor chips. For example, the IBM computer in Figure 12–4 is an industrial computer designed for operation in the harsh production environment. In addition, many vendors provide Intel and RISC chip based computers for manufacturing. These boxes run either the Microsoft NT/2000 operating system or some version of UNIX. The system configuration for cell controllers includes internal hard drive and floppy drive for data and program storage, internal memory (RAM), keyboards, pointing devices, and monitors.

Figure 12–4   IBM Industrial Computer.

Cell Control Software Structure

The operating system software used for cell controllers in most applications is either OS/2, Windows NT/2000, or UNIX. Operating systems such as OS/2, Windows NT/2000, and UNIX that run multiple application programs are called multitasking software. A study of the machining cell in Figure 12–2 provides a good example of the need for this feature. Note that four intelligent or smart machines (TI PLC, bar code concentrator, AB PLC, and CNC machining center) are connected to the cell controller. A machine operator must regularly use the cell controller's direct numerical control (DNC) software to download a new part program to the CNC milling machine. As this DNC function is occurring, the following programs continue to execute in the background: production data from the PLC data registers are uploaded to production control, and bar code data are supplied to automated product tracking software. Work-cell controller applications that demand this level of flexibility require multitasking operating systems.

The cell controller interfaces with other intelligent machines and devices using standard serial data interfaces and networks. The assembly cell's controller in Figure 12–2 has network connectivity to both the local area network (LAN) and the proprietary programmable controller network. In addition, both cell controllers have several serial data interfaces to other smart devices in the cells, such as programmable logic controllers, bar code concentrators, CNC mill, and a robot controller. The serial data interface provides point-to-point communications for high-speed dedicated data transfer required at this level in the communication structure. A typical software configuration in the cell controller to manage these interfaces is illustrated in Figure 12–5. Study the figure until you are familiar with the descriptions in each box.

Figure 12–5   Work-cell Controller Software Layers.

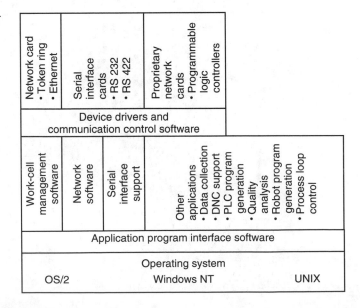

The operating system (OS) is usually one of the three listed at the bottom of the figure. Application program interface (API) and system enabler (SE) software reside between the OS and the applications. The API and SE, like IBM's Distributed Automation Edition (DAE), handle some of the operational overhead common to all applications; in addition, an SE helps manage the resources, like hard drive data storage, across all the computers on a network. For example, text and graphics dumps to the monitor are handled by an API program. Therefore, developers spend less time programming for the operating system and more time on the specific application. The applications can include a wide variety of software, like those listed in Figure 12–5.

The cell controller frequently has various interface cards to support network standards such as Ethernet and token ring. Review the interfaces listed in Figure 12–5. Device drivers and communications control software reside between the interface cards and the application software. In Figure 12–2, for example, the cell controller must have a driver program that satisfies the protocol specifications for the CNC machine before a program to machine a part can be transferred to the mill over the serial data interface. Many vendors supply serial interface cards for use in the controllers in Figure 12–2. These serial data interface cards provide a combination of serial standards such as RS-232 and RS-422. A unique feature of these cards is the presence of a programmable microprocessor on the card that allows the interface to manage data exchange on multiple ports independent of the processor in the main computer.

Work-Cell Management Software

The major application software present in cell controllers is the software used to manage all of the cell activity. The primary role of the cell controller is communications and information processing. For upstream LAN communications, the cell controller is a concentrator of information from the cell and a link to area controllers and other applications, such as MRP II and CAD product design. On the downstream side, the cell controller is a distributor of data and information files to programmable control devices and production equipment. This communications workload, transferring data files ranging from single-bit to megabyte files, often uses over 50 percent of the resources of the processor in the cell controller. Data management from real-time memory resident file storage to support for the CIM relational databases is another cell control function. In addition, management of program libraries for devices in the cell, support for engineering change control, and production tracking represent other cell control responsibilities. The information and communication tasks frequently consume 80 to 90 percent of the resources of the cell controller.

Applications typically loaded into the cell controller include:

- Production monitoring.
- Process monitoring.
- Equipment monitoring.

- Program distribution.
- Alert and alarm management.
- Statistical quality and statistical process control.
- Data and event logging.
- Work dispatching and scheduling.
- Tool tracking and control.
- Inventory tracking and management.
- Report generation on cell activity.
- Problem determination.
- Operator support.
- Off-line programming and system checkout.

12–3 PROPRIETARY VERSUS OPEN SYSTEM INTERCONNECT SOFTWARE

Software to manage the information flow through the cell controller falls into one of three categories: (1) in-house-developed systems, (2) application enablers, and (3) adoption of an open system interconnect (OSI) such as the *manufacturing message specification* (MMS) ISO standard 9506. The first two are proprietary software systems because they are either written for a specific application or developed around a third-party software management shell, called an *application enabler.* The last technique uses an OSI standard such as MMS to develop solutions. A more detailed description of each of these approaches in the following sections illustrates the advantages and disadvantages of each approach.

In-House-Developed Software

Before 1980, most cell control and management software was written by the user using the C programming language. In-house development was necessary because satisfactory third-party software options were not available. The cost of these programming projects was substantial because studies indicate that software development in C cost over $50 per line of code. In addition, the complex and cryptic nature of the code drove up the costs of modifications and updates to existing programs.

The in-house-developed programs were written to address the specific control needs of the cell, and the software interfaced with other in-house-developed software in other enterprise areas. In addition, the lack of available interface software or drivers for equipment in the cell was not a problem because all the code was generated as part of the initial program development. However, these advantages were offset by a major disadvantage in cost and the inability to change the software easily when the cell hardware or configuration changed.

Enabler Software

A common impediment to the development of a fully implemented CIM work cell is the exorbitant cost, time, complexity, and inflexibility of custom-programmed CIM solutions. However, with the introduction of *enabler software,* application development has shifted from software engineers to manufacturing engineers. This shift was possible because enabler software provided a set of software productivity tools to develop control software for the CIM cells. Products such as *Plantworks* from IBM, *Industrial Precision Tool Kit* from Hewlett-Packard, *CELLworks* from FASTech, FIX *DMACS* from Intellution, *Factory Link* from U.S. Data, and In-Touch from Wonderware are available to reduce the difficulty in developing cell control and management applications.

Enablers such as Wonderware and FIX DMACS have a library of driver programs for many commonly used production machines and machine controllers. In addition, they offer LAN and serial data communications support, mathematics and logic functions for internal computations, links to commonly used mainframe and microcomputer relational databases, real-time data logging, real-time generation of graphics and animation of cell processes, alarm and event supervision, statistical process control, batch recipe functions, timed events and intervals, counting functions, and the ability to write custom applications. The screen in Figure 12–6

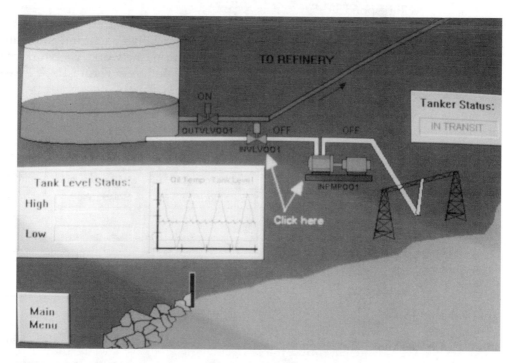

Figure 12–6   Screen from FIX DMACS Showing a Cell Enabler Software.

shows how an enabler dynamically displays a process. The graphic of the process is built using graphic drawing tools supplied with the enabler or from a third-party vendor.

The advantage of enablers is an estimated tenfold improvement in cell control and ease of program development. The primary disadvantage is that the cell control is tied to a third-party software solution, so selecting the best solution is critical. In addition, an enabler is written to satisfy the needs of the average company; therefore, if a specific requirement is not covered, additional Visual Basic or C code programming is necessary to obtain a solution.

OSI Solution

The most frequently used open system interconnect (OSI) solution for cell control is the *manufacturing message specification* (MMS). MMS is a standard (ISO 9506) for communicating information between intelligent devices in a production environment over networks based on the OSI model. The ISO 9506 standard has three parts: (1) service specifications, (2) protocol specifications, and (3) robot interface and protocol specifications. Specifications for other classes of production machines are under development. The MMS standard defines a set of objects that exist within a device; for example, the MMS object could be the location of one axis of the robot's arm. In addition, the MMS defines a set of communication services to access and manipulate the objects and then describes how the devices will respond.

Implementing an MMS solution requires that all the devices support the MMS protocol and that they are linked over a manufacturing automation protocol (MAP) Ethernet or broadband manufacturing network. The work-cell system in Figure 12–2 is reconfigured as an MMS solution in Figure 12–7. Take a few minutes to compare the two implementations.

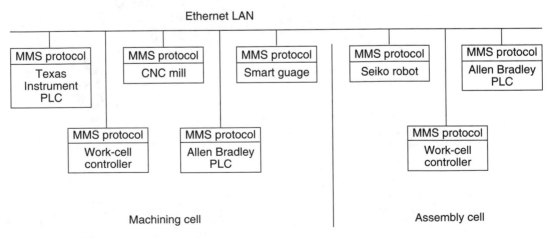

Figure 12–7   MMS Implementation.

Note in Figure 12–7 that all of the smart devices that have MMS support are on the Ethernet network. Also, there is no longer a need for the proprietary PLC network because the PLCs share data directly with each other and the other smart devices over the cell LAN. Any device not on the network would be interfaced to one of the smart devices on the LAN. With the MMS implementation, the work-cell controller no longer serves as the data concentrator because each smart device talks directly with other devices on the LAN. If the area controller in Figure 12–2, for example, needed the value of a memory register in the Allen Bradley PLC, the request would first go the cell controller and then to the PLC. The register information would return by the reverse route, first to the cell controller and then up to the area controller. In the MMS solution, however, the area controller request would go directly to the PLC and the data would be returned directly to the area controller. Each smart device provides real-time data in the MMS configuration without the need for work-cell management software to collect cell information. Many computer vendors and most of the frequently used programmable logic controllers have MMS support available as an option.

The open system interconnect feature of MMS is its primary advantage. Management of the cell is distributed across the network linking all the intelligent devices. Information is requested directly from the target device and delivered directly to the end user with no intermediate computer and software system required. The primary disadvantage is the small current user base and the limited number of equipment vendors who have agreed to support the standard.

12–4 DEVICE CONTROL

Device controllers fall into two categories, *proprietary* and *generic*. Proprietary controllers are generally special-purpose microcomputers built around a common microprocessor chip and programmed to control the target device. For example, the CNC machine tools described in Chapter 10 all have special-purpose computers built into the machine tool for control. In another example, the robot system in Figure 11–17 has a special-purpose controller to move the arm under program control. However, the robot computer controller is located in a remote enclosure and connected to the robot through an electrical interface.

The generic device controllers are *general-purpose* devices designed to interface to different work-cell hardware to provide control. Both proprietary and generic device controllers support *discrete* and *analog* control requirements. Discrete control is used, for example, to turn a motor on or off. Therefore, discrete control implies just two operational states for the device under control.

Analog controllers allow devices to operate over the range from on to off. Analog controllers used in production machines and cells generally control motion or position of an object or the value of a process parameter such as pressure, temperature, level, or flow. For example, an analog temperature controller could hold the temperature of liquid in a tank at any temperature between 20° and 90° C. To achieve this type of control, the energy supplied to the heating element by the controller is set at a level that produces the correct temperature. In contrast, the discrete controller would either turn the heating element on (heat transferred to the liquid) or off (no heat transferred).

Describing all the special-purpose controllers used in manufacturing is difficult in an overview. However, two types of device controllers, *programmable logic controllers* (PLCs) and *computer numerical controllers* (CNCs), are used frequently in production automation, so understanding their operation is important.

12–5 PROGRAMMABLE LOGIC CONTROLLERS

The *programmable logic controller* (PLC) was developed in the early 1970s for General Motors. The early PLCs were special-purpose industrial computers designed to eliminate relay logic from sequential control applications. A simple example of sequential logic control is a good place to start a description of PLCs.

Basic Sequential Logic Control

The simple application illustrated in Figure 12–8 shows the value of PLCs in automation. The overly simplified circuit is designed to control a pump that removes liquid from a tank and a valve that allows liquid to flow into the tank. The following logic is required: (1) the pump can operate only when the input valve to the tank is open, and (2) the input valve can be opened when the pump is either operating or not operating. Study the electrical circuit and verify that the required logic is wired into the circuit.

The circuit works as follows: (1) switch S1 is closed manually and causes the magnetic relay A to be energized; (2) when the relay is energized, the poles on contacts A1 and A2 move from the normally closed (NC) positions to the normally open (NO) positions; (3) the change in contact A1 energizes the input valve and

Figure 12–8 Relay Logic Control Circuit.

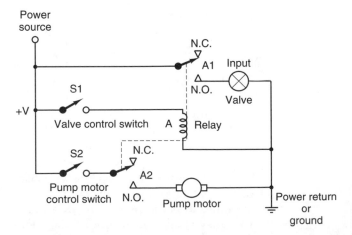

allows liquid to flow into the tank; (4) the change in contact A2 causes no immediate action; and (5) switch S2 is closed manually and causes the pump to operate. Note that the change in relay contact A2 in step 4 to the NO position was necessary for pump operation. If the valve switch S1 is opened manually while the pump is operating, the pump motor stops. It stops because the relay is deenergized and both contacts move back to the NC position. The change in contact A2 opens up the circuit that is energizing the pump, and it stops. Review this section and study Figure 12–8 until operation of this circuit is clear.

Two major problems occur with relay logic circuits. First, relays are mechanical devices, and repeated cycling between the on and off positions causes the contacts to fail. Second, a change in the operational logic or system specifications usually requires major rewiring work on the system. For example, assume that a second pump is necessary but must operate from a second pump switch so that either one or both of the pumps can be operated. This simple change requires a new relay with three sets of contacts because the new pump must use contact A3 to be sure that the second pump operates only when the valve is open. (Implementing this change in a PLC is demostrated later in this chapter.) Replacing relay logic with PLCs makes sequential logic control and system modifications more efficient and less costly

PLC System Components

The *programmable logic controller* (PLC) is a computer designed for control of manufacturing processes, assembly systems, and general automation control. The mechanical configuration of PLCs includes a *rack* into which *modules* are inserted. Figure 12–9 shows the rack of an Allen Bradley PLC with modules. Figure 12–10 shows the *backplane* mounted in the back of the rack with electrical connectors where modules are inserted. The backplane contains the power and signal conductors to interconnect all the modules inserted or installed in the rack. The typical PLC module shown in Figure 12–11 has the electrical connections that make contact with the backplane and connections used to interface the PLC to production equipment. The modules generally fall into one of the following categories: interfaces to production equipment, network and serial communications, special-purpose modules, power supplies, and a microcomputer. A description of systems using each of these modules is included in the following sections.

Basic PLC System Operation

The operation of PLC components for control of automated systems is best described by the block diagram in Figure 12–12. Study the figure and note the components that are part of the PLC (inside the dashed box) and those that are external to the PLC.

The computer, called the *processor* in PLC systems, is at the heart of the PLC operation. *Input modules* accept electrical voltage levels from a wide variety of sensing devices and outputs from other systems. Review the input sources listed in Figure 12–12. The source labeled "Other output" requires some clarification. In

Figure 12–9   Allen Bradley PLC Rack and Modules.

complex automation systems, several PLCs, often manufactured by different vendors, control separate parts of the same automated production process. Frequently information sensed by one PLC must be passed to another PLC because the data affect the part of the process controlled by the second PLC. In that case, the output from one PLC is the input to a second PLC.

The processor performs arithmetic and logical operations on input data and turns outputs in the *PLC output modules* on or off based on the program resident in the processor. The output modules are wired to system components that control the process. The greatest number of PLC automation applications use just these three blocks in the PLC system diagram. Inputs are scanned, and outputs are changed based on input conditions and the logic in the PLC program in the processor.

The *PLC communications modules* are not used as frequently as input and output modules; however, communications is a critical part of every CIM implementation because production data must be available to departments across the enterprise. The *standard networks* box in Figure 12–12 indicates that the PLC can be placed directly on the factory LAN to communicate with other cell or area controllers. The standard network supported by most vendors is Ethernet, with

Figure 12–10   Allen Bradley PLC
Backplane.

protocol standards such as TCP/IP and MMS supported. In addition, proprietary
LAN software from most major vendors is available.

The *proprietary networks* box in Figure 12–12 indicates that most PLC systems
have a proprietary network available to link their PLC processors together in a
local area network. In almost every case, only PLCs from the same vendor are
compatible with the network interface and protocol.

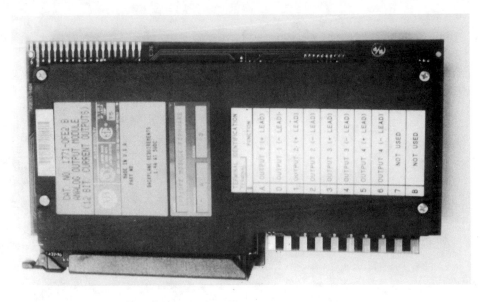

Figure 12–11 Typical PLC Module.

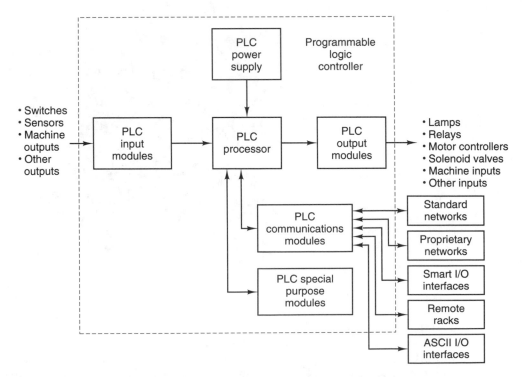

Figure 12–12 PLC System Block Diagram.

The *smart I/O interfaces* box in the figure includes PLC hardware that could revolutionize automation control. The term *smart* implies that the interface includes a microprocessor and can be programmed. Figure 12–13 describes the concept of smart input and output interfaces using an operator panel. Operator panels have switches for control of process machines and devices, and the panels have lights to indicate the condition of process equipment. The operator panels are frequently located in a control room away from the process itself. The traditional approach in building an operator panel (Figure 12–13a) requires a minimum of one wire per switch and lamp plus several return wires between the operator panel and PLC input and output modules. As a result, the wire bundle between these two devices often has hundreds of wires that must be enclosed in conduit over distances of hundreds of feet. In contrast, the smart operator interface (Figure 12–13b) uses the same number of switches and lamps but controls them with a small microcomputer located in the operator panel. The interface between the operator panel and the PLC is just a single coaxial cable that permits the operator panel's microcomputer to communicate with the PLC processor. A large number of smart external devices are available, including motor controllers, process controllers, text readout devices, programmable CRT displays supporting full color and graphics, voice input and output devices, and discrete input and output devices.

In an effort to distribute the control capability across a large automation system, the PLC vendors provide *remote rack* capability (Figure 12–12). A remote rack for the Allen Bradley PLC 5 family of programmable logic controllers is identical to that pictured in Figure 12–9. The rack uses the standard I/O modules for control of machines and processes; however, the processor module is replaced with a remote rack communications module. The remote rack is controlled by the program in the processor in the main PLC rack with communications control between the remote rack and main PLC provided over a single coaxial cable.

The last communications module, illustrated in Figure 12–12, is the ASCII I/O interface. This communication resource is either built into the processor module or comes as a separate module. In both cases, the ASCII interface permits serial data communication using several standard interfaces, such as RS-232 and RS-422. This interface is used to collect data from devices such as the smart gauge in Figure 12–2. In the same figure, the bar code concentrators could have been interfaced to the PLC instead of the cell controller if that connection were more strategic to a successful solution.

Compare the PLC system implementation in the work cells in Figure 12–2 with the PLC model in Figure 12–12 and determine how many of the application interfaces are present in Figure 12–2. The power supply, illustrated in Figure 12–12, is inserted in one of the slots in some PLCs or is a separate component in other vendors' systems. In addition, some vendors incorporate standard communications capability, such as standard and proprietary network connectivity, in the processor so that a separate module is not necessary.

Figure 12–13 Smart I/O Interfaces.

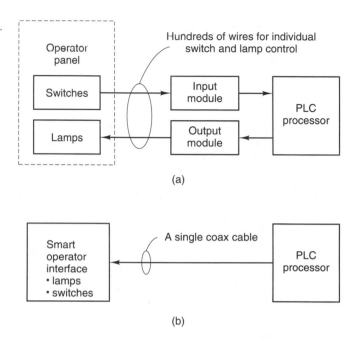

(a)

(b)

PLC Programming

The symbols used to show the logic in PLC programs has changed little since the introduction of the first programmable controller in the 1970s. The process, called *ladder logic programming,* is a variation of the *two-wire diagrams* used to document the wiring of industrial control circuits. The two-wire diagrams have a vertical line at the left (marked L1) and right (marked L2) sides of the paper. The left vertical line usually represents the positive source for power, and the right vertical line represents the power return or ground. All the devices used to control a machine, for example, are drawn between the two vertical lines. The appropriate switches, sensors, and control used to control machine operation are included in the diagram. The circuit in Figure 12–14 illustrates how the pump control circuit in Figure 12–8 would look if it were drawn as a two-wire diagram. A set of standard drawing symbols is used to represent the different input and output devices, such as mechanical switches, sensors, magnetic contactors and relays, and electrical contacts. The component with the "M" identification is a motor contactor or starter used to switch the voltage to the pump motor. Some components, like the motor overloads, are not included in the figure to keep the example simple. Compare the two drawings until all the components in Figure 12–8 are identified in the two-wire diagram. Also, verify that you understand how the logical operation described earlier using the circuit in Figure 12–8 applies to the two-wire diagram in Figure 12–14.

The PLC ladder logic program that would provide the same logical control as circuits in Figures 12–8 and 12–14 is illustrated in Figure 12–15a. The PLC

Figure 12–14 Two-wire Diagram of Pump System.

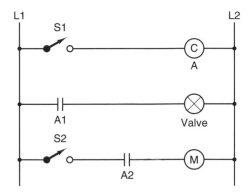

processor *scans* all the inputs and writes the ON versus OFF state of each input into a memory location. The input conditions are applied to the program logic, the appropriate output conditions are determined, and the output terminals are set either high or low accordingly. The example program in Figure 12–15a illustrates how the processor uses input conditions to change output states.

The symbol identified as I1 in ladder rung 01 represents the logical control provided by input 1 on the input module. If a voltage is present at input 1 (the valve control switch S1 is closed), then I1 closes and turns CR1 on. With no voltage at input 1, I1 is open and CR1 is off. CR1, called a control relay, is created in the software operating system in the PLC processor for logical control requirements. The control relay is implemented in the software and does not physically exist. In most PLCs the processor permits the programmers to use large numbers of control relays for logical control needs.

When CR1 is turned to the logical ON state, all the contacts labeled CR1 change state. Therefore, in the ladder logic diagram in Figure 12–15a, both CR1 contacts close. When the CR1 contact in rung 02 closes, the PLC processor causes the output module terminal 1 to switch to a voltage state that causes the solenoid valve connected to the output to energize. This action causes liquid to enter the tank.

When the pump motor control switch S2 closes, input 2 of the input module has voltage applied and the I2 contacts close. The two contacts in rung 03, CR1 and I2, are now both closed, so the 02 output is energized. With 02 active, output 2 changes from high to low, the motor starter is energized, and the pump motor is running. The pump motor cannot be turned on if the valve switch is open because the contacts CR1 would be open. Spend time studying the operation of the program and PLC interface in Figure 12–15 until you are familiar with the notation and operation. The logical operations and control performed by PLC processors include *AND, OR, INVERSION, timers, counters,* and *comparitors,* to name a few.

The advantages of PLCs over conventional relay logic include:

■ *Easy interface modifications.* Consider the change described earlier for the pump system. The PLC solution makes the change to a second pump and

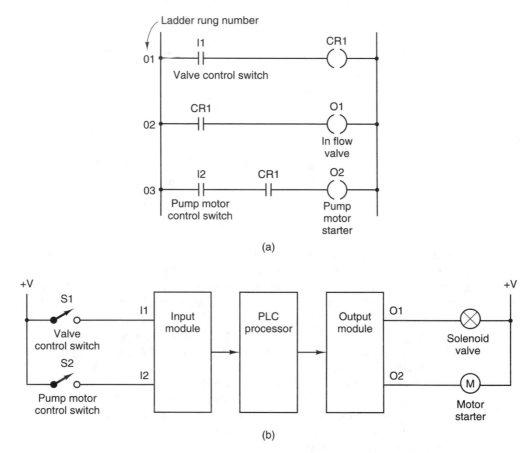

Figure 12–15 PLC Program and I/O Interface: (a) Ladder Logic Program; (b) PLC Wiring Interface.

pump switch relatively easy. In Figure 12–16b, the second pump control switch is wired into an available terminal (I3) on the input module. The second magnetic starter for the added pump is wired into an available terminal (03) on the output module. The change in the program for the second pump is illustrated in Figure 12–16a. In most system modifications, the major changes occur in the PLC program, which is easy to modify. Wiring changes little because the interface connects inputs to input modules and outputs to output modules. How the inputs and outputs interact is a function of the PLC program.

■ *Improved maintenance and troubleshooting.* In general, PLCs are reliable and have few maintenance problems. If a problem does occur in any module or in the processor, the module or processor can be changed in a matter of seconds without any changes in wiring. The PLC also makes troubleshooting the entire control system less difficult because a technician can check the

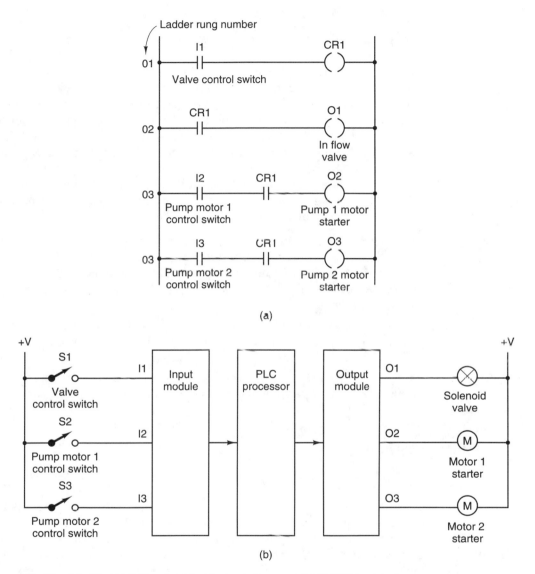

Figure 12–16 Modified PLC Circuit: (a) Ladder Logic Program; (b) PLC Wiring Interface.

status of each input and force a change in each output to identify the input or output device causing the problem.

■ *Off-line programming.* In the past, PLCs could be programmed only with special programming terminals supplied by the vendor; however, the present systems use microcomputers running special PLC programming software as programming terminals. The microcomputer either is attached directly to the PLC processor or is a node on the PLC network. In both situations, most of the application programming is performed using micro-computer hardware and software resources and does not require the PLC to be in the *program mode*. This process, called *off-line programming*, allows new program development and current program modifications without taking the PLCs out of the production process. After the development or modification of a program, the PLC is put in the program mode and the code for the new application or modification is downloaded to the processor. This operation takes less than a second, so the downtime in production is minimal. Some program modifications, such as changing variable values or setpoints, are performed while the processor is in the *run mode*, and no production time is lost. In contrast, installing and modifying relay logic circuits often takes days or weeks with considerable lost production as the control circuits are interfaced to the production system.

■ *Broad application base.* PLC software supports a broad range of discrete and analog applications in various industries. With just program and interface module changes, a PLC can be moved from sequential control of discrete actions in an assembly cell application to control of the liquid level and temperature in a process.

■ *Low cost.* The cost of PLCs has dropped significantly in the last ten years. For example, powerful PLCs designed for machine control applications are available for less than $500.

■ *Broad range of PLCs and peripheral devices.* Most PLC vendors offer a wide range of PLCs and peripheral devices that interface to the processor or PLC network.

Other Programming Options

Programming PLCs using ladder logic diagrams has existed from the beginning. While the use of ladder logic offers a familiar programming environment for industrial applications, the process has some disadvantages. For example, the unstructured nature of the process results in multiple ladder logic solutions for the same control problem. In addition, interlocks are required to eliminate the undesired interaction between two different parts of the ladder logic program. Also, large ladder logic programs are difficult to troubleshoot because contacts of a control relay could appear anywhere on the ladder. PLC vendors have developed various solution to these problems.

A European standard called *Grafcet* and a similar U.S. process called sequential function charts are used to overcome many of the problems associated with conventional ladder logic programming. Both are graphical methods of functional analysis that represent the functions of sequentially automated machines as a sequence of steps and transitions. Each step includes commands or actions for control of the machine or system that are either active or inactive. If a step is active, the commands present in the step are executed. Using the pump system in Figure 12–15, for example, valve control could be a step. If that step were active, S1 would control the valve. The flow of control in Grafcet passes from one step to another through conditional transitions that are either *true* or *false.* If the transition is *true,* control passes from the current step to the next. After the transition, the previous step is inactive and the current step is active. Each control function in a sequential process is represented by a group of steps and transitions called a *function chart.* A function chart for a semiautomatic punch press is illustrated in Figure 12–17. The system starts with step 1 active, which is a wait state. Transition from step 1 to step 2 is true when the *start button* is pressed and the punch is in the fully opened position. When the transition to step 2 occurs, step 1 becomes inactive and the press starts descending. When the press reaches the fully closed position, the transition to step 3 occurs. Step 2 becomes inactive, step 3 becomes active, and the press starts to rise. When the press reaches the fully open position, a transition from step 3 back to step 1 occurs and the cycle is read to be repeated the next time the start button is pressed. Several companies have adopted programming standards that emulate the Grafcet standard; for example, TI/Siemens has *stage programming.*

12-6 COMPUTER NUMERICAL CONTROL

The definition of computer numerical control (CNC) was restricted initially to numerical control (NC) machines that incorporated internal computers to handle machine control and program execution. Currently, CNC represents all production machines that use on-board computers to control the movement of tooling with production programs. The basic robot system described in Figure 11–1 is an example

Figure 12–17 Grafcet Function Chart.

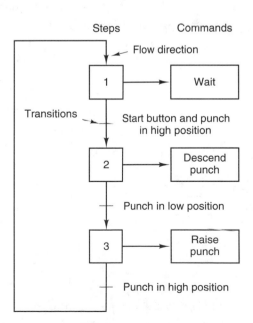

of a machine that would fit the definition of CNC. A generic block diagram for CNC machines is provided in Figure 12–18. Compare the robot controller in Figure 11–1 with the CNC block diagram and verify that the functions are similar.

The input to the CNC system comes from either an operator or from another machine. Frequently, the input for CNC machines is another computer; for example, many of the CNC machines, described in Chapter 10, received the program they used to cut parts from a computer that translated the part drawing from a CAD vector file to a machine code program file used by the CNC machine to cut the part. The processing that converts part geometry to an intermediate machine code is either performed by an NC software module inside the CAD software or by a separate computer-aided manufacturing (CAM) software program such as SmartCAM or MasterCAM. The translation from the intermediate machine code to a program for a specific brand and model of CNC machine is performed by a program called a *postprocessor*. The postprocessor produces a program with the tooling and move commands required for a specific CNC machine. When this CNC machine code program is loaded into the memory in Figure 12–18, the control and arithmetic elements use the program to drive the tooling through the desired motions. A large number of production machines use CNC. Included in the group are robots, coordinate measuring machines (CMMs), and most material processing machines. One of the first and still a common CNC programming application includes numerical control (NC) of lathes and mills. An overview of the programming techniques used on these types of process machines follows.

CNC Programming

CNC programming of mills and lathes is like any other machine programming process; it requires a complete understanding of the programming language. The language used for NC of mills and lathes is often referred to as *G codes*. The

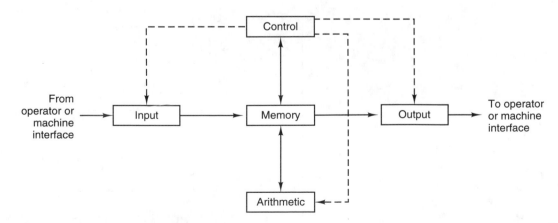

Figure 12–18   Five Major Functional Units of a CNC.

process used for milling machines and machining centers provides a good example of G-code use because it encompasses about 75 percent of all NC operations. The following five categories of programming commands and techniques are used for mill NC programming.

Basic Commands.   The following command codes and their functions are used to write an NC program for a noncomplex part.

- Motion commands (G00, G01, G02, G03).
- Plane selection (G17, G18, G19).
- Positioning system selection (G90, G91).
- Unit selection (G70 or G20, G71 or G21).
- Work coordinate setting (G92).
- Reference point return (G28, G29, G30).
- Tool selection and change (Txx M06).
- Feed selection and input (Fxxx.xx, G94, G95).
- Spindle speed selection and control (Sxxxx, M03, M04, M05).
- Miscellaneous functions (M00, M01, M02, M07, M08, M09, M30).

Compensation and Offset.   The following commands are used to define work coordinate systems, perform tool diameter compensations, and compensate for tool length differences.

- Work coordinate compensation (G54–G59).
- Tool diameter (radius) compensation (G40, G41, G42).
- Tool length offset (G43, G44, G49).

Fixed Cycles.   These commands in the following three categories provide a method of executing a series of repetitive machining operations with a single code block:

- Standard fixed cycles (G80–G89).
- Special fixed cycles.
- User-defined fixed cycles.

Macro and Subroutine Programming.   Modern NC controllers support basic computer programming syntax that includes defined variables, arithmetic operations, and execution of logical decisions to define part geometry mathematically.

Advanced Programming Features.   These are generally considered to be user-defined controls.

A simple example of G-code programming illustrates how NC programming codes are used.

Example 12–1

Write a G-code program to cut the path illustrated in Figure 12–19 with a cutter speed of 9 inches per minute (IPM).

Solution:

N05	G90	Absolute coordinate system.
N10	G00 X 1.0 Y1.0	Rapid mode move to A (1.0, 1.0).
N15	G01 Y6.0 F9.0	Linear interpolation, cut from A (1.0, 1.0) to B (1.0, 5.0) at a rate of 9 inches per minute.
N20	X7.0	Cut to C (7.0, 5.0).
N25	X1.0 Y1.0	Cut to A (1.0, 1.0).

Note that some commands in the previous example remain in effect after they are used. For example, X1.0 did not have to be repeated in line N15 if the X distance did not change, and the Y distance was not needed in line N20 for the same reason. Also the feed rate was not repeated for lines N20 and N25 because no change in the cutting rate was desired.

Like all programming languages, a full programming course is necessary to use all of the NC G-code commands for mills and lathes. Complete coverage of G-code programming is beyond the scope of this book. However, a list of

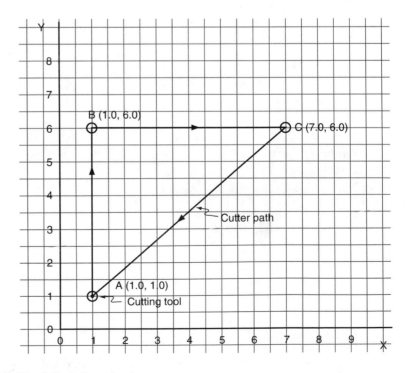

Figure 12–19   G-Code Programming.

G codes for a Fanuc lathe controller is provided in Appendix 12–1 at the end of the chapter.

Manual Versus Automated Programming

Programming of NC machines takes two forms: manual programming and code generation with the support of CAM software. Example 12–1 is an example of manual programming. Starting with a drawing of the milled part, the programmer creates a G-code sequence that drives the machine's cutting tool over the desired path. CAM-generated NC code uses a postprocessor for the target machine tool to convert the drawing of the part directly to the G-code program that will run on the machine selected. The CAM software and postprocessors, described in Chapters 4 and 5, fell into two catagories. One type, MasterCAM and SmartCAM, is stand-alone and works with drawing files from all the major CAM vendors. The second type, developed by the CAD vendor, is integrated with the CAD program and runs as part of the integrated CAD/CAM design software.

12–7 AUTOMATIC TRACKING

Tracking the movement of parts and tools through production is a major task. The primary technique used to follow movement of parts is *bar codes*. The concept of giving products a unique label started in the grocery industry in the early 1900s. The first bar code, a circular design, was patented in 1949; however, over fifty unique bar code symbols have followed that first circular design.

Bar Code Symbols

Bar code scanning is rapidly replacing the keyboard and other data-entry devices for recording manufacturing information and data. Sophisticated automation on the shop floor uses bar code technology to provide the information for material resource planning, statistical process control, production program selection, and other CIM system applications.

A bar code is a symbol composed of parallel bars and spaces with varying widths. The symbol is used as a graphical code to represent a sequence of alphanumeric characters that includes punctuation. For example, the three bar codes pictured in Figure 12–20 represent the numbers written below the codes. The codes can also represent combinations of letters and numbers. More than fifty codes were developed and each offers one or more of the following code characteristics: robust character set, structural simplicity, generous tolerances for printing and reading, and a high density-to-size ratio. The two most popular codes used in industry are the *interleaved two-of-five* and *code 39.*

Interleaved Two-of-Five

The interleaved two-of-five is a numeric code, which means that only numbers can be represented by the code. A sample interleaved two-of-five code is illustrated

Figure 12–20   Sample Bar Codes.

in Figure 12–21. The code was designed with two key features: (1) the code has a high density, so that a large amount of information can be encoded in a short space; and (2) the code has a wide tolerance for printing and scanning. This code is a continuous code, which means that both the bars and spaces are coded. The odd-number digits are represented by the bars, and the even-number digits are represented by the spaces. The specification requires that the code represent an even number of digits. To satisfy this requirement, a zero is placed in front of all numbers with an odd number of digits before converting to the interleave code.

The interleaved two-of-five is effective for coding information on corrugated boxes, where bars must be wide to permit automatic scanning while the boxes are moving on conveyors; however, space on the boxes is usually limited. The code is used widely in the automotive industry and in warehousing and heavy industrial applications.

Code 39

Code 39, the most widely used and accepted industrial bar code symbol, was developed in 1975 by Dr. David Allais and Ray Stevens for what is now Intermec, Inc. The code, illustrated in Figure 12–22, has become the *de facto* industrial standard for item identification. Code 39, also called a three-of-nine code, is a complete code system. The code can represent the twenty-six letters of the alphabet, the ten digits, and seven additional characters. The code has unique start and stop

Figure 12–21 Interleave Two-of-Five Bar Code.

Code table	
Character	Code
0	00110
1	10001
2	01001
3	11000
4	00101
5	10100
6	01100
7	00011
8	10010
9	01010

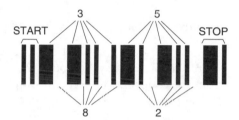

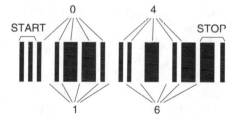

Figure 12–22 Code 39 Bar Code.

bits and can vary in length. All characters are self-checking, and as a discrete code, the intercharacter space is not part of the code.

Major advantages of this code are the high level of data security provided by the self-checking feature and the wide tolerance for printing and scanning. For example, with good dot matrix bar code printers and properly selected bar code scanners, the error rate is less than one character substitution in 3 million characters scanned. If high-quality printers and scanners are used, the value drops to one substitution error in 70 million characters scanned. For even greater data security, an additional check character is added to the code. As a result of this high security, code 39 was added to standards developed by health care and automotive industries and by the Department of Defense.

Reading Bar Codes

The scanners used to read bar codes have three major components: (1) a light source to illuminate the code, (2) a photodetector to sample the light reflected from the code, and (3) a microcomputer to convert the photodetector output into the series of letters and numbers represented by the code. The operation is straightforward. A source of light from a laser, light-emitting diode (LED), or incandescent lamp is aimed at the bar code symbol. The dark bars absorb the light and the spaces reflect the light back to the photodetector. From these data, the width of the bars and spacing are determined by a microcomputer. In addition, the computer decodes the bar code and provides a serial data output of the code in ASCII format. This output is passed to another computer or programmable logic controller to use in a host of production applications.

Scanners fall into three categories: *handheld contact type, handheld noncontact type,* and *fixed-station systems.* Figure 12–23 shows a handheld type scanner used in industrial applications. Handheld scanners are connected to portable data storage devices that are connected to a computer, or transmit the data over radio frequency links to a host computer. The simplest form of scanners used in a large number of applications are the handheld type with a *light wand.* In the contact type of scanner, the wand has a light source (LED or incandescent lamp) and photodetector in the pencil-like wand and two fiber-optic cables connecting these devices to the tip of the wand. As the wand is dragged over the bar code, the width of bars and spaces is read from the reflections of the light through these fiber links. Handheld scanners using laser light sources are used to make noncontact readings. The laser's pinpoint light beam permits scans with the reader some distance from the bar code. The principle of operation is the same as that of the contact-type scanner. Light reflected back to the photodetector provides information on bar and space width and orientation.

Two other types of scanners are the *moving-* and *fixed-beam* devices. The moving beam typically uses a helium–neon laser as a light source and sweeps the laser beam across the object at a rate of 1440 passes per second in search of the bar code. The advantage of this type of scanner is that bar code symbol placement is not critical as long as the code falls into the field of view of the scanner. The fixed-beam type of scanner using a laser, LED, or incandescent

Figure 12-23   Bar Code Scanner.

lamp illuminates a large area of the bar code without moving the beam. However, the fixed beam requires that the bar code be moved past the scanner to generate signals required by the photodetector. Again, reflected light is used to decode the information in the code.

Other Tracking Mechanisms

In many applications where parts or tools need to be tracked, the environment does not permit bar codes to be applied or read. For example, the cutting tools used on CNC machines are costly and their location in the production area must be tracked. The environment does not support bar codes, however, because of the cutting fluids and the general handling the tools receive. A common tracking technique for cutting tools and other devices that cannot use bar codes is the use of *radio-frequency tags.* The tags are passive electronic circuits that transmit a code when subjected to radio-frequency (RF) energy. The passive electronic circuit, shaped like a small computer chip, is held in a slot in the tool base by epoxy. Each circuit transmits a unique code when the tracking device focuses RF energy at the tool. The tracking device receives the tool code and records the number. In addition to cutting tools, the RF tag is useful for many other harsh application environments.

Another tracking technique, frequently used on production pallets, utilizes a binary code to track objects. The pallets have a sequence of holes placed somewhere on the surface. At preselected points along the production process, the holes are aligned with proximity sensors or limit switches with a sensor or switch over each hole. To give a pallet a unique code, some combination of holes is filled with pins. When the pallet passes under the sensors or switches, the holes with pins trip the sensor or switch. With four holes, sixteen unique codes or pallet identification numbers are available.

Tracking of products using one of the standard bar codes or other types of tracking techniques is a common practice in automated work cells. In addition to providing data on the location of products, parts, and tools, the unique part code can also identify the part to the machine. In an application at Allen Bradley, for example, a bar code placed on the base of a blank motor contactor tells the production system what type of contact to build. At every step in the totally automated process, machines add the correct parts to the contactor based on the information read from the bar code. Material, tools, and product tracking play a major role in most world-class production facilities.

12–8 NETWORK COMMUNICATIONS

A key requisite for any CIM installation is access to a single product database from all departments in the enterprise. The machine, cell, and area controllers described in the last sections must be linked to departments across the enterprise for successful implementation of a common database CIM system. Developing the database and network to achieve this goal is not a trivial task; as a result, most organizations view the process as evolutionary and spread the work over three or more years.

Identifying the requirements for the enterprise network is a prerequisite for the design of the physical configuration and selection of network hardware and software. Important considerations are identified and discussed below.

Enterprise Data

The type of data files communicated across the network and stored in the central data repository dictates network characteristics and configuration. The term *data file* is often used to describe the information stored in a computer, and an appreciation for the difference between data files is requisite for understanding network characteristics.

The storage of manufacturing information or data in a computer is analogous to storing the same data in a file cabinet. In the file cabinet, the information or data printed on paper is placed in folders, which are then placed into drawers in a predetermined order. The type of manufacturing information placed in the folders can be words or numbers, pictures, and line drawings. The CIM central database has similar data storage requirements. In the computer, the storage device is divided into sections similar to the drawers of the file cabinet, and the data are stored electronically in a predetermined order. The major difference between the paper and computer storage process is the conversion of the information and data from paper to the electronic format.

The computer data files generated across the enterprise fall into three classifications: *text* (words and numbers), *vector* (line drawings), and *image* (pictures). The data type produced and the size of the data file required for storage are functions of the process; for example, 2-D product drawings developed with CAD software are stored as vector files. Text files use the least amount of computer memory or storage space, and image files use the greatest. A comparison of the size for each file type illustrates the complex problem surrounding the development of a central CIM data repository for all enterprise data. A standard page of text would take no more than 2500 storage locations in a central data repository. A moderately complex 2-D CAD drawing that would fit on the same size page could require 50,000 storage locations, and an image file for the same size page could be over a million storage locations. These numbers will fall with technology innovations; however, the relative position of each file type with respect to size will not change.

In general, the product and process element where product design and documentation occur generates all three types of data files in the design and documentation of products. However, the majority of the files are vector type with large data files. The manufacturing planning and control (MPC) area generates text files the majority of the time; however, the MPC function works with drawing files (vector type) from the design area. The shop floor area generates text files in most applications but uses files in all three formats.

The required data transfer rate for the network is higher if large data files must be exchanged between users in the enterprise. Also, the network design is affected by the requirement for a large number of concurrent users requesting large data files.

12–9 SUMMARY

In this chapter, we described the hierarchy used by the CIM enterprise for control of devices, cells, areas, and corporatewide systems. The cell control hardware is classified as either intelligent or nonintelligent, with intelligent systems including an on-board computer. Most cells are organized with a cell controller at the top of the control hierarchy, with the remaining cell hardware interfaced to the controller directly or through other intelligent devices. The cell controller is usually either a microcomputer or a minicomputer running one of the following operating systems: OS/2, Windows NT/2000 or UNIX. The major application present in most cells is the cell control management software that manages the bidirectional data flow between production equipment and the remainder of the enterprise. The software falls into two categories: proprietary and open system interconnect. The proprietary group includes in-house-developed software and application enablers provided by third-party software developers. Internal system compatibility is a major benefit of the software written specifically for the cell control application. However, the cost of internally developed packages is forcing companies to switch to enabler software. Enabler software, developed by third-party vendors, is a control shell that is tailored for each company's specific control application. The open system interconnect cell control solution most frequently selected is manufacturing message specification (MMS). MMS is a standard for communicating information between intelligent devices in a production environment over networks based on the OSI model. The primary advantage of MMS is that intelligent devices are nodes on a standard network, and the communication among all the devices is device to device without any need for data concentrators such as a cell controller.

Device control uses a combination of proprietary and generic control devices. The proprietary controllers are built specifically to control a single type of machine, such as a CNC machine tool. Generic controllers are general-purpose devices that are applied to the control of various automation machines. One frequently used generic type of controller is a programmable logic controller (PLC). PLCs were designed initially to replace mechanical relays in sequential control logic circuits in industrial applications. PLC technology has advanced significantly in the last twenty five years, and PLCs are now used for a broad range of discrete and analog control applications. The applications range from a single PLC for machine control to networks of PLCs that run entire production systems. Although the representation of PLC programs in the United States continues to use a ladder logic format, the European community has moved to a more structured format called Grafcet. In addition, the method of programming has changed from special-purpose programming units to microcomputers connected to the PLC or on the PLC network. The advantages of PLCs include easy interface modifications, improved maintenance and troubleshooting, off-line programming, a broad application base, low cost, and a broad range of PLC types and peripheral devices.

Another type of special-purpose controller frequently found in industry is the computer numerical controller. This type of controller, most often associated with metal-cutting machine tools, is used for several production machines that require high-level control of operation or motion. The robot, for example, uses a type of CNC controller to drive the arm through the programmed motions required for a production operation.

Automatic tracking is a requirement present in every manufacturing operation. Bar codes, the most frequently used tracking technology, are symbols composed of parallel bars and spaces with varying widths. Over fifty different bar codes are used to represent letters, numbers, and punctuation, with interleaved two-of-five and code 39 the two most commonly used codes. Bar codes are read with a wide variety of scanners that fall into three categories: handheld contact type, handheld noncontact type, and fixed-station systems. In addition to bar codes, the use of radio-frequency tags and mechanically generated binary-coded tracking systems are also used.

A communications network for data and information is necessary for successful operation of a central database in a CIM implementation. All the data stored in the CIM database fall into one of the following three categories: text (words and numbers), vector (line drawings), and image (pictures).

REFERENCES

GRAHAM, G. A., *Automation Encyclopedia*. Dearborn, MI: Society of Manufacturing Engineers, 1988.

GROOVER, M. P., *Automation, Production Systems, and Computer-Integrated Manufacturing.* Upper Saddle River, NJ: Prentice Hall, 1987.

LIM, S. C. J., *Computer Numerical Control.* Albany, NY: Delmar Publisher, 1994.

LLOYD, M., "Graphical Function Chart Programming for Programmable Controllers," *Control Engineering,* October 1985, pp. 37–41.

LUGGEN, W. W., *Flexible Manufacturing Cells and Systems.* Upper Saddle River, NJ: Prentice Hall, 1991.

REHG, J. A., *Introduction to Robotics in CIM Systems.* Upper Saddle River, NJ: Prentice Hall, 1992.

SOBCZAK, T. V., *A Glossary of Terms for Computer Integrated Manufacturing.* Dearborn, MI: CASA of SME, 1984.

WEBB, J. W., *Programmable Controllers: Principles and Applications.* Columbus, OH: Charles E. Merrill, 1988.

QUESTIONS

1. Describe the interface between the enterprise network in Figure 12–1 and the factory LAN in Figure 12–2.
2. Why are point-to-point communication systems and fast response necessary for the data interfaces at the cell control level?
3. Describe the difference between intelligent and nonintelligent cell control devices and list the nonintelligent devices in Figure 12–2.
4. Describe the two different types of cell controllers most frequently used and the operating systems associated with cell control applications.
5. What limits DOS from being an effective cell control operating system?
6. Describe the different types of interface cards frequently found in cell controllers.
7. What is the major type of application software found in most cell controllers, and what is the major function served by the software?
8. Describe the difference between proprietary and open system interconnect software.
9. What are the advantages and disadvantages of in-house-developed cell control software?
10. Describe application enabler software and list the advantages and disadvantages.
11. Describe the most frequently used OSI software for cell control applications.
12. What are the advantages of using MMS for cell control applications?
13. Describe the difference between proprietary and generic device controllers.
14. What is the difference between discrete and analog control requirements?
15. What are the characteristics of a basic sequential logic control application in industry?

16. List and describe the system components for a programmable logic controller.
17. Describe the basic PLC operation.
18. Describe the data communication options available in PLC applications.
19. Describe how smart I/O interfaces operate and how they can significantly affect automation applications.
20. Describe the process for programming a PLC for an industrial application.
21. Why is the Grafcet programming technique better than ladder logic?
22. What advantages do programmable controllers offer in cell and area control applications?
23. Describe the characteristics of CNC.
24. Describe three different types of automatic tracking techniques used in industry.
25. Compare and contrast the interleaved two-of-five and code 39 bar codes.
26. What are the major components in a bar code scanner?
27. Describe the different types of bar code scanners and the accuracy possible with different configurations.
28. Describe the three different types of data files stored in the CIM central database.

PROJECTS

1. Using the list of companies developed in Project 1 in Chapter 1, determine which companies are using cell controllers, programmable controllers, automatic tracking, and factory automation networks.
2. Select one of the companies in Project 1 using cell controllers and describe in detail the hardware and software used in the system.
3. Select one of the companies in Project 1 that uses automatic tracking and describe in detail the type of system implemented and the operational characteristics present in the system.
4. Select one of the companies in Project 1 that has a factory network and describe in detail the type of system implemented and the operational characteristics present in the system.

APPENDIX 12–1: TURNING G CODES

G-codes are sometimes called preparatory functions and are specified by the G address followed by a two-digit identification number. Two types of G codes are commonly used: modal codes and nonmodal codes. A modal code is effective until it is overridden by another G code in the same group; a nonmodal code is effective only for the block in which it is used. The code and function of some G codes used in turning are the same for milling; however, others have different assigned functions. The following list shows the Fanuc G codes used in the United States for turning with controllers:

G00   Rapid traverse (positioning).

G01   Linear interpolation.

G02*   Circular interpolation (clockwise).

G03*   Circular interpolation (counterclockwise).

G04   Dwell (temporary stop).

G20*   Inch data input (G70 for some systems).

G21   Metric data input (G71 for some systems).

G27   Reference point return check.

G28   Reference point return.

G30   Second, third, fourth reference return.

G33   Thread cutting.

G34   Variable-lead thread cutting.

G40   Tool nose radius compensation cancel.

G41   Tool nose radius compensation right.

G42   Tool nose radius compensation left.

G65   Macro call command.

G70   Finishing cycle.

G71   Stock removal in turning.

G72   Stock removal in facing.

G73   Pattern repeating.

G74   Peck drilling in Z axis.

G75   Grooving on X axis.

G76   Multiple threading cycle.

G77   Diameter cutting cycle.

G78   Thread cutting cycle.

G79   End face turning cycle.

G92   Coordinate system setting, or maximum spindle speed setting.

G94   Feed rate per minute.

G95   Feed rate per revolution.

G96   Constant surface speed control.

G97*   Constant surface speed control cancel.

*Modal operation.

Quality and Human Resource Issues in Manufacturing

Many manufacturers in Europe and the United States experienced a loss in market share in the mid-1970s; even producers who were able to maintain their market share felt the pressure of foreign competition. At the same time, manufacturing technology started to mature with the introduction of sophisticated and reliable hardware such as robots, computer numerical machine tools, programmable logic controllers, material-handling systems, coordinate-measuring machines, and automatic storage and retrieval systems. Computers became smaller and less expensive, and distributed computing became a reality. In addition, the number of off-the-shelf software solutions for manufacturing and production control increased dramatically. Many manufacturers enthusiastically invested in manufacturing resource planning II (MRP II) technology solutions, assuming that greater data capture-and-control power would automatically produce the efficiencies necessary to restore the enterprise to world-class performance. Thus it appeared that technology had matured just in time to save the day.

Although the introduction of technology caused some improvement in performance, many potentially first-rate U.S. manufacturing companies failed to soar as a result of advanced manufacturing systems. In some cases, manufacturing efficiencies dropped when technology was introduced. The lesson learned by manufacturing in the mid-1970s is that technology alone cannot improve business performance, and a new emphasis on *quality* and the development of the *workforce* was required. The delivery of higher performance is achieved only through people operating in a quality-rich environment on good production processes, applications that use technology effectively. In this chapter we provide an overview of the quality concept and the development of human resources.

13–1 QUALITY

The move to *total quality* means different things to different people. In some industries, the term *quality* refers to the processes and tools used to measure and record the degrees of quality achieved in production. A technique, such as *statistical process control* (SPC) that records the degree of variation in a product on charts with statistically determined limits is an example of a process. The emphasis on SPC in the early 1980s was a reaction to the superior quality of products received from offshore competition. U.S. industries recognized the need for quality products in the early 1970s, and isolated fixes such as SPC were adopted in an attempt to match the quality of offshore competition.

The quality focus in the enterprise is not just the adoption of technology such as SPC, smart gauges, coordinate-measuring machines, and vision systems. The quality focus is a broad view of all operations that blends the effort of all employees and the necessary technology into a comprehensive plan to approach a fault-free production environment. This quality process includes four major elements: (1) appropriate technology at all levels in the enterprise to control and measure critical quality parameters, (2) good design that permits manufacturing and assembly with near zero defects, (3) management understanding of all quality-related issues and the commitment to strive for a defect-free operation within the enterprise, and (4) broad workforce training and involvement of every employee in the effort to reach the quality goals. This broad view of quality is addressed by the process called *total quality management*, and development of these elements is the focus for the next sections.

13-2 TOTAL QUALITY MANAGEMENT

Total quality management (TQM) has two components, the principles of the TQM process and tools to get the job done. The principles or philosophy associated with TQM allow an organization to overcome the traditional barriers and impediments that prevented the management group from utilizing the potential, skills, and knowledge of every employee. Following the TQM philosophy, the institution sets higher goals, recognizes and eliminates barriers to change, and solicits the opinions and ideas of every employee. TQM helps everyone see more of the big organizational picture by focusing on the common quality goal to "get it right the first time."

In addition to a guiding philosophy or principle, TQM is also driven by tools. These qualitative and quantitative quality tools are not new; however, the method of application is new when used with the TQM philosophy. The tools help an organization to understand better the way they do business and to measure the degree of improvement along the TQM journey. The tools help individuals, departments, and the total enterprise recognize when the goals for improved *productivity, performance, efficiency, customer satisfaction, work life,* and *quality* are reached.

The domain for TQM is defined in the illustration in Figure 13-1. The cycle illustrated in the figure shows the product moving from customer demand through the many phases of manufacturing until a finished product is shipped This illustration is also a TQM cycle because a concern for the quality at every step in the sequence is a focus of the total quality management process.

Definition of TQM

Total quality management (TQM) is a term coined to describe Japanese-style management approaches to quality improvement. Total quality management has taken on many meanings. Simply put, TQM is a management approach to long-term success through customer satisfaction. TQM is based on the

Figure 13–1 Production and TQM System Cycle.

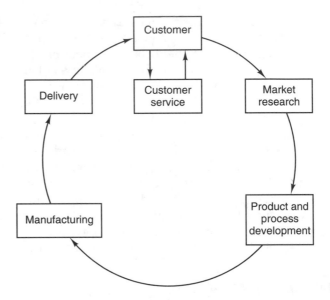

participation of all members of an organization in improving processes, goods, services, and the culture in which they work. The methods for implementing this approach are found in the teachings of quality leaders such as Philip B. Crosby, W. Edwards Deming, Armand V. Feigenbaum, Karu Ishikawa, J. M. Juran, and Genichi Taguchi. (*APICS Dictionary*, 9th Ed., 1998)

Study the definition and identify the three dominant themes present.

The three critical components for the TQM process to flourish are (1) a management group willing to allow everyone in the organization to participate in the decision process, (2) a process of continuous improvement for *all* elements in the enterprise, and (3) the use of multifunctional teams. The first element, participative management, is an evolutionary process that requires trust from both management and labor. The management must look for examples of labor's ability to participate positively in the management process and must recognize the capabilities and contributions that employees make to improve the production process and enhance the enterprise business. Management must take the first step to lower the barriers between themselves and labor by giving employees the opportunity to participate in decision making. It is an act of faith that management must take to launch the TQM process in the slow journey toward full participative management.

The continuous improvement process is an admission that every activity or process in the enterprise could be improved. This commitment never changes, regardless of the length of time that continuous improvement is practiced. No process will ever be perfect because some improvement will always be possible.

The TQM teams are a cross-section of members from inside and outside the enterprise who represent some part of the process under study. The team frequently includes enterprise employees associated with the process; vendors who supply raw materials, services, or production equipment; and the customers who use the products. Two types of team processes occur in the CIM enterprise. The first, described above, are cross-functional teams with representation from a broad cross-section of the enterprise and external sources. These teams are primarily process troubleshooting teams that are formed to work on an identified improvement and then are disbanded after the process improvement is implemented and successful. The second type of team, a self-directed work team, is formed to provide for self-governance of a production area. These self-directed teams have members only from the specific process area, and their responsibility includes the full authority to run the process area or production cell. A full description of self-directed work teams is provided in a later section.

TQM Principles for Implementation

Total quality management includes six principles that must be considered in an implementation: customer focus, process focus, prevention focus, workforce mobilization focus, fact-based decision-making focus, and continuous feedback focus. Each of these implementation principles is described below.

Customer Focus.   Most employees know that customers are important to the survival of the enterprise; however, few know the full definition of the term *customer* from the TQM perspective. In TQM, the term *customer* refers to the external customers who place the orders and receive the finished goods. In addition, everyone in an organization has internal customers. For example, the internal customers for the heat-treating department are all the other departments that send work to heat treat as part of the production process. Every person in the organization has internal customers who must be served just as well as the external ones who receive the finished products. To implement TQM successfully, every employee must know and serve all the needs of their external and internal customers.

Process Focus.   In the past the focus of the enterprise was on the results and not on the process used to get the results. In a TQM implementation, the focus is on the process; the results take care of themselves. Improvement in the process is triggered by results that do not meet or exceed the internal and external customers' expectations. This improvement is continuous in the well-oiled TQM system.

Prevention Focus.   Too often, correction of quality problems of internal and external customers occurs after the defective items are identified by inspection. In the worst case, the inspection is a final inspection, and many defective parts are identified as scrap or requiring rework. TQM focuses on prevention of problems, not on inspection of final results; as a result, more defects are prevented before they happen, or problems are identified early in the production cycle before many parts are affected.

Workforce Mobilization Focus.   "Checking your brain at the time clock" was a common practice in organizations that did not recognize and use the expertise present in the workforce. The contribution of every employee toward meeting the enterprise goals changes significantly under TQM. Therefore, the TQM implementation must include a change in the way management views the capability of the workforce and a mobilization of the total workforce for the solution of common enterprise problems.

Fact-Based Decision-Making Focus.   Finger-pointing over problems in processes and products has no place in a TQM implementation. Responsibility for fixing the problem falls on everyone who is associated with the defective process or product. The TQM implementation uses a continuous improvement process and multifunctional teams to analyze the problems carefully, document the facts, propose and implement solutions to the problems, and verify that the solution is permanent.

Continuous Feedback Focus.   The term *feedback* is synonymous with good communication. The success of a TQM implementation is ensured only when open honest communication occurs from the top of the enterprise to the bottom. Equal access to good information and open communications across the enterprise are necessary for TQM.

An organization moving toward a goal of total quality management must work to have all six of these principles in place for the TQM process to be successful and lasting.

TQM Implementation Process

TQM implementation includes a five-step process and an implementation schedule. The five steps are (1) preparation, (2) planning, (3) assessment, (4) implementation, and (5) evangelization. Each of the five steps is described briefly below.

1. The first step prepares the organization for the TQM implementation by starting with the CEO and other top management. The key point at this time is *getting top management commitment* to the principles and changes in the organization required by TQM. Top management must develop a vision statement, set corporate goals, draft a policy to support the strategic plan, and commit the resources to implement TQM successfully.

2. The *corporate council,* the initial planning group, lays the foundation for the TQM process. The vision statements, goals, and strategic plan from step 1 are used to develop the implementation plan and start to commit resources to the TQM process.

3. In the assessment step, the strengths and weaknesses of the organization are studied and documented. The process includes the exchange of information through the use of surveys, evaluations, questionnaires, and interviews from individuals across the enterprise and at all operational levels. This self-assessment provides both individual and group perceptions of the organization and operations.

4. The implementation stage starts execution of the TQM process. All the effort from the previous three stages starts to pay off with a focused training plan

for managers and the workforce. The improvement process starts in this step with the formation of *process action teams* (PATs) to evaluate processes and implement new processes and changes as necessary.

5. Evangelization is the process of taking the TQM process that was implemented successfully in the initial area to other departments and elements in the enterprise. In the first four steps, much process and organization information and knowledge were acquired. In addition, the PATs have recommended and made process improvements on the initial TQM project area. This success record is used as the launching pad for the introduction of TQM into other areas within the enterprise.

The TQM process is vital to the success of a CIM implementation because CIM requires a quality product and quality in the organization's operation. TQM supports the development of both product and organizational quality. In addition, TQM and CIM work well together because both require a change in the management style, the use of teamwork, improved communications, continuous improvement, and more self-management at all levels in the organization.

The quality tools used to measure the current status of the enterprise are critically important to the successful TQM implementation. Their importance becomes apparent when the PATs start to gather data for improvement of a process. The tools used in this process are described in the next section.

13–3 QUALITY TOOLS AND PROCESSES

The goal of every manufacturer is to produce items correctly the first time. The traditional quality assurance and quality control techniques worked to improve product quality but only after the mistakes were made. This traditional approach to quality has a major impact on productivity and product cost. For example, research performed by the American Society of Quality Control indicates that 15 to 30 cents of every sales dollar of products manufactured is lost due to the traditional approach of fixing poor production. In the service industries, the cost is closer to 35 cents for every dollar of service delivered. World-class enterprises that work to correct quality problems before the product is manufactured spend less than 5 cents for the same dollar of sales. One of the most frequently used tools in measuring and improving production is *statistical process control.*

Statistical Process Control

The name *statistical process control* (SPC) implies that *statistical* analysis of manufactured parts will be used to *control* and *improve* the manufacturing *process.* SPC is a group of statistical methods used to *measure, analyze,* and *regulate* a production process to reduce *defects.* A defect is defined as any variation of a required characteristic of the product or part that deviates sufficiently from the nominal value to prevent the product or part from fulfilling the physical and functional requirements of the customer. Product or part variations are grouped into two categories: variable and attribute characteristics. Variable characteristics such as length or weight are measured in physical units. Attribute characteristics such as surface finish are either good or bad. Product and part variations of these two characteristics are monitored and controlled through SPC.

The basis for SPC is the natural tendency for most manufacturing processes to produce products that conform to a *normal distribution.* Although space does not permit an in-depth development of this concept, an overview of the general process is necessary for an understanding of the SPC process. Variable characteristics are specified by the designer as a nominal value (target) and a tolerance about the target (variability). For example, a drawing of a shaft would show the nominal length of the shaft and the upper and lower variation (tolerance) that is acceptable. The manufacturing process attempts to produce the shaft at the nominal value of the length characteristic. No process is perfect, however, and some variation about the nominal length will occur. Some shafts will be longer than the nominal value, and some will be shorter. When the length of each shaft produced by the manufacturing process is plotted on a bar chart, called a *histogram,* the variation usually approximates a normal distribution. Study the normal distribution illustrated in Figure 13–2. The center of the curve represents all the shafts that are at the *mean* (μ) or average value for all shafts produced. The number of parts larger than the mean are plotted to the right of the mean, and the parts smaller than the mean are plotted on the left. Note that the vertical axis represents the number of shafts at the different lengths, and the horizontal axis represents the mean length and positive and negative variations from the mean.

The variability from the mean, represented by a statistical term called *standard deviation* (σ), is illustrated in Figure 13–2. The standard deviation is a measure of dispersion or variability and indicates how far away shaft values lie from the mean value of all shafts. The statistical formula for standard deviation relates the mean value for all shafts, the number of shafts, and the variation from the mean for specific shafts. Therefore, the standard deviation formula determines the number of shafts that lie at some variation from the mean.

The critical property of the normal distribution is that the percentage of parts that vary from the mean is predictable. The curve in Figure 13–3 illustrates this concept. Note that 68.26 percent of the parts are within 1 standard deviation of the mean, and that only 0.26 percent fall outside ±3 standard deviations. Therefore, knowing that a process output conforms to a normal distribution, you can predict the number of parts that will fall outside set variation limits. These concepts are used to produce *control charts* that track production variation and lead to process improvement.

Control Charts

Control charts, derived from statistical analysis of the process, are important SPC tools because they indicate whether a process is statistically predictable. The four basic functions provided by control charts include (1) defining the parameters of the process, (2) predicting when a process is changing due to a special cause outside

Figure 13–2   Normal Distribution Curve.

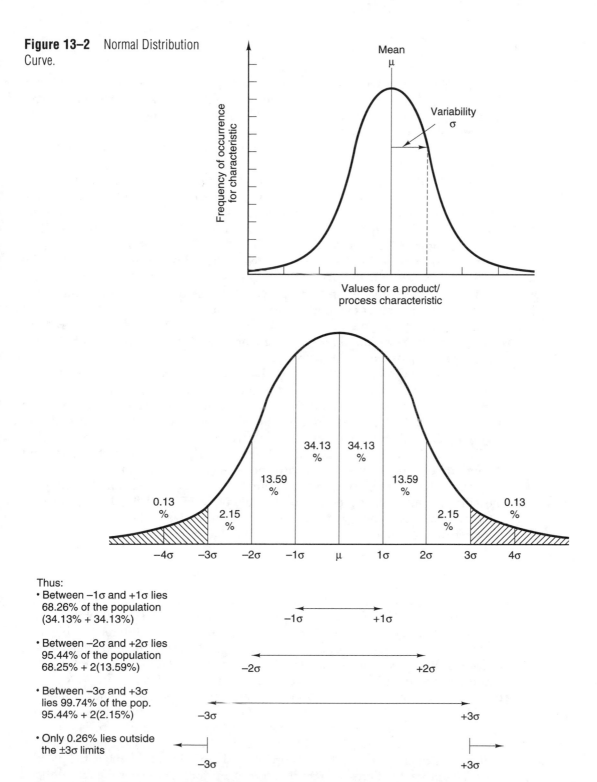

Mean
μ

Variability
σ

Frequency of occurrence for characteristic

Values for a product/ process characteristic

34.13 %   34.13 %

13.59 %   13.59 %

0.13 %   0.13 %

2.15 %   2.15 %

−4σ   −3σ   −2σ   −1σ   μ   1σ   2σ   3σ   4σ

Thus:
• Between −1σ and +1σ lies 68.26% of the population (34.13% + 34.13%)

−1σ   +1σ

• Between −2σ and +2σ lies 95.44% of the population 68.25% + 2(13.59%)

−2σ   +2σ

• Between −3σ and +3σ lies 99.74% of the pop. 95.44% + 2(2.15%)

−3σ   +3σ

• Only 0.26% lies outside the ±3σ limits

−3σ   +3σ

Figure 13–3   Variation from the Mean on a Normal Curve.

the normal process variability, (3) showing when the state of statistical control for the process has been reached, and (4) monitoring a statistically stable process and alerting operators when a change takes place that could affect cost or quality. All control charts have a centerline and *upper* and *lower* control limits (UCL and LCL) determined from statistical data. These control limits are usually spaced 3 standard deviations above and below the centerline.

Four control charts are commonly used to track variable and attribute characteristics. The *X-bar* and *R charts*, the most frequently used control charts, track and identify causes of variation in variable data from parts and products. The charts, usually plotted as a pair, have the X-bar chart above the R chart as illustrated in Figure 13–4. The X-bar chart plots the averages of a sample group of parts measured over time intervals. The result indicates the center value that the process follows. The R chart plots the range for the samples measured and indicates the variability that exists in the process. For a stable process, the mean or average and the standard deviation are used to establish estimates for the center and width of the process. The results are then compared to the desired process center (the nominal value required for the design) and the desired process width (the tolerance established in the design process).

The two control charts that are used for attribute data are the p-chart and the c-chart. The *p-chart*, also called the *percentage chart*, indicates the percentage of parts in the measured group, often called a *subgroup*, that are defective due to a problem associated with an attribute characteristic. A sample of the p-chart is provided in Figure 13–5. The *c-chart* is similar to the p-chart because it focuses on defects associated with attribute data. The p-chart data count the number of defective parts and do not differentiate between parts with one defect and those with multiple defects. In contrast, the c-chart graphs the total number of defects present in the subgroup measured, so multiple defects in a part would be included in the data. As a result, the c-chart process is effective only when the sample size of the measured parts is the same in each inspection. An example of a c-chart is provided in Figure 13–6.

Other Quality Tools

Several additional techniques are used in concert with SPC to solve production quality problems. Two frequently used tools are Pareto analysis and Ishikawa diagrams. *Pareto analysis* uses the frequency of occurrence of the different problems to set the priority for quality control analysis. Pareto diagrams, named after Vilfredo Pareto, an Italian economist, display the causes of a problem in the relative order of importance. An example of a Pareto diagram is shown in Figure 13–7. Note that the vertical axis has the number of occurrences of defects, and the horizontal axis has the groups of different problems identified. A Pareto diagram gives a clear picture of the relative importance of each grouping. For example, in Figure 13–7, problems 5, 8, and 1 are present 71 percent of the time when the quality problem occurs. The diagram suggests that correcting these three problems is the place to start. Therefore, the Pareto diagram becomes a prioritization for the efficient and effective use of resources in the solution of a quality problem. From this process,

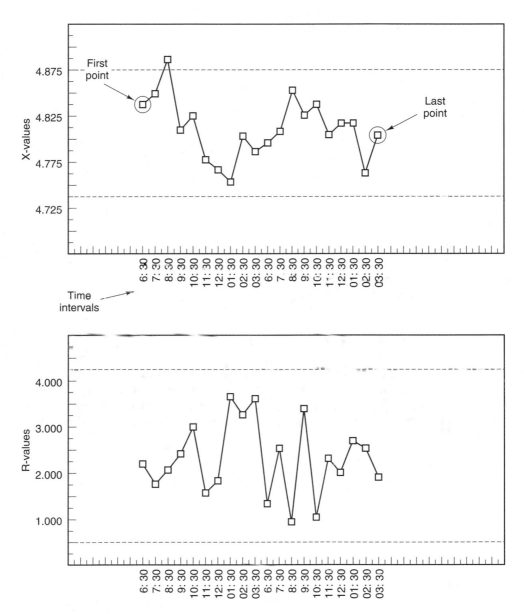

Figure 13–4 X-bar and R Charts.

the 80–20 rule evolved. Generally, 80 percent of the results come from 20 percent of the causes. The Pareto diagram helps to focus on the area where the greatest opportunities for improvement exist.

The *cause-and-effect* diagram, also called the *fishbone diagram* or *Ishikawa diagram*, was developed by Karou Ishikawa to identify the cause of problems in

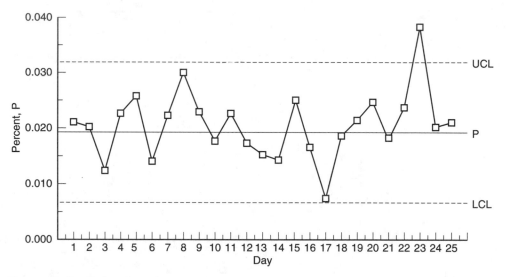

Figure 13–5 P-chart Example.

Japanese manufacturing. The cause-and-effect diagram allows a manufacturer to illustrate on a single diagram the relationship between the causes (machine adjustment) and the effect (poor quality). There are three basic types of cause-and-effect diagrams used by industry: *dispersion analysis* (Figure 13–8), *cause enumeration* (Figure 13–9), and *production process classification* (Figure 13–10). Study the diagrams in these figures and note similarities and differences. The first two

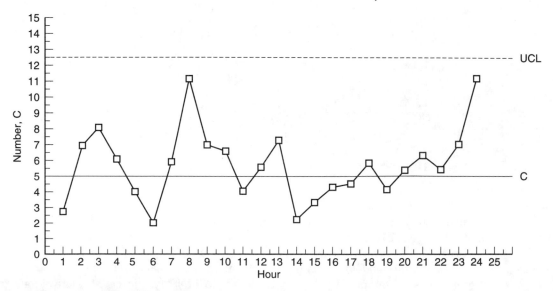

Figure 13–6 C-chart Example.

Figure 13–7 Pareto Diagram.

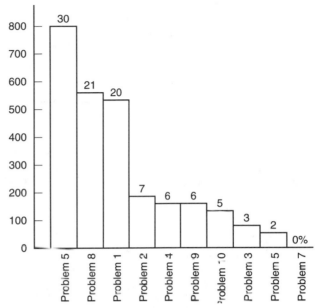

diagrams have the same basic format with four branches labeled *workers, materials, inspection,* and *tools* and the effect identified in the box on the right. The majority of the causes fit into one of these four categories.

Dispersion analysis (Figure 13–8) is the most frequently used diagram and allows problem solvers to focus on why variations occur. When a control chart

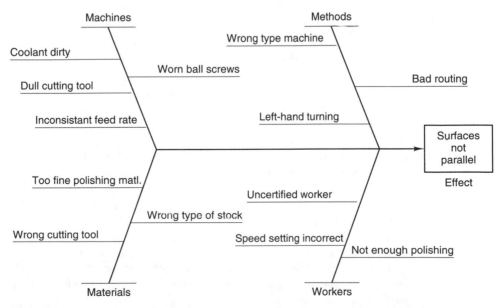

Figure 13–8 Dispersion Analysis Chart.

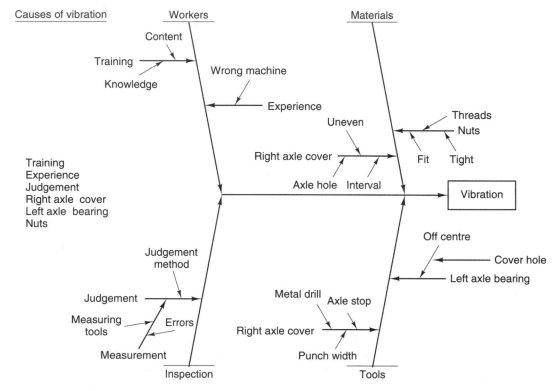

Figure 13–9   Cause Enumeration Chart.

indicates that a process is out of statistical control, a dispersion analysis diagram is used to determine the cause.

The cause enumeration type of fishbone diagram (Figure 13–9) is used to list all of the causes of variation that can affect a process. The cause enumeration diagram helps to identify the sets of common causes that inhibit a process from meeting specifications.

The final type of diagram, production process classification, is used to study an entire process. Figure 13–10 shows all machines used in a sequential process and the possible causes at each machine that could result in a defective product.

The graphic diagrams shown in Figures 13–7 through 13–10 are among the most effective problem-solving tools used in manufacturing. A team of problem solvers from different functional areas frequently creates these diagrams. The graphic representations help people visualize the problem and the factors that contribute to it. Working with people involved with a problem to create the analysis graphics is an effective way to encourage everyone's sincere participation.

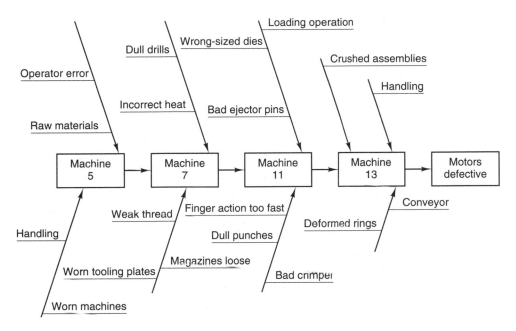

Figure 13–10   Product Process Classification

Solving SPC Problems

The charts are part of the four basic steps used to solve SPC problems. The four steps include:

1. *Problem identification.* Problems in quality are detected by the operator, in downstream operations, or at inspection. Techniques used in problem identification include *Pareto analysis, control charts,* and *process capability studies.*

2. *Identification of cause(s).* The cause(s) of the problem usually results from a process element, such as people, tools, machines, material, or process methods. Investigation of the problems often includes *cause-and-effect analysis* using an *Ishikawa diagram.* If industrial experiments are needed to help differentiate among interrelated influencing factors, *Taguchi techniques* for experiment design are used.

3. *Problem correction.* The range of possible corrective actions, from a simple machine adjustment to redesign of a complete process, is broad. In addition to the corrective action, a verification that the fix has corrected the problem must be performed.

4. *Maintaining control.* After the problems are corrected, process control is achieved using either statistical methods or physical systems. The statistical technique uses control charts and the physical system uses instructions embedded in hardware and software. Constant improvement in the system is possible only if the corrections to the system are maintained.

Quality of Design Versus Quality of Conformance

The importance of the quality of the design has been emphasized throughout the product and process section of this book. A process developed in Japan offers a quantitative approach to analyzing the quality aspects of the design. The technique, named after its inventor Taguchi, addresses the quality problems that are built into the product and process during their design.

Figure 13–11 shows the TQM and manufacturing cycle from Figure 13–1 with two areas highlighted. The *market research* activity is where customer needs and expectations are defined, and the *product and process development* activity is where standards and specifications for the products and processes are developed. Taguchi calls these two areas the *off-line quality system*, and Juran calls them the *quality of design* area. The quality of design area is critical because 80 percent of the quality problems that surface during production of the product are built into the product through bad design and poor process selection and development.

The remaining half of the product and TQM cycle in Figure 13–11 is called the *on-line quality system* by Taguchi. In this area, *quality of conformance* is practiced by the production and quality control departments. The primary quality tools to check conformance to specifications include testing, inspection, supplier certification, and SPC. However, only 20 percent of the quality problems encountered during production can be eliminated because the balance of the problems are in the design.

The impact of design on quality is well documented, and the following technique illustrates how a well-developed product and process design produces products with near zero defects.

Figure 13–11 Quality of Design and Quality of Conformance in TQM Cycle.

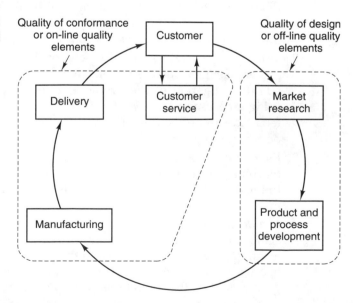

13–4 DEFECT-FREE DESIGN PHILOSOPHY

In 1798 the U.S. government needed a large supply of muskets, and production at the federal arsenal could not meet the demand. Muskets could not be purchased from Europe because war with the French was imminent. Eli Whitney offered to produce 10,000 muskets for the U.S. government and to meet a 28-month delivery requirement. It took 10½ years to complete the project because the musket parts produced in separate batches had a high degree of variability. As a result, musket parts selected at random for assembly would not fit together; therefore, each musket had to have the parts specially selected and assembled individually. After several years of work to reduce the variability of musket parts, Whitney's guns could be built with parts selected randomly from a pile, and parts from different guns were interchangeable. These same issues dominate the manufacture of products today, and the need for a process that produces near zero defects during production is necessary. Such a process is called the *six-sigma design process*.

Six-Sigma Design

Review the normal distribution curve in Figure 13–2 and the data associated with the curve in Figure 13–3. The products from most manufacturing processes have a normal distribution, which means that 99.74 percent of all production falls within plus or minus three standard deviations (±3 sigma) of the average center value. The relationship between the normal curve and the product and process design determines how the process variation will affect the final product assembly. The affect of the normal variation is reduced if the range of produced part values, ±3 sigma, falls comfortably within the tolerance specified for the part. Figure 13–12 illustrates this concept. The ±3 sigma range for the process

Figure 13–12 Process Variation and Design Tolerance.

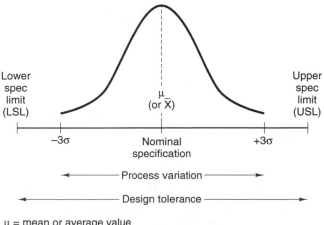

μ = mean or average value
σ = standard deviation

variation is called the *process capability,* and defects occur when this process capability is wider than the design tolerance or *specification width* marked by the lower specification limit (LSL) and upper specification limit (USL). Anything that causes the process capability to exceed either specification limit (LSL or USL) causes defects in the product.

The *capability index* (Cp), a measure of the inherent capability of the process to produce parts that meet the design requirement, is found by dividing the total specification width by the process capability. The larger the capability index, the more likely that the parts will satisfy the production requirement. One method of ensuring defect-free parts would be to make the specification width as wide as possible and the process capability narrow.

Defects-per-Unit Benchmark

A benchmark used to judge a superior manufacturer, one that is world class, is the number of defects per unit in complex products. The superior manufacturer has a defects per unit that approaches zero, while the more typical manufacturer has a defect level of five or higher. The automotive industry can provide a good example of this concept. Many of the Japanese car manufacturers have defects per car in the range of 1.5 to 2, while their American counterparts have levels of 8 or higher. If you consider the production of hundreds of cars per day with each car having thousands of parts, the concept of less than 2 bad parts per car appears impossible to reach. The key to achieving this level of quality lies in the capability index for the critical design tolerances on each of the parts used in the complex assembly. The superior company enforces a design margin of Cp equal to or greater than 2, while the typical company uses a value of Cp equal to 1.33 or less. The following examples illustrate this important difference.

The typical company with a Cp equal to 1.33 sets the LSL and USL at ±4 sigma while the process capability or variation remains at ±3 sigma. This situation is illustrated in Figure 13–13. This capability index appears to be satisfactory for quality production; however, it is not uncommon for the average value or center point of a normal distribution to shift ±1.5 sigma. The effect of such a shift is illustrated in Figure 13–14. The area outside the specification limit

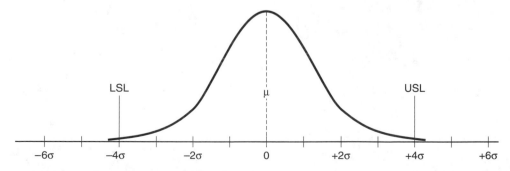

Figure 13–13   Process Capability and Specification Width with a Capability Index of 1.33.

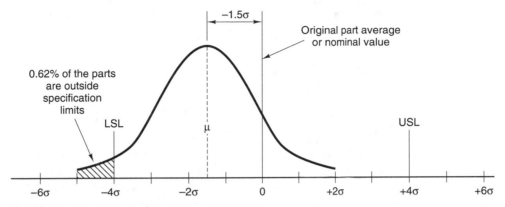

Figure 13–14 Shift of 1.5σ in a Normal Distribution.

represents 0.62 percent of the production. Although 0.62 percent is small, it means that for every 1 million parts produced, 6200 will be defective in the specification measured. This number of defective parts is not acceptable by today's quality standards.

Changing the specification width to ±5 sigma (Cp = 1.66) reduces the defects to 200 parts per million. This level of defects is too high for a company with a world-class quality goal; as a result, a specification width of ±6 sigma is used to ensure a near-zero-defect operation. The normal distribution with a ±6 sigma specification width is illustrated in Figure 13–15. Note that a shift of ±1.5 sigma in the critical specification has a minimal effect on the product quality, and the number of defects is just 3.4 per million parts produced. There are many more issues associated with achieving this level of quality production; however, for this overview the key point is that specification limits must be a minimum of two times the process variation. This limit permits the process to shift ±1.5 sigma, a

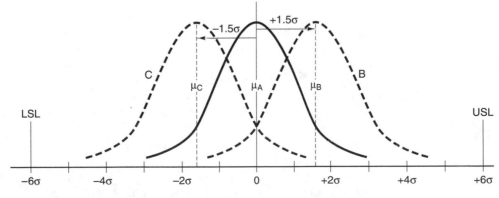

Figure 13–15 Six-sigma Specification Limits.

common variation for processes that follow the normal distribution, with little effect on product quality.

Seven Steps to Six-Sigma Capability

Many world-class manufacturers set a goal of zero defects in critical product and part parameters. To achieve that goal, many use the following seven-step process.

1. Identify the product or part characteristics critical to satisfying the physical and functional requirements of the customer and any standards agencies.
2. Identify the specific critical dimensional and functional characteristics of the products associated with the critical characteristics identified in step 1.
3. Identify and/or develop the process sequence necessary to control the critical dimensional and functional characteristics.
4. Determine the nominal design value and the maximum dimensional and functional tolerances that permit assembly and operation of the product at a level that guarantees customer satisfaction.
5. Determine the process capability for the processes selected/developed for the critical dimensional and functional characteristics in step 3.
6. Check the capability index, Cp, for each of the critical dimensional and functional characteristics.
7. If Cp is not equal to or greater than 2, change the design of the product and/or the process to achieve a Cp equal to or greater than 2. Another alternative is to institute process control measures that narrow the process capability sufficiently to achieve the necessary Cp for high quality manufacturing.

The process just described permits a part to be run using standard production processes with the guarantee that 999,997 parts out of every million produced can be used in the assembly process. When this process is applied across all critical dimensions on all parts in the product, defects associated with part compliance and poorly manufactured parts are less than three bad products out of a million produced.

Given the current capability of the production process, the six-sigma program attempts to make the specification width on critical dimensions and functions twice the process capability. Achieving this 2:1 ratio ensures a near-zero-defects production operation. Design of component parts and processes is a critical part of achieving the capability index of 2 or greater. When economically possible, new process machines capable of holding tighter tolerances on critical dimensions are purchased. Frequently, however, the burden falls on the product design group to produce a design with wider specification widths. In that case, the principles associated with design for manufacture and assembly (DFMA) are used to produce parts with specifications consistent with the six-sigma program needs. In addition, many of the design tools associated with the new CAD software help to create designs that support this production philosophy.

13–5 THE CHANGING WORKFORCE

Management of the manufacturing workforce in a highly automated facility is changing as a result of internal and external factors. Two key internal issues that affect management of the workforce emerge as a result of automation: (1) the interaction of the operator with the machine changed significantly when manual machines were replaced with computer-controlled processes, and (2) the highly integrated nature of the automated production systems and the enterprise requires a new set of skills.

Many external factors are linked to changes in workforce management. The two most significant factors are the *expanding global marketplace* and the *evolution of the workforce*. The issues surrounding these internal and external factors are addressed in the following sections.

Management Change: Internal Factors

The internal factors that forced a change in the management of workers resulted from the introduction of manufacturing automation. In general, automation created an artificial distance between the craftsperson or worker and the finished part or product. For example, when the processes were manual, worker action directly affected the size, shape, look, and feel of the finished product. They handled the product and were directly responsible for the quality of the product and the results achieved. Their manual skills as craftspeople were reflected in every finished part.

As the production process was automated, many of the skills of the craftsperson were replaced with a different set of skills as a machine operator. Responsibility for the dimensional integrity of the part shifted from the craftsperson to the automated machine, and the operator's responsibility shifted to control of the automated process. In highly automated work cells, the operator may not even touch the raw material or the finished product. This change in the production process changed the role played by both the machine operators and the first-line supervisors for the area.

In the manual operation, the first-line supervisor frequently was the master craftsperson who brought employees up the productivity curve by teaching and demonstrating techniques on the various machines used in production. The supervisor's information base was a mixture of learned skill and experience. The introduction of automated machines forced the supervisor into the new role of resource provider and coach. With the automated machine in control of the process, the supervisor becomes a manager of the resources required by machine operators. This task includes, for example, acting as an interface to the other departments responsible for machine programming and resolving a more complex set of production problems. The supervisor's production skills must be augmented with skills in communications, negotiation, and technical problem solving, and the ability to resolve human resource and personnel issues, teach, and coach in a team environment. The supervisor and all the production employees are required to have a broader view of the enterprise and the ability to use the wide range of information available in the computer-integrated enterprise database to improve the process constantly and increase productivity.

Management Change: External Factors

The two most significant factors are the *expanding global marketplace* and the *evolution of the workforce.* While the first challenges the enterprise's ability to survive, the second offers a mechanism to meet market demands that result from worldwide competition. Together, these two external factors demand a change in management at all levels in the enterprise.

The traditional vertical management structure, used by almost every manufacturing organization for the last seventy-five years, is under pressure from top to bottom. Top management is under pressure to create an organization with few management layers so that response is quicker and overhead cost is reduced. As a result, middle-management positions are eliminated and the associated tasks are absorbed by those managers retained, assumed by automation, or just abolished. Decision making is less centralized; as a result, the authority and responsibility for production decisions are pushed closer to the manufacturing floor. A changing management environment from the top and an evolving workforce at the production level puts first-line supervisors between forces of change that have never been experienced in the history of manufacturing. Global competition is forcing a different management style in U.S. companies, and the evolution that has occurred in the workforce over the last 100 years makes a new style possible. The evolution in the U.S. workforce opens the door for *self-directed work teams* and a major change in the way that manufacturing management functions.

13–6 SELF-DIRECTED WORK TEAMS

The concept behind *self-directed work teams* has been evolving for the last fifty years. The dramatic changes in the global market and the increase in the use of technology in manufacturing since the 1970s has accelerated the interest in team-based management. Global market pressures require manufacturers to respond rapidly to customer needs, frequent new product introductions, increased productivity at all levels in the organization, lower production cost, and higher quality. Achieving these *order-winning criteria* with the workforce management policies of the 1950s and 1960s is not possible. As a result, manufacturers interested in achieving world-class status are ready to consider a change in management style and to embrace the concept of manufacturing teams. The following sections cover a short history of self-directed work teams and a description of their operation and implementation requirements.

Development of Self-Directed Work Teams

Management style in the United States has been moving away from autocratic management toward worker empowerment for many years. A brief overview of this transition puts into perspective the shift to self-directed work teams.

The chart in Figure 13–16 describes the development of the management style used since the eighteenth century. Study the chart until you are familiar with the data. Manufacturing started in the United States as *cottage industries*, where families produced products for distribution within the town or village. The radius of impact was usually less than thirty miles, and the management style was autocratic or paternalistic. The head of the household, usually the eldest male, was both the dominant leader and typically the master craftsman for the product produced. For example, if the industry was ironwork, the father or grandfather was usually the master blacksmith. It was not uncommon for everyone in the family to have some job or responsibility in the business; however, all control and power resided with the master craftsman.

In the mid-nineteenth century, several events caused a change in manufacturing: (1) the population of the nation grew, and town markets became regional markets spanning hundreds of miles; (2) the implementation of the power spinning shaft, powered by water wheels, provided the energy to run all the machines in the factory; and (3) the development of the steam engine permitted wider distribution of finished products and mobile power so that factories could be located anywhere. This stage in the transition was marked by a move away from single-family operations to company affiliations. However, the management style did not change. The autocratic leadership moved from the family cottage industry to a position as plant manager or factory owner. The central source of authority was still present with just a larger family of unrelated workers.

The current industrial mode, industrial automation, started with electrification of the country and continues today with the rapid development of high technology. The markets are worldwide, with distances marked in thousands of miles. National companies are now multinational, with raw material sources, labor, and distribution coming from across the globe. The management style necessary for survival in this climate is *worker managed.*

The transition in the management style over the last fifty years started with the industrial buildup for World War II. The first stage in the empowerment process was the move to *delegation.* Delegation is not empowerment; rather, it is the granting of permission to carry out a task. The next step was the establishment of *quality circle* groups to perform group problem analysis. In the 1970s the transition to empowerment started with quality circle groups given greater authority to work on their own and the initiation of cross-functional corrective action teams. However, neither of these initial efforts experienced the empowerment of the self-directed work teams used today.

Work system mode	Economic impact area	Management style
Cottage industry	Town or village	Autocratic and paternalistic
Production plant	Region	Autocratic and paternalistic
Automated manufacturing	World	Worker-managed

Figure 13–16   Development of Management Styles.

Establishing Self-Directed Work Teams

There are many misconceptions about what constitutes a self-directed work team. A brief discussion of the major misconceptions helps to bring the concept into sharper focus. The most common is that the process is *not about power.* The entire issue is the authority to act and the power to carry the actions to completion. The major variable that affects the implementation of self-directed teams is the importance of power to management. If power is important, the transition to self-directed teams is difficult. Supervisors and workers have a new role when self-directed teams are implemented, and neither group risks a loss of jobs. Job security is a problem only for supervisors who cannot coach without power and for workers who want management to think for them.

Other misconceptions about self-directed teams are that they are easy, take a limited amount of time, are management driven, and are not imperative. Companies are not implementing self-directed teams because management wants to give up power. Teams are installed because management recognizes that the workforce has changed and that to be competitive and survive, the enterprise must have every employee working on the solution to manufacturing problems. In world-class companies, the important issue is how many people are thinking about the enterprise problems. Self-directed work teams are a significant change in management style that provides a real transfer of power from supervision to the work team.

The Transition to Work Teams

One of the first questions deals with the *readiness* of the workforce. No amount of training can change a workforce not ready to accept the responsibility placed on a self-directed team. Stated in the simplest terms, the readiness test identifies workers who understand the operation of the process and want ownership of it. Workers were asked, "Could you do your job better if something were changed?" Those ready for self-directed work teams know specifically what needs to be changed to improve their performance. On the other hand, workers who don't know would not operate well in the team environment. The evolution toward self-directed teams is driven by a work force that is generally more intelligent because of media and information access. In addition, workers are process competent and know how the production system functions. Finally, they have a higher level of motivation and sense of ownership about their jobs. If these conditions are not present, the group is not ready for the self-directed work team concept.

The transition to self-directed work teams is smooth when four rules are followed: (1) management must understand the issues and transfer of power required for functional teams, (2) a plan must be developed for the specific plant location based on local culture and existing readiness, (3) functional teams must be trained in skills required for self-management, and (4) the transition through the three stages of empowerment must be spread over five to eight years. Functional self-directed work teams have been implemented in many companies following these four rules.

Description of a Self-Directed Work Team

The common characteristics for the generic work team include the following:

- Frequently, a self-directed work team is formed from a function work group currently in a process area. The term *functional* means that all members of the team come from a single process area. Cross-functional work groups composed of employees from several different areas cannot operate in the self-directed mode.
- A self-directed work team meets as a functional group to manage the process area and resolve problems that reduce the effectiveness and profitability of the process.
- Team members are trained to work in self-managed mode and are managed by natural leadership from within the group.
- The team improves the production process in their area using the following six-step process: (1) identification of a process problem, (2) development of a process improvement plan, (3) implementation of the process improvement, (4) tracking of plan effectiveness through data collection, (5) evaluation of process data to determine performance, and (6) participation in both personnel and financial rewards as a team.

A successful team is identified by the existence of interdependence between the functional team members and exercising *true power* through the ability to take corrective action.

Installing Self-Directed Work Teams

The installation process passes through the following three stages.

1. *Process focus.* In this stage, functional work teams are formed, training is carried out in critical interpersonal skills (communication, problem solving, consensus building, and conflict resolution), technical and process training is provided, relationships within the group are established, and natural leadership elements are developed. However, the primary focus of the team is on gaining ownership of the process, and this stage takes from eighteen months to two years to complete.
2. *Self-directed work teams with supervision.* In this stage, the work teams begin to perform the duties for the functional area but with the help of supervision. The supervisors start to assume the role of *coach* but retain the right to final approval for work team decisions. The help provided by the supervisor includes budget development and presentation, peer assessment techniques, and vendor interfacing and communications. The time required to work through all of the elements in this stage is usually two to three years.
3. *Full self-managed work teams.* In the final stage, the supervisor is totally removed from the group and acts only as a coach or advisor. The team assumes full responsibility for operation of the process area. The team is

empowered to perform all of the tasks previously performed by the supervisor, which include setting schedule priorities, planning working hours and duties, disciplinary action for team members, selection and hiring of new team members, and termination of current team members. It often takes one to three years to perfect the self-management process.

After the implementation is complete, management becomes the "what" team to decide what and how much to make; the self-directed work team becomes the "how" team that decides how best to produce it. The critical job for management is the development of clear objectives and goals with an openness and patience to resolve the disagreements that occur along the implementation process. For the teams, the critical work includes establishing a climate of support and trust, development of problem-solving skills and leadership, and regular reviews of tasks and participants' progress.

The implementation process normally moves through the following steps:

- Executive training.
- Planning and establishment of the work team method.
- Work team training on a team-by-team basis.
- Implementation of work team control over production processes as the employees are trained and implemented into teams.
- Work team beginning to perform the following tasks: participate in pre-shift meetings, participate in formal scheduled team meetings, examine the production process, plan corrective action for process problems, and take responsibility for operation of the process.

Self-directed work teams have two types of authority: sovereign and negotiated. In the production area, a team has the power to change anything viewed as an improvement in the manufacturing process. However, any change that affects functions outside the process area must be negotiated with the external team before any changes are made. Problems that cross into other areas fall into three general categories: *operational* problems, *broad-based* problems, and *enterprise-wide* problems. In practice, operational problems that affect another team are solved by sending two team members to a team meeting of the other work team. For example, the speed of loading of parts into a machine is slowed by the way the parts are oriented on the pallets at an upstream work center. An improved pallet orientation is developed and presented to the work team by two members at the upstream work center.

When the problem falls into the broad-based category, a joint work team meeting is planned to solve the problem. For example, the plating process has an increase in parts rejected due to a nonuniform finish. The team isolates the problem to handling and processing in another work center. Because the source of the problem is not defined clearly and may affect other parts produced with the same routings, a joint team meeting between the plating department work team and the process center work team is held to define and propose a solution to the problem.

Major problems affect a range of enterprise departments and require a cross-functional meeting with representatives from several work teams. For example, a work-center performing parts assembly routinely has a problem assembling mating parts. The cross-functional solution team would have representatives from assembly, material processing, design, and production engineering.

Companies that implement self-directed work teams have an advantage over manufacturers who choose to stay with the traditional supervisor/worker relationship. Consider the following two scenarios. In organizations where the workers don't have to think after they arrive for work, only a handful of employees are working on the solution to enterprise problems. However, the company that adopts the work team concept successfully has everyone working on the solution to enterprise problems. Over time, the company with self-directed work teams will win over the "mindless" competition.

13–7 SUMMARY

The lessons learned by manufacturing in the 1970s is that technology alone cannot improve business performance, and a new emphasis on quality and development of the workforce was required. The move to quality has two major elements: a new philosophy on how the business must function and the use of quality tools to achieve the desired results. This broader view of quality is included in a process called total quality management (TQM). TQM has two components: principles and tools. The principles allow management changes that lower the barriers to a successful TQM implementation. The tools permit quantitative and qualitative measurement of the system to determine how well the process is meeting organizational goals.

The definition of TQM has three dominant themes: (1) participative management for everyone in the enterprise, (2) implementation of a successful continuous improvement process, and (3) efficient use of multifunctional teams. The six important principles that must be considered in a TQM implementation include (1) a focus on customer needs and satisfaction; (2) a focus on the process used to produce a product as opposed to a result or product focus; (3) a focus on the prevention of problems in areas such as quality, production, machine operation, and engineering design changes; (4) a focus on using the brain power of every employee in the organization; (5) a focus on basing decisions on facts about manufacturing and design without finger-pointing and blame when problems occur; and (6) a focus on developing an enterprise where communication channels are always open so that product data and process information flow freely among all levels and all employees.

Implementing TQM is a five-step process: preparation, planning, assessment, implementation, and evangelization. The first step sets the goal for the organization, and in the second step the corporate council builds plans around these enterprise goals and guidelines. The third step identifies the strengths and weaknesses of the enterprise using a broad-based assessment process. TQM is implemented in the fourth step using process action teams to look at the weaknesses in the processes and to implement changes to make improvements. In the fifth and last step, the successful TQM processes are moved to other parts in the enterprise for implementation.

The quality tools used to support the quantitative and qualitative needs in TQM include statistical process control (SPC), Pareto analysis, Ishikawa diagrams, and Taguchi techniques. The SPC process uses four different types of control charts (X-bar, R, p, and c) to analyze the performance of a process and suggest changes to reduce the variability present in the process.

The goal of producing with near zero defects is reached through the adoption of a six-sigma design and production process. The basis for six-sigma design and production is a focused effort on making the maximum limits on critical dimensional tolerances of parts twice the width of the process variability. If machines produce parts with the standard variation found in a normal distribution, most of the parts will fall between ±3 standard deviations of the average value. If the allowable specification width for critical part dimensions is ±6 standard diviations, even with normal process variation fewer than four defective parts are produced in every batch of 1 million. This process leads to near-zero-defect production.

Management of the automated workforce is changing as a result of two key internal factors: the relationship between the operator and the machine, and the highly integrated nature of automated production systems. External factors affecting workforce management are the expanding global marketplace and evolution of the workforce. The most significant event resulting from evolution of the workforce is the move to self-directed work teams. The move to these teams is a result of two factors: first, for the organization to remain competitive, global competition requires problem solving from everyone in the enterprise; and second, the workforce has evolved to a state where employees are ready to take the responsibility to control the production processes assigned to them by management. After self-directed work teams are implemented, management becomes the "what" team to decide what and how much to make; the teams deal with the "how" issues, where they focus on how to produce the product most efficiently.

REFERENCES

ARNSDORF, D., *Technology Application Guide: Quality and Inspection.* Ann Arbor, MI: Industrial Technology Institute, 1989.

BYHAM, W. C., *Zapp! The Lightning of Empowerment.* Pittsburgh, PA: Development Dimensions International Press, 1989.

COX, J. F., and J. H. BLACKSTONE, ed. *APICS Dictionary,* 9th Ed. Falls Church, VA: APICS—The Educational Society for Resource Management, 1998.

CROSBY, P. B., *Quality Is Free.* New York: McGraw-Hill, 1979.

EVANS, J. R., and W. M. LINDSAY, *The Management and Control of Quality,* 3rd ed. Minneapolis/St. Paul, MN: West Publishing Co, 1996.

JABLONSKI, J. R., *Implementing Total Quality Management.* Albuquerque, NM, Technical Management Consortium, 1991.

JURAN, J. M., and F. M. GRYNA, *Quality Planning and Analysis,* 3rd ed. New York: McGraw-Hill, 1993.

LOCHNER, R. H., *Designing for Quality.* Milwaukee, WI: Statpower Associates, 1991.

MOTOROLA, *Design for Manufacturability.* Tucson, AZ: Motorola, Inc., 1988.

SOBCZAK, T. V., *A Glossary of Terms for Computer-Integrated Manufacturing.* Dearborn, MI: CASA of SME, 1984.

QUESTIONS

1. Describe the two different views of quality in the area of manufacturing.
2. Describe the two components present in TQM.
3. Define *total quality management.*
4. Describe the three dominant themes present in the definition for TQM.
5. Name the six principles that are critical for a successful TQM implementation and describe each briefly.
6. Describe the TQM implementation process.
7. Define *SPC.*
8. Describe the normal distribution and the significance of 3 standard deviations.
9. What are control charts, and how are they used to control a process?
10. Compare and contrast the four types of control charts with respect to their process control function.
11. Describe how Pareto charts are used to solve process problems.
12. Describe how Ishikawa diagrams are used to solve process problems.
13. Name and describe three types of cause-and-effect diagrams.
14. Describe the solution process for problems identified with SPC.
15. Compare and contrast quality of design versus quality of conformance.
16. Describe the six-sigma design process.
17. Why is a specification width of 4 standard deviations inadequate for high-quality production?
18. What is the relationship between DFMA and near-zero-defects production?
19. Why is defects per unit a good method for measuring the quality of a product?
20. Briefly describe the seven steps required for near-zero-defects manufacturing.
21. What are the key internal and external issues affecting the management of the workforce?
22. Describe the evolution of self-directed work teams.
23. Describe the four rules used in the transition to self-directed work teams.
24. What are the common characteristics that describe a self-directed work team?
25. Describe the four stages in the implementation of self-directed work teams.
26. Describe the authority vested in self-directed work teams.

PROJECTS

1. Using the list of companies developed in Project 1 in Chapter 1, create a matrix that illustrates which company is using TQM, SPC, other quality tools, six-sigma design process, and self-directed work teams.

2. Select one company from the list in Project 1 above that uses TQM, and describe how the process was implemented.

3. Select one company from the list in Project 1 that uses SPC, and describe how they implemented the process, what control charts are used, and how they solve problems identified through SPC.

4. Select one company from the list in Project 1 that uses self-directed work teams, and describe how they implemented the process, the current stage of implementation, and how they determine when a team is ready for self-governance.

APPENDIX 13–1: DEMING'S FOURTEEN POINTS

1. Create constancy of purpose toward improvement of the product and service to all customers.

2. Adopt the new philosophy that comes from the new economic age.

3. Cease dependence on mass inspection for quality products by building quality into the product and process.

4. End the practice of awarding business on the basis of price and change to a meaningful measure of quality.

5. Find problems and improve continuously the system of production and service.

6. Institute modern methods of training on the job.

7. Institute leadership at all levels of management.

8. Drive out the fear of suggesting changes to company policy and operations.

9. Break down the barriers between departments.

10. Eliminate numerical goals, posters, and slogans for the workforce that work for higher levels of productivity without improving the production methods.

11. Eliminate work standards that prescribe numerical quotas.

12. Remove barriers that reduce pride in workmanship for hourly workers, engineers, and managers.

13. Institute a vigorous program of education and self-improvement.

14. Make everyone in the company responsible for achieving the transformation.

Index